T0295496

# Interactions and Interdependence in Plant Communities

# Interactions and Interdependence in Plant Communities

Edited by Jason Eady

New York

Published by Syrawood Publishing House,
750 Third Avenue, 9th Floor,
New York, NY 10017, USA
www.syrawoodpublishinghouse.com

**Interactions and Interdependence in Plant Communities**
Edited by Jason Eady



International Standard Book Number: 978-1-64740-348-5 (Hardback)



**Cataloging-in-publication Data**

Interactions and interdependence in plant communities / edited by Jason Eady.
p. cm.
Includes bibliographical references and index.
ISBN 978-1-64740-348-5
1. Plant communities. 2. Plant ecology. 3. Biotic communities. I. Eady, Jason.
QK911 .I58 2023
581.524 7--dc23

# TABLE OF CONTENTS

# PREFACE

The collection of plant species which are present in a particular geographical area, and which forms a relatively uniform patch that can be differentiated from neighboring patches of different vegetation types, is known as a plant community. The study of these communities of plants is integral towards developing an understanding of the impact of dispersal, response to disturbance and tolerance to environmental conditions in different varieties of plant species. Interactions and interdependence in plant communities involve the study of the influence of facilitation and positive interactions among species in plant communities. It includes species specificity in facilitative interactions, indirect facilitative interactions, and potential evolutionary aspects of positive interactions. This book includes some of the vital pieces of work being conducted across the world, on various topics related to interactions and interdependence in plant communities. It attempts to assist those with a goal of delving into the field of botany. This book is a resource guide for experts as well as students.

The information shared in this book is based on empirical researches made by veterans in this field of study. The elaborative information provided in this book will help the readers further their scope of knowledge leading to advancements in this field.

Finally, I would like to thank my fellow researchers who gave constructive feedback and my family members who supported me at every step of my research.

**Editor**

1

# Phenological Plasticity of Wild and Cultivated Plants

*Amber L. Hauvermale and Marwa N.M.E. Sanad*

**Abstract**

The future survival of wild and cultivated plant species will depend on their ability to adapt to environmental changes caused by climate change. Phenological plasticity describes physiological, developmental, cellular, and epigenetic mechanisms that contribute to genetic diversity and adaptability. Many studies evaluating plasticity using trees, cereals (barley, wheat, and rice), pulses, and weeds have discovered that plasticity mechanisms differ between wild and cultivated plant populations. Major findings indicated by these studies are: (1) invasiveness and adaptability in wild and/or "weedy" plant species may be controlled by specific plasticity genes, (2) adaptability is directly connected to adaptive responses and fitness, and (3) domestication and cultivation have altered plasticity mechanisms. Therefore, selective breeding requires a holistic understanding of plant plasticity. Breeding strategies should consider differences in plasticity mechanisms between wild and cultivated plant populations to reintroduce genetic diversity of plasticity from wild relatives.

**Keywords:** cellular plasticity, climate change, developmental plasticity, drought response, epigenetic plasticity, germination, hormone signaling, physiological plasticity, phenological plasticity, seed dormancy, selective breeding

## 1. Introduction

Global climate changes undoubtedly impact adaptability in plants by altering mechanisms of physiological plasticity [1]. Modifications in mechanism occurring at the morphological, anatomical, and physiological level are regulated by the capacity of a plant to adjust to abiotic and biotic stresses [1–4]. The resulting survival response and survival capacity may vary depending on plant life stages [1–4]. Plasticity mechanisms discovered in plants are like those described in animals and humans, illustrating the conserved connection between environmental selection and adaptive response [2, 3, 5–11]. Research into the connection between environmental stress, environmental selection, and plant plasticity has also identified both general and unique plasticity mechanisms that differ between wild, i.e., non-cultivated, and cultivated plant species [1, 12–15]. However, a review analyzing the contribution of key traits responsible for varied plasticity mechanisms in wild and cultivated plants has not occurred. Thus, the range of plasticity occurring in wild plants will be compared with plasticity mechanisms in cultivated plants. Similarities and differences in plasticity responses will be highlighted between the two groups, with a specific focus on climate imposed global abiotic stresses like drought [14].

All plants have evolved unique life cycle characteristics that enhance survival and adaptation to diverse short and long-term climatic events that limit resources. Phenotypic responses occur at every stage of plant development, and influence overall plasticity from one generation to the next. Understanding and tracking phenotypic plasticity of wild plants in cultivated plants first requires defining biological reaction norms and their alternatives to clearly illustrate the differences between biological plasticity and non-plastic responses. Examples of phenotypic responses include: (1) rapid seedling growth (2) a short vegetative phase, (3) deep root systems, (4) high seed output, (5) discontinuous or extensive seed dormancy, (6) efficient cellular defense machinery, and (7) environmental plasticity. Although all plants exhibit phenotypic responses, the level of response is largely influenced by the degree of cultivation. Several species of trees and weeds are exceptional models for defining and tracking the range of both short and long-term heritable charac-teristics of wild plasticity [1, 4, 12, 16–21]. Drought response studies in agronomi-cally important, and highly cultivated crops like wheat, add perspective about the contributions of selective breeding programs; how increased cultivation results in gains or losses in adaptive responses and plasticity [7]. Transitional plant models, such as *Chenopodium quinoa* (quinoa) and *Hordeum vulgare* (barley) will be used to illustrate the evolutionary path from wild plasticity to cultivated plasticity.

## 2. Environmental changes impact phenotypic plasticity

Climatic events trigger heterogeneous responses in plants. Plant responses occurring from biotic or abiotic factors drive two distinct adaptation mechanisms, natural selection and phenotypic plasticity. Both mechanisms reveal the full genetic capacity of plants [22, 23]. The genetic makeup or genotype of each plant species determines how a plant will react in new environments [24]. Accumulated exposure to novel environmental stresses over many generations may increase selection toward the frequency of favorable alleles versus a reduction of unfavorable alleles, and results in less genetic diversity [22]. Otherwise, in natural selection, any change in plant phenotype is defined as phenotypic plasticity [25]. Changes in phenotypic plasticity impact individual fitness without changing genetic diversity [22, 26]. Sometimes a novel genotypic response does not deviate from a normal range of reactions, i.e., the reaction norm, and sometimes it does [27]. Thus, plants have a wide array of genotypic responses that impact phenotype. Non-cultivated plant species like trees acquired wild plasticity through the combination of both the long-term accumulation of genetic changes and the conservation of favorable survival strategies through time [24]. Adaptive responses result in phenotypic plasticity [22, 26]. Adaptive responses also maximize phenotypic fitness, or the ability to respond and survive in chang-ing environments [27]. Breeding programs have accelerated the adaptive process to abiotic stresses, like drought, in domesticated plant species by selecting for tolerance to drought or increased resource-use efficiency [28]. This approach has allowed breeders to select for favorable plant responses based on flexibility to varied environmental changes. A broad understanding of wild plasticity in non-domesticated plant species will enhance and extend our current understanding of the range of plasticity mechanisms in cultivated plants [2, 3, 5–11].

## 3. The plasticity spectrum

All terrestrial plants are stationary and adjust phenotypic responses to survive in fluctuating environments [22, 26]. A wide spectrum of adaptive variation occurs with a specific phenotypic response, and which is defined as phenotypic plastic-ity [14]. Three recognizable outcomes associated with a phenotypic response, as illustrated in **Figure 1**, are: (1) a neutral response, (2) an adaptive response, or (3) a maladaptive response [13, 22, 27]. Each panel illustrates the relationship between a phenotypic response and a change in environment. Red, green, or blue colored lines represent different genotypes or individuals [13, 22, 27].

A neutral response occurs when there is no observable change in plant fitness or plasticity after exposure to novel environmental stress (**Figure 1a**). Canalisation and developmental stability are components of neutral responses that create some confusion in understanding and mapping phenotypes [29]. Canalisation describes the occurrence of a constant phenotype in a given population that is not influenced by environmental or genetic regulation [29]. Developmental stability describes the degree to which organisms withstand environmental changes or genetic perturbations during development [29]. Canalisation measures gene rigidity or the resistance of genes to altered function during environmental changes [29, 30]. Canalisation is a useful measure of genetic robustness and is more frequently described than adaptive plasticity in plants [29, 30].

Adaptive responses occur in new environments and may or may not occur as a direct result of genetic variation [29]. Adaptive responses result in beneficial changes that maximize phenotypic fitness (**Figure 1b–d**) [27]. Not all phenotypic changes occur because of beneficial adaptive responses [27, 29, 30]. Individuals within a population may experience random passive phenotypic changes that are limited to specific phenotypic traits or that act more broadly impacting adaptive performance at all stages of plant development [29, 30]. Plasticity may be controlled by a single gene or many genes [31, 32]. The plasticity threshold of a plant is a func-tion of individual, pleiotropic, and collective responses within a population. This mosaic of responses influences genotypic selection [33, 34].

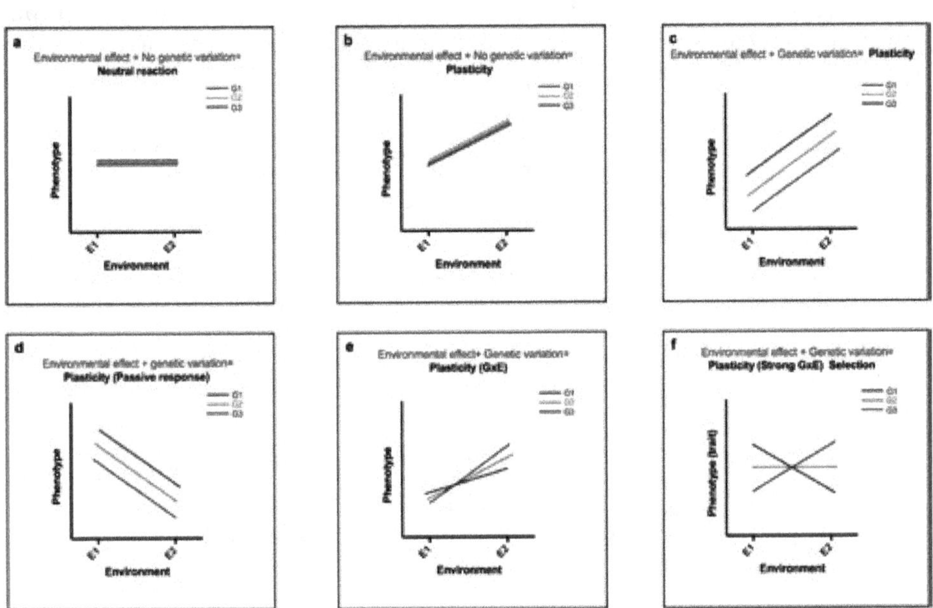


**Figure 1.**
*Recognition of the different reaction norms. The three major responses; neutral, adaptive, and maladaptive, which occur within the plasticity spectrum [13, 22, 27, 34, 35]. A neutral response (a). An adaptive response without genetic variation (b), or with genetic variation (c, d). A non-plastic or maladaptive response (e). All three phenotypic responses occurring simultaneously (f).*

Not all adaptive strategies are beneficial for plants and often result in decreased fitness or yield [35]. A maladaptive response describes a phenomenon which reflects the absence of plasticity (**Figure 1e**) [34]. Maladaptive responses are not easy to distinguish from neutral responses because the average response of the population may mask any decline in response by individuals within the population over a long period of time [35]. Maladaptive responses are often misinterpreted as adaptive responses and difficult to study genetically [34, 35].

All phenotypic responses, neutral, positive, and negative, may occur simultaneously within an individual or across a population (**Figure 1f**) [13]. Changes in plasticity may be measured by examining the relationship between a specific genotype (G) in a specific environment (E) [13]. A genotype-by-environment (GXE) study tracks genetic plasticity and is a powerful tool for targeted genotypic selection [13, 33, 34].

## 4. Characteristics of wild plasticity: examples in trees and weeds

Phenotypic plasticity, especially within wild plant populations, is a mechanism that enhances plant invasion and survival [12]. The invasiveness of a plant species is influenced by many phenotypic characteristics and responses [12]. The three major phenotypic characteristics that impact plasticity in wild plant populations are plant development, plant morphology, and plant physiology (**Figure 2**) [36]. Phenotypic responses associated with each characteristic occur at every stage of plant development, influencing the overall plasticity from generation to generation (**Figure 2**). Common phenotypic responses known to be associated with plant development, plant morphology, and plant physiology include: (1) rapid seedling growth allowing maximum capture of light, water, and nutrients [37–41], (2) a short vegetative phase allowing life cycle completion in various growing seasons and conditions [42–47], (3) deep root systems allowing plants to survive through drought conditions [47, 48], (4) high seed output ensuring spatial and temporal dispersal, (5) discontinuous or extensive seed dormancy ensuring germination only in favorable conditions [49–51], (6) efficient

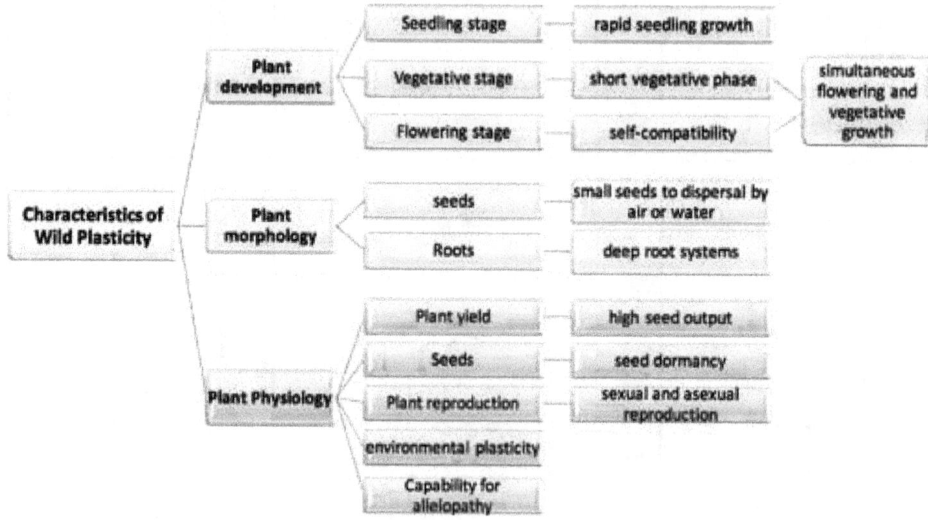


**Figure 2.**
*The key characteristics and responses of wild plant plasticity including plant development (green), plant morphology (yellow), and plant physiology (blue) [1, 13, 36–53].*

cellular machinery for scavenging reactive oxygen species (ROS) [52, 53], and (7) environmental plasticity, or the ability to respond to changing biotic or abiotic environmental factors [1, 13].

Phenotypic plasticity was first described for non-cultivated plants species including trees and weeds [1, 4, 12, 17–20, 54]. Trees are excellent models for studying phenotypic plasticity due to their longevity [12]. Trees have developed a diverse set of plasticity mechanisms that are specific for both short and long development programs occurring in different developmental tissues at the same time [12]. Simultaneous root and leaf canopy development are an example of parallel programing [12, 54]. Phenotypic plasticity in trees occurs through a diverse collection of physiological, anatomical, and morphological responses [12, 54]. Many studies exploring global warming have investigated the possibility of using physiological or morphological indicators of beneficial adaptive responses as predictors of species survival [1]. Adaptive mechanisms in trees, as well as other plants, are important for mitigating the stress that is associated with fluctuations in native environments or, after new colonization, for rapid adaptation to novel environments [54–57]. Studies investigating drought stress in trees have shown that by reducing the leaf canopy and increasing root proliferation, trees become more drought tolerant because both phenotypic responses limit water loss [58]. The occurrence of phe-notypic responses occurring in parallel suggests that there may be a coordinated regulation of these traits [59, 60]. Other traits indicative of drought responses and plasticity in trees include leaf area, leaf dry mass, leaf mass per area (LMA), leaf tissue density, net photosynthesis, stomatal conductance, leaf respiration, water use efficiency, leaf water potential at midday, total chlorophyll content, relative water content, gross photosynthesis, leaf transpiration, and the ratio between leaf respiration and net photosynthesis [58]. Drought avoidance may also be viewed as a strategy for drought tolerance by altering the timing of growth and reproduction [14]. By maximizing the adaptive response of traits related to drought response, the overall fitness of an existing population of trees has the potential to adapt to a new environment [58]. However, if a given climatic event exceeds the limit of adaptive capacity, the same population of trees may also be replaced by a new, more adapted species [14].

Plasticity in weeds, as with trees, is governed by adaptive responses that impact physiology, morphology, and anatomy [36]. However, unlike trees, many weed species have relatively short life spans and must make rapid and frequent adjustments to environmental changes to ensure survival [16]. In addition to the wild characteristics for plasticity listed above, other characteristics in weeds that demonstrate enhanced phenotypic plasticity include: discontinuous or extensive seed dormancy ensuring germination only in favorable conditions, indeterminant or simultaneous flowering and vegetative growth, self-compatibility allowing genetic divergence from previous generations without requiring special pollinators to ensure seed viability, long-distance seed dispersal by air or water; competition with crop plants resulting in reduced crop yield, sexual and asexual reproduction strategies; and allelopathy, or the ability to produce chemicals that retard or kill other plants (**Figure 2**) [16, 36, 61]. Adaptive responses in weeds occur throughout development [16, 36]. Sometimes adaptive responses are more apparent in plant architecture than in signaling responses, are more pronounced at certain developmental stages or in specific populations, or involve the same tissue types during different developmental phases [1, 16, 36, 61]. The invasiveness of weeds is thought to be associated with several phenotypic plasticity traits including plant height, flower development, flowering, and light quality, [62, 63]. A direct correlation between plant height and invasiveness remains unclear. However, there may be an association between tall plant phenotypes, increased phenotypic diversity, and higher plant abundance in unfavorable environments [63]. Associations have not been observed with flowering phenology among native and non-native plant populations, but this may be because flowering time is dependent on the environment [62–64]. Flower development and invasiveness in Purple loosestrife (*Lythrum salicaria*) demonstrate that both anther and stigma respond to changes in soil moisture during either of vegetative and reproductive development [65]. Tufted knotweed (*Polygonum cespitosum*) has been enhanced through adaptive responses to drought and high temperature without any observable decrease in fit-ness when grown in the shade [66]. Narrow-leaf plantain (Plantago lanceolata) is very sensitive to changes in light quality and modulates seed germination and leaf size as a mechanism for shade avoidance [67]. It is also the case in this weed species that leaf size and germination patterns share common physiological mechanisms where the short leaf phenotype is more plastic than the long leaf phenotype in shady conditions [20]. This discovery illustrates that wild plasticity is a dynamic network of processes that work synergistically to enhance the likelihood of sur-vival [20]. Both trees and weeds demonstrate how, through the process of natural selection, non-cultivated plants have adopted very different and dynamic strate-gies that ensure reproduction and survival [1, 4, 12, 16–21, 24].

## 5. Genetic regulation of plasticity

There is not enough data about the genes, promoters, and regulatory elements that control "invasive" or "weedy" phenotypes commonly observed in wild plant popula-tions. However, the phenotypes provide key insights into potential gene families and signaling pathways. Adaptive phenotypes also provide evidence that plasticity responses are controlled genetically and by specific plasticity genes [6]. The accu-mulation of genetic modifications associated with adaptive responses can be tracked through time and are genetically controlled [6]. Two models have been proposed to explore how changes in adaptive response occur. The first model proposes that the expression of structural genes varies as the environment changes [68]. Genetic plasticity is not regulated by plasticity genes, rather by changes in gene expression of structural genes resulting in phenotypic changes or plasticity [68]. The second model proposes that specific regulatory genes, i.e., plasticity genes mediate responses for structural genes [69, 70]. The resulting change in expression of the regulatory genes in response to environmental changes is what ultimately controls the pattern of plasticity [69, 70].

## 6. Plasticity mechanisms

There are four primary mechanisms in wild plant populations that regulate plasticity through adaptive responses [6]. The four plasticity mechanisms are physiological, developmental, cellular, and epigenetic responses [6]. Physiological plasticity describes all physiological responses associated with phenotypic traits and signaling networks [40, 71]. Developmental plasticity is associated with human or animal neural developmental, and plant embryonic development, in response to stress [6, 40]. Cellular plasticity describes adaptive responses within cells that are often associated with reducing reactive oxygen species accumulation through redox mechanisms [72]. Epigenetic plasticity describes changes to molecular mechanisms in response to abiotic stresses resulting in altered gene expression and function without changes in the DNA [71].

## 6.1 Physiological plasticity

Physiological plasticity is the most dynamic of plasticity mechanisms and is often involved in all other mechanisms of plasticity [40, 71]. Novel and emerging environments trigger many physiological responses such as carbon dioxide ($CO_2$) assimilation, changes in chlorophyll content, water use efficiency, sugar sensing and photosynthesis [73]. Physiological changes correlate directly to plant fitness, and changes in plasticity determine how a plant responds to environmental stresses [73, 74]. Studying the association between a physiological phenotype and changes in gene expression within wild populations will make it is possible to identify and target genes that are responsible for adaptive responses, i.e., plasticitygenes [24, 73, 74]. In this way plasticity genes and gene variants become a selective tool for understanding plasticity heritability dynamics, as well as identifying positively adapted populations [24, 73–75].

Seed dormancy is an excellent example of physiological plasticity [73, 74] Seed dormancy prevents germination out of season, even under favorable conditions, and ensures species survival of natural catastrophes [16, 76]. Environmental cues such as light, temperature, and moisture impact the depth of seed dormancy and the length of time required for dormancy release [76]. In weeds, discontinuous or extensive seed dormancy ensures germination only in favorable conditions and confers environmental plasticity, or the ability to respond to changing biotic or abiotic environmental factors [16].

Discontinuous or extensive seed dormancy impacts environmental plasticity through variable emergence timing throughout a growing season [76]. Discontinuous seed dormancy is likely a major "weedy" characteristic contributing to physiological plasticity in many wild plants and weed populations [76]. Downy brome (*Bromus tectorum L.*) is an invasive grass weed in both natural and agricultural environments which produces seeds with discontinuous seed dormancy [77–80]. New downy brome seedlings have the capacity to emerge in any season; early and late in the fall, before and after cool season crops or native grasses emerge, and even in the spring [80]. Differences in emergence timing in downy brome may be due to differences in dormancy status and may occur because of phenotypic and genotypic variation within a single population cluster, the presence of multiple population clusters within a single location, and the viability of seed in the seed bank [80]. The successful invasion of wild plant populations is measured by the number of individuals in a population, reproductive output, the range of habitats occupied, and the ability for survival and adapt in new environments through time [16]. Therefore, downy brome is an excellent example of a model colonizing species as it allocates most of the developmental time to seed production [16]. Downy brome increases the chances for survival of future generations, by maximizing contributions to seed banks [76]. Physiological plasticity mechanisms like prolific seed production, discontinuous seed dormancy, and variable germination in weeds increase the likelihood of outcompeting wild and cultivated plant species in native and non-native environments [76].

There is currently very little information about the specific genes or molecular mechanisms regulating dormancy or dormancy loss in many weeds or wild plant species [77]. Gaps in molecular information slow the progress for understanding the impact of wild plasticity on adaptability [1, 16]. However, detailed physiological observations and translational research are useful tools. These are powerful tools for studying the mechanisms that drive physiological plasticity in the seed and throughout all plant life stages, in natural and agricultural environments, and in both wild and cultivated plant populations [1, 16, 82–86].

Basic research has established that a seed's transition from dormancy to germination is controlled by the plant hormones, abscisic acid (ABA) and gibberellin (GA) [81, 82]. ABA establishes seed dormancy during embryo maturation and maintains dormancy in mature seeds, whereas GA stimulates seed germination [34, 36]. Dormancy studies in model systems including *Arabidopsis thaliana*, *Brachypodium distachyon*, *Hordeum vulgare*, and *Triticum aestivum* draw a clear connection between ABA, GA signaling mechanisms, seed dormancy and dormancy loss [81–84]. These studies also provide a framework for comparing the similarities, differences of mechanisms regulating physiological plasticity, and the degree of conservation within wild plant populations [50, 76, 83–100]. Carefully documenting development from seedling to seed in wild plant populations including weeds, provides a wealth of information about phenotypic plasticity in varied environments, and demonstrates the value in using wild species as models for understanding the full capacity of phenotypic plasticity in nature [85].

### 6.2 Developmental plasticity

Developmental plasticity was first identified in, and is most often associated with, human and animal development [40]. Developmental plasticity refers to the impact of environmental stimuli on embryonic development [6]. Within the plant biology community, there remains some skepticism surrounding the existence of plant developmental plasticity mechanisms, and how to best identify and characterize them [5–7]. Despite these challenges, recent paradigm shifts in conventional thought have resulted in significant efforts toward studying the impact of developmental plasticity in cultivated plant species [5–7, 40, 72].

Developmental plasticity directly impacts phenotypic plasticity and is characterized using GXE experiments that investigate the interactions of genotype in a given environment [7, 101–104]. Developmental plasticity occurs commonly within plant populations when a given population inhabits moderate environments [2, 18]. Abiotic stresses, like drought, trigger physiological and developmental plasticity in plants [7]. The degree of developmental plasticity observed in plants resulting from abiotic stress is directly connected to a plant's development phase [7]. Some phases of development are more responsive to environmental changes and display a more plastic response than others [7]. It was found in spring wheat that the early developmental stages til-lering and heading (after spike formation) show more morphological and physiologi-cal plasticity than other developmental phases [7]. Cold tolerance in quinoa is also based on developmental plasticity, and associated with grain formation [105, 106]. Flowering time is another trait associated with developmental plasticity across plant species [6]. A shift in flowering time in response to drought allows for accelerated seed set, thus ensuring species survival, even in non-ideal growing conditions [6].

### 6.3 Cellular plasticity

Plant cellular plasticity allows cells to respond to the negative impacts of biotic and abiotic stresses. Cellular plasticity occurs through long-range signaling via hydraulic, electrical, and chemical signaling mechanisms [107]. One example of chemical signaling directly connected with plant cellular plasticity occurs when plants experience oxidative stress. Environmental stresses stimulate the production of toxic chemicals known as reactive oxygen species (ROS) [108]. The function of scavenging enzymes is to quench the flux of ROS [108–114]. When ROS levels are elevated due to environmental stress, the activity of scavenging enzymes, including ascorbate peroxidase, superoxide dismutase (SOD), and catalase (CAT) increases [108–114].

Cellular plasticity is a very dynamic process whereby ROS scavengers are acting simultaneously in different cellular compartments including the cell wall membrane, cytoplasm, chloroplast, mitochondria, peroxisomes, and the apoplast [115–117]. The peroxisomes are the most important indicators of environmental stress, ROS-scavenger activity, and cellular plasticity [115]. Peroxisomes proliferate in response to an array of environmental stresses including light, ozone, metal, and salt [119]. Peroxisome number may fluctuate depending on cultivar or genotype [118–122]. An emerging hypothesis about cellular plasticity is that relative peroxisome abundance may be a good predictor for cellular plasticity mechanisms [123, 124]. Peroxisomal proliferation occurs because of environmental stress, and, any change in a phenological trait occurring from a change in environment is defined as cellular plastic-ity [123, 124]. Investigations of peroxisome proliferation in response to drought tolerance demonstrate that peroxisome abundance is correlated with abiotic stress response and impacts GXE interactions [123, 124]. A negative correlation also exists between peroxisome abundance and several phenological traits including plant biomass, root dry weight, and grain yield [123, 124]. Therefore, peroxisome abundance is an emerging tool for measuring cellular plasticity mechanisms of adaptation, and ROS homeostasis [123, 124].

### 6.4 Epigenetic plasticity

Plasticity responses exist as both the inherent genetic machinery (past regulatory events), and as part of regulation occurring outside of the genetic code (epigenetically) [71]. Epigenetic mechanisms include DNA methylation, non-coding RNA, chromatin remodeling, and histone modifications [71]. Changes in the environment trigger heritable changes in gene expression which result in stable phenotypes [71]. DNA methylation is the most common, and perhaps best understood mechanisms controlling epigenetic plasticity in plants [71]. Studies using Arabidopsis epigenetic recombinant inbred lines (epiRILS), i.e., lines with nearly identical genomes but contrasting DNA methylation patterns, demonstrated that plasticity to water availability and nutrient loss is controlled through changes in DNA methylation [80]. Epigenetic changes rather than genetic changes contribute to changes impacting phe-notypic plasticity [71]. Other research has demonstrated that epigenetic regulation impacts heritability in specific phenological traits like plant height, plant biomass, seed/fruit production, the root-to-shoot ratio, and flowering time [71, 125–127]. Heritable traits are very important in breeding programs, and the role of epigenetics in regulating these traits is now only being characterized and understood [128].

## 7. The path to domestication: learning from transitional models

Decreased genetic diversity in plants populations is often associated with increased cultivation [7, 129]. Less cultivated plant populations tend to have more genetic diversity or "wildness" than plants that have been domesticated [129]. Wild characteristics broaden genetic responses and are valuable for maintaining phenotypic plasticity [129–131]. Leveraging broad genetic responses to enhance plasticity is especially important for the survival of plant species in unpredictable and changing climates [74, 75].

Since the dawn of agriculture, farmers have used selective breeding techniques for cultivating and domesticating wild plants for food [132]. Seeds from wild plant populations are smaller, an adaptation thought to enhance dispersal [132]. From an agricultural perspective, increased domestication is useful for reliable germination, uniform emergence, uniform stand establishment, larger seed size, increased yield, and improved nutrition [132]. Domestication of wild maize, soybean, and barley has resulted in significant increases in seed size [86]. However, there has also been a negative cost associated with domestication [86]. In maize, soybean, and rice, domestication and intensive cultivation have resulted in the elimination of genetic loci in modern crop cultivars [86, 133–136]. Breeding strategies that do not address adaptation and plasticity decrease trait diversity and may limit the development of new crop varieties with the ability to adapt to insects and extreme environmental fluctuations [133–136].

A reduction in heritability of favorable traits within breeding populations has been one of the main reasons plant breeders have explored the possibility of integrating genetic diversity from wild populations (landraces) back into selective breeding programs [133–137]. Two wild plant models that have been very instrumental in the effort to introduce diversity back into breeding population are: (1) barley (*Hordeum vulgare*); a standard model for monocots, and (2) quinoa (*Chenopodium quinoa* Wild); a model for dicots [137–145]. These two models are very powerful because they highlight a clear transition from wild populations to domesticated cultivars. They also provide tools for understanding plasticity by comparing characteristics that have remained constant, changed, or been lost through a history of domestication [137–145].

Barley was domesticated very early in history from the wild grass relative, *Hordeum spontaneum* [137]. Barley, along with einkorn (*T. monococcum*, genomes AmAm) and emmer (*T. turgidum* ssp. *dicoccoides*, genomes BBAA), marked the beginning of domestication in cereals [137]. Barley is often used as a model to improve crops like wheat (*Triticum aestivum*) [138]. Barley demonstrates a wide range of plasticity including superior growth in nutrient-limited environments, and adapted root architecture [139, 140]. Although there are evolutionary similarities between barley and other monocots like wheat, the orthologous genomic regions between the two species have a completely diverged [141–143]. However, genomic similarities between barley and wheat have enriched the comparative studies of plasticity and provide new information about horizontal gene transfer [141–143].

Quinoa, like barley, was recognized as a valuable food resource, and was domes-ticated very long ago [144]. Although quinoa has been highly domesticated, it retains vast genetic variability and plasticity with a wide range of resistance to many abiotic and biotic stresses [144, 145]. Quinoa thrives in extreme environmental con-ditions including in regions with high salinity soils, areas of extremely low precipi-tation, and environments with extremely cold temperatures [105, 146]. Moreover, quinoa grain is resistant to starch degradation in environments susceptible to extreme temperature and moisture fluctuations [147]. The differences in plasticity discovered between wild and domesticated quinoa species illustrate the importance of continued studies identifying physiological and genetic mechanisms regulating plasticity [147]. These discoveries also highlight the feasibility and importance of selectively breeding for gene targets that improve adaptability and fitness [133–136]. Additionally, because quinoa is a polyploid, it is a rich resource for studying how complex genomes contribute new dimensions of genetic regulation to phenological plasticity [147]. Recent studies investigating modern cultivated varieties of quinoa show that cellular plasticity mechanisms, and more specifically ROS homeostasis, are dependent on both genotype and type of stress [123]. The emerging discoveries in quinoa are important because they provide a model for how plasticity mecha-nisms present in other polyploid crop species may be regulated [123].

## 8. Discussion

The discovery and utilization of improved traits that enhance the adaptability of crops to increasingly variable environments will help to ensure long-term crop stability in changing climates [74, 128, 129]. Knowledge of phenological plasticity in wild populations will continue to benefit breeding programs [28]. Although wild genomes increase genetic complexity and may impact plasticity and fitness in unpre-dictable ways through changes in development, morphology, or physiology, one of the discovered benefits of increased diversity is increased adaptability [71, 129]. Over the last decade, advancements in genetics, molecular biology, systems biol-ogy, and statistical modeling have removed many of the barriers for understanding the regulation of complex plasticity networks in plants [13]. Association mapping, next generation sequencing, and genotype-by-phenotype (GWAS) approaches have greatly improved our comprehensive understanding of plasticity and the impacts of genomic selection [141–143]. Additionally, translational approaches utilizing a wealth of genomic information from both model plant systems and non-domesticated relatives have provided a framework for parallel studies in a wide range of plant populations. These studies have helped to uncover the developmental, cellular, and epigenetic mechanisms that regulate plasticity in all plants [6, 13, 71, 74, 142, 143].

## 9. Conclusions

One of the benefits of increasing genetic diversity in domesticated populations, from a long-term agricultural perspective, is the increased likelihood of plant population survival in unpredictable environments. In the past, evaluating the con-tributions of specific traits on phenological plasticity in plants was challenging due to experimental limitations and gaps in knowledge. However, emerging research continues to be extended from model systems directly to wild and cultivated plant populations to uncover the full potential of plasticity. New areas of research will need to investigate plasticity using a systems biology approach. Work should con-tinue to explore the degree of conservation of plasticity existing between monocots and dicot crops, as well as comparing the contributions of ploidy on diversity. Other areas of research should address how DNA methylation and epigenetic mechanisms contribute to plant plasticity and may be fully utilized in plant improvement programs. Additional work should focus on how the simultaneous deployment of multiple plasticity mechanisms during plant developmental shift in changing envi-ronments using newly identified plasticity markers like the peroxisomes. Continued plasticity research will be is critical for understanding how to maximize the benefits of both domestication and wild genetic diversity to maximize adaptation and fit-ness in a new area of climate diversity.

## Author details

Amber L. Hauvermale[1] and Marwa N.M.E. Sanad[2*]

1 Department of Crop and Soil Sciences, Washington State University, Pullman, WA, USA

2 Department of Genetics and Cytology, National Research Centre, Giza, Egypt

*Address all correspondence to: mn.sanad@nrc.sci.eg and marwa.sanad@wsu.edu

## References

[1] Gratani L. Plant phenotypic plasticity in response to environmental factors. Advances in Botany. 2014;**2014**:1-17

[2] Sultan SE. Commentary: The promise of ecological developmental biology. Journal of Experimental Zoological Biology and Molecular and Developmental Evolution. 2003; **296**:1-7

[3] West-Eberhard MJ. Developmental Plasticity and Evolution. New York: Oxford University Press; 2003. pp. 1-794

[4] Pigliucci M, Murren CJ, Schlichting CD. Phenotypic plasticity and evolution by genetic assimilation. The Journal of Experimental Biology. 2006;**209**:2362-2367

[5] Palmer CM, Bush SM, Maloof JN, et al. Phenotypic and Developmental Plasticity in Plants. Chichester, UK: John Wiley & Sons, Ltd; 2011

[6] De Jong M, Leyser O. Developmental plasticity in plants. Cold Spring Harbor Symposia on Quantitative Biology. 2012;**77**:63-73

[7] Sanad MNME, Campbell KG, Gill KS. Developmental program impacts phenological plasticity of spring wheat under drought. Botanical Studies. 2016;**57**:1-35

[8] Horton TH. Fetal origins of developmental plasticity: Animal models of induced life history variation. American Journal of Human Biology. 2005;**17**:34-43

[9] Minelli A, Fusco G. Developmental plasticity and the evolution of animal complex life cycles. Philosophical Transactions of the Royal Society, B: Biological Sciences. 2010;**365**:631-640

[10] Bateson P, Barker D, Clutton-Brock T, Deb D, D'Udine B, Foley RA, et al. Developmental plasticity and human health. Nature. 2004;**430**:419-421

[11] Nettle D, Bateson M. Adaptive developmental plasticity: What is it, how can we recognize it and when can it evolve? Proceedings of the Royal Society B: Biological Sciences. 2015;**282**:1005

[12] Funk JL. Differences in plasticity between invasive and native plants from a low resource environment. Journal of Ecology. 2008;**96**:1162-1173

[13] Nicotra AB, Davidson A. Adaptive phenotypic plasticity and plant water use. Functional Plant Biology. 2010;**37**:117-127

[14] Richter S, Kipfer T, Wohlgemuth T, Moser B. Phenotypic plasticity facilitates resistance to climate change in a highly variable environment. Oceologia. 2012;**1**(69):269-279

[15] Brachi B, Aimé C, Glorieux C, Cuguen J, Roux F. Adaptive value of phenological traits in stressful environments: Predictions based on seed production and laboratory natural selection. PLoS One. 2012;**7**:P32069

[16] Baker H. Characteristics and Modes of origin of weeds. In: Genetics of Colonizing Species. New York: Academic Press Inc.; 1965. pp. 147-172

[17] Marshall DR, Jain SK. Phenotypic plasticity of *Avena fatua* and *A. barbata*. The American Naturalist. 1968;**102**:457-467

[18] Sultan SE. Phenotypic plasticity for plant development, function and life history. Trends in Plant Science. 2000;**5**:537-542

[19] Callaway RM, Pennings SC, Richards CL. Phenotypic plasticity and interactions among plants. Ecology. 2003;**84**:1115-1128

[20] Daehler CC. Performance comparisons of co-occurring native and alien invasive plants: Implications for conservation and restoration. Annual Review of Ecology, Evolution, and Systematics. 2003;**34**:183-211

[21] Rejmánek M, Richardson DM, Pylek P. Plant invasions and invasibility of plant communities. In: Maarel E, Franklin J, editors. Vegetative Ecology. 2nd ed. Wiley; 2013. pp. P387-P424

[22] Grenier S, Barre P, Litrico I. Phenotypic plasticity and selection: Nonexclusive mechanisms of adaptation. Scientifica. 2016;**7021701**:1-9

[23] Fay JC, Wittkopp PJ. Evaluating the role of natural selection in the evolution of gene regulation. Heredity. 2008;**100**:191-199

[24] Anderson JT, Willis JH, Mitchell-Olds T. Evolutionary genetics of plant adaptation. Trends in Genetics. 2011;**27**:258-266

[25] Przybylo R, Sheldon BC, Merila J. Climatic effects on breeding and morphology: Evidence for phenotypic plasticity. The Journal of Animal Ecology. 2000;**69**:395-403

[26] Reed TE, Schindler DE, Waples RS. Interacting effects of phenotypic plasticity and evolution on population persistence in a changing climate. Conservation Biology. 2011;**25**:56-63

[27] Ghalambor CK, McKay JK, Carroll SP, Reznick DN. Adaptive versus non-adaptive phenotypic plasticity and the potential for contemporary adaptation in new environments. Functional Ecology. 2007;**21**:394-407

[28] Fritsche-Neto R, DoVale C. Breeding for stress-tolerance or resource-use efficiency? In: Fritsche-Neto R, Borém A, editors. Plant Breeding for Abiotic Stress Tolerance. Springer; 2012. pp. 13-19

[29] Waddington CH. Canalization of development and the inheritance of acquired characters. Nature. 1942;**150**:563-565

[30] Debat V, David P. Mapping phenotypes: Canalization, plasticity and developmental stability. Trends in Ecology & Evolution. 2001;**16**:555-561

[31] Van Gestel J, Weissing FJ. Is plasticity caused by single genes? Nature. 2018;**555**:19-20

[32] Belsky J, Jonassaint C, Pluess M, Stanton M, Brummett B, Williams R. Vulnerability genes or plasticity genes? Molecular Psychiatry. 2009;**14**:746-754

[33] Pigliucci M, Schmitt J. Genes affecting phenotypic plasticity in *Arabidopsis*: Pleiotropic effects and reproductive fitness of photomorphogenic mutants. Journal of Evolutionary Biology. 1999;**12**:551-562

[34] Des Marais DL, Juenger TE. Pleiotropy, plasticity, and the evolution of plant abiotic stress tolerance. Annals of the New York Academy of Sciences. 2010;**1206**:56-79

[35] Dewitt TJ, Sih A, Wilson DS. Costs and limits of phenotypic plasticity. Trends of Ecology & Evolution. 1998;**13**:77-81

[36] Sutherland S. What makes a weed a weed: Life history traits of native and exotic plants in the USA. Oecologia. 2004;**141**:24-39

[37] West-Eberhard MJ. Phenotypic plasticity and the origins of diversity. Annual Review of Ecology and Systematics. 1989;**20**:249-278

[38] Sultan SE. What has survived of Darwin's theory? Phenotypic plasticity and the neo-Darwinian legacy. Evolution of Trend Plant. 1992;**6**:61-71

[39] Sánchez-Gómez D, Valladares F, Zavala MA. Functional traits and plasticity in response to light in seedlings of four Iberian forest tree species. Tree Physiology. 2006;**26**:1425-1433

[40] Borges RM. Plasticity comparisons between plants and animals: Concepts and mechanisms. Plant Signaling & Behavior. 2008;**3**:367-375

[41] Chapin FS III. The mineral nutrition of wild plants. Annual Review of Ecology and Systematics. 1980;**11**:233-260

[42] Lambers H, Poorter H. Inherent variation in growth rate between higher plants: A search for physiological causes and ecological consequences. Advances in Ecological Research. 1992;**23**:187-261

[43] Aerts R, Peijl M. A simple model to explain the dominance of low-productive perennials in nutrient-poor habitats. Oikos. 1993;**66**:144-147

[44] Chapin FS III, Autumn K, Pugnaire F. Evolution of suites of traits in response to environmental stress. The American Naturalist. 1993;**142**:78-92

[45] Valladares F, Martinez-Ferri E, Balaguer L, Perez-Corona E, Manrique E. Low leaf level response to light and nutrients in mediterranean evergreen oaks: A conservative resource-use strategy? The New Phytologist. 2000;**148**:79-91

[46] Pearson T, Burslem D, Goeriz R, Dalling J. Regeneration niche partitioning in neotropical pioneers: Effects of gap size, seasonal drought and herbivory on growth and survival. Oecologia. 2003;**137**:456-465

[47] Sapkota TB, Askegaard M, Laegdsmand M, Olesen JE. Effects of catch crop type and root depth on nitrogen leaching and yield of spring barley. Field Crops Research. 2012;**125**:129-138

[48] Alvarez-Flores R, Nguyen-Thi-Truc A, Peredo-Parada S, Joffre R, Winkel T. Rooting plasticity in wild and cultivated Andean Chenopodium species. Plant and Soil. 2018;**425**:479-492

[49] Finch-Savage WE, Cadman CS, Troorop PE, Lynn JE, Hilhorst HW. Seed dormancy release in *Arabidopsis Cvi* by dry after-ripening, low temperature, nitrate and light shows common quantitative patterns of gene expression directed by environmentally specific sensing. The Plant Journal. 2017;**51**:60-78

[50] Finch-Savage WE, Leubner-Metzger G. Seed dormancy and the control of germination. The New Phytologist. 2006;**171**:501-523

[51] Simpson GM, editor. Seed Dormancy in Grasses. New York Cambridge University Press; 2007

[52] Miller G, Suzuki N, Ciftci-Yilmaz S, Mittler R. Reactive oxygen species homeostasis and signaling during drought and salinity stresses. Plant, Cell & Environment. 2010;**33**:453-467

[53] Sharma P, Jha AB, Dubey RS, Pessarakli M. Reactive oxygen species, oxidative damage, and antioxidative defense mechanism in plants under stressful conditions. Journal of Botany. 2012;**26**:1-27

[54] Vitasse Y, Bresson CC, Kremer A, Michalet R, Delzon S. Quantifying phenological plasticity to temperature in two temperate tree species. Functional Ecology. 2010;**24**:1211-1218

[55] Williams DG, Mack RN, Black RA. Ecophysiology of introduced *Pennisetum setaceum* on Hawaii: The role of phenotypic plasticity. Ecology. 1995;**76**:1569-1580

[56] Yeh PJ, Price TD. Adaptive phenotypic plasticity and the successful colonization of a novel environment. The American Naturalist. 2004;**164**:531-542

[57] Atkin OK, Loveys BR, Atkinson LJ, Pons TL. Phenotypic plasticity and growth temperature: Understanding interspecific variability. Journal of Experimnetal of Botany. 2006;**57**:267-281

[58] Aroca R, editor. Plant Responses to Drought Stress: From Morphological to Molecular Features. Verlag Berlin Heidelberg: Springer; 2012. 413 p

[59] DeLucia E, Maherali H, Carey E. Climate-driven changes in biomass allocation in pines. Global Change Biology. 2000;**6**:587-593

[60] Markesteijn L, Poorter L. Seedling root morphology and biomass allocation of 62 tropical tree species in relation to drought-and shade-tolerance. Journal of Ecology. 2009;**97**:311-325

[61] Baker HG. Weeds-native and introduced. Journal of California Horticulture Society. 1962;**23**:97-104

[62] Pyšek P, Richardson DM, editors. Traits associated with invasiveness in alien plants: Where do we stand? In: Biological Invasions. Berlin Heidelberg: Springer Verlag; 2008. pp. 97-125

[63] Goodwin BJ, Allister AJMC, Fahrig L. Predicting invasiveness of plant species based on biological information. Conservation Biology. 1999;**13**:422-426

[64] Pyšek P, Brock JH, Bímová K, Mandák B, Jarošík V, Koukolíková I, et al. Vegetative regeneration in invasive *Reynoutria* (Polygonaceae) taxa: The determinant of invisibility at the genotype level. American Journal of Botany. 2003;**90**:1487-1495

[65] Mal TK, Lovett-Doust J. Phenotypic plasticity in vegetative and reproductive traits in an invasive weed, *Lythrum salicaria* (Lythraceae), in response to soil moisture. American Journal of Botany. 2005;**92**:819-825

[66] Sultan SE, Matesanz S. An ideal weed: Plasticity and invasiveness in *Polygonum cespitosum*. Annals of the New York Academy of Sciences. 2015;**1360**:101-119

[67] Van Hinsberg A. Morphological variation in *Plantago lanceolata* L.: Effects of light quality and growth regulators on sun and shade populations. Journal of Evolutionary Biology. 1997;**10**:687-701

[68] Via S. Adaptive phenotypic plasticity: Target or by-product of selection in a variable environment? The American Naturalist. 1993;**1**(42):352-365

[69] Schlichting CD. The evolution of phenotypic plasticity in plants. Annual Review of Ecology and Systematics. 1986;**17**:667-693

[70] Schlichting CD, Pigliucci M. Gene regulation, quantitative genetics and the evolution of reaction norms. Evolutionary Ecology. 1995;**9**:154-168

[71] Zhang L, Lu X, Lu J, Liang H, Dai Q, Xu G-L, et al. Thymine DNA glycosylase specifically recognizes 5-carboxylcytosine-modified DNA. Nature Chemical Biology. 2012;**8**:328-330

[72] Kuwabara A, Nagata T. Cellular basis of developmental plasticity observed in heterophyllous leaf formation of *Ludwigia arcuata* (Onagraceae). Planta. 2006;**224**:761-770

[73] Kimball S, Gremer JR, Angert AL, Huxman TE, Venable DL. Fitness and physiology in a variable environment. Oecologia. 2012;**169**:319-329

[74] Becklin KM, Anderson JT, Gerhart LM, Wadgymar SM,

Wessinger CA, Ward JK. Examining plant physiological responses to climate change through an evolutionary lens. Plant Physiology. 2016;**172**:635-649

[75] Mitchell-Olds T, Willis JH, Goldstein DB. Which evolutionary processes influence natural genetic variation for phenotypic traits? Nature Reviews Genetics. 2007;**8**:845-856

[76] Bewley JD, Bradford KJ, Hilhorst HWM. In: Nonogaki H, editor. Seeds: Physiology of Development, Germination and Dormancy. Springer; 2013. pp. 1-15

[77] Rydrych D, Muzik T. Downy brome competition and control in dryland wheat. Agronomy Journal. 1968;**60**:279-280

[78] Young JA, Evans RA, Eckert RE. Population dynamics of downy brome. Weed Science. 1969;**17**:20-26

[79] Stahlman P, Miller S. Downy brome (*Bromus tectorum*) interference and economic thresholds in winter wheat (*Triticum aestivum*). Weed Science. 1990;**38**:224-228

[80] Blackshaw R. Downy brome (*Bromus tectorum*) control in winter wheat and winter rye. Canadian Journal of Plant Science. 1994;**74**:185-191

[81] Koornneef M, Jorna ML, der Brinkhorst-van Swan DLC, Karssen CM. The isolation of abscisic acid (ABA) deficient mutants by selection of induced revertants in non-germinating gibberellin sensitive lines of *Arabidopsis thaliana* (L.) heynh. Theoretical and Applied Genetics. 1982;**61**:385-393

[82] Finkelstein RR, Reeves W, Ariizumi T, Steber CM. Molecular aspects of seed dormancy. Annual Review of Plant Biology. 2008;**59**:387-415

[83] Karssen CM, Laçka E. A revision of the hormone balance theory of seed dormancy: Studies on gibberellin and/or abscisic acid-deficient mutants of *Arabidopsis thaliana*. Plant Growth Substances. Verlag Berlin Heidelberg: Springer; 1986. pp. 315-323

[84] Walker-Simmons M. ABA levels and sensitivity in developing wheat embryos of sprouting resistant and susceptible cultivars. Plant Physiology. 1987;**84**:61-66

[85] Morris CF, Moffatt JM, Sears RG, Paulsen GM. Seed dormancy and responses of caryopses, embryos, and calli to abscisic acid in wheat. Plant Physiology. 1989;**90**:643-647

[86] Taylor IB, Burbidge A, Thompson AJ. Control of abscisic acid synthesis. Journal of Experimental Botany. 2000;**51**:1563-1574

[87] Kushiro T, Okamoto M, Nakabayashi K, Yamagishi K, Kitamura S, Asami T, et al. The *Arabidopsis* cytochrome P450 CYP707A encodes ABA 8′-hydroxylases: Key enzymes in ABA catabolism. The EMBO Journal. 2004;**23**:1647-1656

[88] Chono M, Honda I, Shinoda S, Kushiro T, Kamiya Y, Nambara E, et al. Field studies on the regulation of abscisic acid content and germinability during grain development of barley: Molecular and chemical analysis of pre-harvest sprouting. Journal of Experimental Botany. 2006;**57**:2421-2434

[89] Millar AA, Jacobsen JV, Ross JJ, Helliwell CA, Poole AT, Scofield G, et al. Seed dormancy and ABA metabolism in arabidopsis and barley: The role of ABA 8′-hydroxylase. The Plant Journal. 2006;**45**:942-954

[90] Okamoto M, Kuwahara A, Seo M, Kushiro T, Asami T, Hirai N, et al. CYP707A1 and CYP707A2, which

encode abscisic acid 8′-hydroxylases, are indispensable for proper control of seed dormancy and germination in *Arabidopsis*. Plant Physiology. 2006;**141**:97-107

[91] Barrero JM, Talbot MJ, White RG, Jacobsen JV, Gubler F. Anatomical and transcriptomic studies of the coleorhiza reveal the importance of this tissue in regulating dormancy in barley. Plant Physiology. 2009;**150**:1006-1021

[92] Schramm EC, Abellera JC, Strader LC, Campbell KG, Steber CM. Isolation of ABA-responsive mutants in allohexaploid bread wheat (*Triticum aestivum* L.): Drawing connections to grain dormancy, preharvest sprouting, and drought tolerance. Plant Science. 2010;**179**:620-629

[93] Barrero JM, Jacobsen JV, Talbot MJ, White RG, Swain SM, Garvin DF, et al. Grain dormancy and light quality effects on germination in the model grass *Brachypodium distachyon*. New Phytologist. 2012;**193**:376-386

[94] Ariizumi T, Hauvermale AL, Nelson SK, Hanada A, Yamaguchi S, Steber CM. Lifting DELLA repression of *Arabidopsis* seed germination by non-proteolytic gibberellin signaling. Plant Physiology. 2013;**162**:2125-2139

[95] Chono M, Matsunaka H, Seki M, Fujita M, Kiribuchi-Otobe C, Oda S, et al. Isolation of a wheat (*Triticum aestivum* L.) mutant in ABA 8′-hydroxylase gene: Effect of reduced ABA catabolism on germination inhibition under field condition. Breeding Science. 2013;**63**:104-115

[96] Schramm EC, Nelson SK, Kidwell KK, Steber CM. Increased ABA sensitivity results in higher seed dormancy in soft white spring wheat cultivar 'Zak'. Theoretical and Applied Genetics. 2013;**126**:791-803

[97] Barrero JM, Downie AB, Xu Q , Gubler F. A role for barley CRYPTOCHROME1 in light regulation of grain dormancy and germination. The Plant Cell. 2014;**26**:1094-1104

[98] Hauvermale AL, Tuttle KM, Takebayashi Y, Seo M, Steber CM. Loss of *Arabidopsis thaliana* seed dormancy is associated with increased accumulation of the GID1 GA hormone receptors. Plant & Cell Physiology. 2015;**56**:1773-1785

[99] Tuttle KM, Martinez SA, Schramm EC, Takebayashi Y, Seo M, Steber CM. Grain dormancy loss is associated with changes in ABA and GA sensitivity and hormone accumulation in bread wheat, *Triticum aestivum* (L.). Seed Science Research. 2015;**25**:179-193

[100] Lawrence NC, Hauvermale AL, Dhingra A, Burke IC. Population structure and genetic diversity of *Bromus tectorum* within the small grain production region of the Pacific Northwest. Ecology and Evolution. 2017;**7**:8316-8328

[101] Chaves MM, Maroco JP, Pereira JS. Understanding plant responses to drought from genes to the whole plant. Functional Plant Biology. 2003;**30**:239-264

[102] Rizza F, Badeck FW, Cattivelli L, Lidestri O, Di Fonzo N, Stanca AM. Use of a water stress index to identify barley genotypes adapted to rainfed and irrigated conditions. Crop Science. 2004;**44**:2127-2137

[103] De Leonardis AM, Marone D, Mazzucotelli E, Neffar F, Rizza F, Di Fonzo N, et al. Durum wheat genes up-regulated in the early phases of cold stress are modulated by drought in a developmental and genotype dependent manner. Plant Science. 2007;**172**:1005-1016

[104] Milad SI, Wahba LE, Barakat MN. Identification of RAPD and ISSR markers associated with flag leaf senescence under water-stressed conditions in wheat (*Triticum aestivum* L.). Australian Journal of Crop Science. 2011;**5**:334-340

[105] Espindola G. Respuestas fisiológicas, morfológicas y agronómicas de la quinoa al déficit hídrico [thesis]. Chapingo, México: These de maitrise, Colegio de Postgraduados Institución de Enseñanza e Investigaciér': Ciencias Agrícolas; 1986

[106] Rea J, Tapia M, Mujica A. Practicas agronomicas. In: Tapia M, Gandarillas H, Alandia S, Cardozo A, Mujica A, editors. Quinua y Kaiiiwa. Cultivos Andinos. Rome, Italy: FAO; 1997

[107] Huber AE, Bauerle TL. Long-distance plant signaling pathways in response to multiple stressors: The gap in knowledge. Journal of Experimental Botany. 2016;**67**:2063-2079

[108] Gamble PE, Burke JJ. Effect of water stress on the chloroplast antioxidant system I. Alterations in glutathione reductase activity. Plant Physiology. 1984;**76**:615-621

[109] Smirnoff N. The role of active oxygen in the response of plants to water deficit and desiccation. New Phytologist. 1993;**125**:27-58

[110] Noctor G, Foyer CH. Ascorbate and glutathione: Keeping active under control. Annual Review of Plant Physiology and Plant Molecular Biology. 1998;**49**:249-279

[111] Rubio MC, González EM, Minchin FR, Webb KJ, Arrese-Igor C, Ramos J, et al. Effects of water stress on antioxidant enzymes of leaves and nodules of transgenic alfalfa overexpressing superoxide dismutases. Physiologia Plantarum. 2002;**115**:531-540

[112] Jiang M, Zhang J. Water stress-induced abscisic acid accumulation triggers the increased generation of reactive oxygen species and up-regulates the activities of antioxidant enzymes in maize leaves. Journal of Experimental Botany. 2002;**53**:2401-2410

[113] Guo Z, Ou W, Lu S, Zhong Q. Differential responses of antioxidative system to chilling and drought in four rice cultivars differing in sensitivity. Plant Physiology and Biochemistry. 2006;**44**:828-836

[114] Møller IM, Jensen PE, Hansson A. Oxidative modifications to cellular components in plants. Annual Review of Plant Biology. 2007;**58**:459-481

[115] Foyer CH, Noctor G. Redox sensing and signaling associated with reactive oxygen in chloroplasts, peroxisomes and mitochondria. Physiologia Plantarum. 2003;**119**:355-364

[116] Apel K, Hirt H. Reactive oxygen species: Metabolism, oxidative stress, and signal transduction. Annual Review of Plant Biology. 2004;**55**:373-399

[117] Nyathi Y, Baker A. Plant peroxisomes as a source of signaling molecules. Biochimica et Biophysica Acta (BBA)—Molecular Cell Research. 2006;**1763**:1478-1495

[118] Ferreira RMB, Bird B, Davies DD. The effect of light on the structure and organization of lemna peroxisomes. Journal of Experimental Botany. 1989;**40**:1029-1035

[119] Morre DJ, Sellden G, Ojanpera K, Sandelius AS, Egger A, Morre DM, et al. Peroxisome proliferation in Norway spruce induced by ozone. Protoplasma. 1990;**155**:58-65

[120] Romero-Puertas MC, McCarthy I, Sandalio LM, Palma JM, Corpas FJ, Gómez M, et al. Cadmium toxicity and oxidative metabolism of pea leaf

peroxisomes. Free Radical Research. 1999;**31**:25-31

[121] Oksanen E, Häikiö E, Sober J, Karnosky DF. Ozone-induced $H_2O_2$ accumulation in field-grown aspen and birch is linked to foliar ultrastructure and peroxisomal activity. New Phytologist. 2004;**161**:791-799

[122] Mitsuya S, El-Shami M, Sparkes IA, Charlton WL, Lousa CDM, Johnson B, et al. Salt stress causes peroxisome proliferation, but inducing peroxisome proliferation does not improve NaCl tolerance in *Arabidopsis thaliana*. PLoS One. 2010;**5**:9408

[123] Fahy D, Sanad MNME, Duscha K, et al. Impact of salt stress, cell death, and autophagy on peroxisomes: Quantitative and morphological analyses using small fluorescent probe N-BODIPY. Scientific Reports. 2017;**7, 39069**

[124] Marwa NM, Sanad E, Andrei S, Kimberley A. Garland-Campbell. Differential dynamic changes of reduced trait model for analyzing the plastic response to drought: A case study in spring wheat. Frontiers in Plant Science. 2019. DOI: 10.3389/fpls.2019.00504

[125] Lynch M, Walsh B. Genetics and analysis of quantitative traits. Oxford University Press; 1998. pp. 1-980

[126] Johannes F, Porcher E, Teixeira FK, et al. Assessing the impact of transgenerational epigenetic variation on complex traits. PLoS Genetics. 2009;**5**:e1000530

[127] Roux F, Colomé-Tatché M, Edelist C, Wardenaar R, Guerche P, Hospital F, et al. Genome-wide epigenetic perturbation jump-starts patterns of heritable variation found in nature. Genetics. 2011;**188**:1015-1017

[128] Gallusci P, Dai Z, Génard M, Gauffretau A, Leblanc-Fournier N, Richard-Molard C, et al. Epigenetics for plant improvement: Current knowledge and modeling avenues. Trends in Plant Science. 2017;**22**:610-623

[129] Zhang H, Mittal N, Leamy LJ, Barazani O, Song B-H. Back into the wild; applying untapped genetic diversity of wild relatives for crop improvement. Evolutionary Applications. 2017;**10**:5-24

[130] Pigliucci M, Kolodynska A. Phenotypic plasticity to light intensity in *Arabidopsis thaliana*: Invariance of reaction norms and phenotypic integration. Evolutionary Ecology. 2002;**16**:27-47

[131] Bossdorf O, Pigliucci M. Plasticity to wind is modular and genetically variable in *Arabidopsis thaliana*. Evolutionary Ecology. 2009;**23**:669-685

[132] Osborne C. The conversation, an academic rigour, journalistic flair [Internet]. 2017. Did the first farmers deliberately domesticate wild plants? Available from: http://theconversation.com/did-the-first-farmers-deliberately-domesticate-wild-plants-77434

[133] Wright SI, Bi IV, Schroeder SG, Yamasaki M, Doebley JF, MD MM, et al. The effects of artificial selection on the maize genome. Science. 2005;**308**:1310-1314

[134] Hyten DL, Song Q , Zhu Y, Choi I-Y, Nelson RL, Costa JM, et al. Impacts of genetic bottlenecks on soybean genome diversity. Proceedings of the National Academy of Sciences. 2003;**103**:16666-16671

[135] Xu X, Liu X, Ge S, et al. Resequencing 50 accessions of cultivated and wild rice yields markers for identifying agronomically

important genes. Nature Biotechnology. 2012;**30**:105-111

[136] Zhou Z, Jiang Y, Wang Z, et al. Resequencing 302 wild and cultivated accessions identifies genes related to domestication and improvement in soybean. Nature Biotechnology. 2015;**33**:408-414

[137] Harlan JR, Zohary D. Distribution of wild wheats and barley. Science. 1996;**153**:1074-1080

[138] Kartha KK, Nehra NS, Chibbar RN. Genetic engineering of wheat and barley. In: Robert J. Henry, John A. Ronalds, editors. Improvement of Cereal Quality by Genetic Engineering. New York: Springer; 1994:21-30

[139] Elberse IAM, van Damme JMM, van Tienderen PH. Plasticity of growth characteristics in wild barley (*Hordeum spontaneum*) in response to nutrient limitation. Journal of Ecology. 2003;**91**:371-382

[140] Bingham IJ, Bengough AG. Morphological plasticity of wheat and barley roots in response to spatial variation in soil strength. Plant and Soil. 2003;**250**(2):73-282

[141] Ramakrishna W, Dubcovsky J, Park Y-J, Busso C, Emberton J, Sanmiguel P, et al. Different types and rates of genome evolution detected by comparative sequence analysis of orthologous segments from four cereal genomes. Genetics. 2002;**162**:1389-1400

[142] SanMiguel PJ, Ramakrishna W, Bennetzen JL, Busso CS, Dubcovsky J. Transposable elements, genes and recombination in a 215-kb contig from wheat chromosome 5Am. Functional and Integrative Genomics. 2002;**2**:70-80

[143] Dubcovsky J, Dvorak J. Genome plasticity a key factor in the success of polyploid wheat under domestication. Science. 2007;**29**:316-393

[144] Del Castillo C, Winkel T, Mahy G, Bizoux J-P. Genetic structure of quinoa (*Chenopodium quinoa* Willd.) from the Bolivian altiplano as revealed by RAPD markers. Genetic Resources and Crop Evolution. 2007;**54**:897-905

[145] Rojas W, Mamani E, Pinto M, Alanoca C, and Ortuño T. Identificación taxonómica de parientes silvestres de quinua del Banco de Germoplasma de Granos Altoandinos. En Revista de Agricultura. Revista de Agricultura-Año 60, Nro. 44. Cochabamba, Bolivia. 2008. pp. 56-65

[146] Jacobsen SE. The worldwide potential for quinoa (*Chenopodium quinoa* Willd.). Food Reviews International. 2003;**19**:167-177

[147] Ahamed NT, Singhal RS, Kulkarni PR, Pal M. A lesser-known grain, *Chenopodium Quinoa*: Review of the chemical composition of its edible parts. Food and Nutrition Bulletin. 1998;**19**:61-70

# 2

# Aphid-Plant Interactions: Implications for Pest Management

*Sarwan Kumar*

**Abstract**

Aphids are important herbivores and important pest of many field and forest crops. They have specialized long and flexible stylets which are adapted to feeding on phloem sap. To establish successful feeding on host plant, they need to counter a range of both physical and chemical defenses. The defenses employed by plants can have direct effect on the aphid species through difficulty in establishing successful feeding due to the presence of trichomes, thick cell wall, etc. or effect on their biology with lethal consequences in extreme cases (direct defenses). In contrast to this, plants can attract natural enemies of aphids through the release of volatile compounds (the so-called "cry or call for help") (indirect defense). The information on different defense strategies employed by plants can be utilized to enhance the level of resistance (R) to develop sustainable pest management strategies.

**Keywords:** Aphidoidea, insect-plant interactions, phloem feeding, plant defense, sieve elements

## 1. Introduction

Aphids constitute a major group of crop pests that limit productivity of many crops and cause serious damage to plants both by direct feeding and indirectly as vectors of many diseases. Despite being a relatively small insect group (about 5000 known species) compared to 10,000 species of grasshoppers, 12,000 species of geometrid moths, and 60,000 species of weevils, aphids are a serious problem for agriculture [1–3]. Of the 5000 known species in family Aphididae, 450 are endemic on crop plants, and 100 have successfully exploited the agricultural environment to the extent that they are of significant economic importance [3]. They are the specialized phloem sap feeders resulting in significant yield losses in many crops. It is their ability to rapidly exploit the ephemeral habitats that makes them seri-ous pests, and this ability results from (i) their high reproductive potential, (ii) their dispersal capacities, and (iii) their adaptability to local survival [2]. Unlike majority of insects, aphids exhibit parthenogenetic viviparity—phenomenon that limits the need for males to fertilize females and eliminates egg stage from their life cycle. Thus, aphids reproduce clonally and give birth to young ones, and embryonic development of an aphid begins before its mother's birth leading to telescoping of generations. All these traits allow aphids to exploit the periods of rapid plant growth, conserve energy, and allow for short generation times; nymphs of certain aphid species can reach maturity in as little as 5 days [4].

The well-known parthenogenesis exhibited by aphids sets them apart from other Hemiptera and has a great influence on their biology. In addition to parthenogenesis, many species of aphids also exhibit alternation of generations. The system of alternating one bisexual generation with a succession of parthenogenetic, all-female generation evolved as far back as the Triassic [3] which was later coupled with evolution of viviparity. All these led to reduction in their development period allowing them to multiply at a faster rate. Further, to conserve energy and to invest it in maximizing their reproduction and survival, aphid colonies exhibit wing dimorphism to produce highly fecund wingless morphs or less prolific winged progeny that can disperse to new host plant.

## 2. Aphid biology and behavior

Aphids are specialized phloem sap feeders and chemists *par excellence*. In most of cases, they exhibit passive feeding by high pressure within the sieve elements (SEs) and feed on virtually all plant families. While most of the species are specialists on a single host plant, some of them are generalists with relatively broad host range [5]. The aphid life cycles involve sexual and asexual morphs, and most of the species have relatively complicated life cycles with morphs that specialize in reproduction, dispersal, and survival under adverse conditions. Based on host utilization, aphids have two different types of life cycle: heteroecious or host alternating and monoecious/autoecious or nonhost alternating. Heteroecious species live on one plant species (primary host) in winter and migrate to another taxonomically unrelated plant species (secondary host) in summer and again migrate to primary host in autumn. While oviparity is exhibited on the primary host, on the secondary host, they reproduce parthenogenetically. These changes in sexual fate and reproductive mode are condition dependent and explain the extraordinary plasticity in development in response to environmental cues. Aphid species that interrupt parthenogenetic reproduction with sexual reproduction are termed as holocyclic. In contrast to host-alternating aphids, nonhost-alternating aphids remain either on the same or closely related host species throughout the year. They complete both sexual life cycle as well as parthenogenetic life cycle on the same host species. In contrast to this, there are species which do not produce eggs and are known as anholocyclic. Some species, particularly those having cosmopolitan distribution, exhibit both holocyclic and anholocyclic life, both at the same time in different geographical areas [6] but rarely both monoecy and heteroecy [7]. The presence of both biparental sexual and asexual life cycle ensures that aphids take advantage of both genetic recombination that help them to evolve and parthenogenesis (very convenient to exploit short-lived hosts).

## 3. Aphid mouthparts

The beak-like modification of mouthparts (labium, labrum, maxillae, and mandibles) is a distinct character of members of order Hemiptera. Generally the labium (and rarely labrum) is modified into rostrum, into the groove of which needlelike mandibular and maxillary stylets rest when not in use [8]. These needle-like mouthparts enable insects to penetrate the plant tissue and feed on the plant sap. Mandibles constitute the outer stylets and are important in physical penetration of cell walls, while maxillae form the inner ones [9] and form major role in selection of host plant [10]. Since the stylets can penetrate the individual cells due to their microstructure, this enables the aphids to puncture the symplast without wounding. This behavior is important for phloem-feeding insects which helps them to inoculate viruses into vascular and nonvascular plant cells. Recently, Uzest et al. [11] reported the existence of distinct anatomical structure called "acrostyle" on the tips of maxillary stylets of aphids which is an expanded part of cuticle visible in the common duct of all aphid species.

The presence of four- or five-segmented rostrum (labium) is the characteristic of the family Aphididae [12], and five-segmented labium does not occur in the other groups of Hemiptera. The four-segmented labium has been confirmed in members of Aphidinae, e.g., *Aphis fabae* [13], *Myzus persicae* [14], and *Schizaphis graminum* [15], and the five-segmented labium is confirmed only in Lachninae, e.g.,*Lachnus roboris* (L.), which has resulted from the secondary division of the apical segment [16]. However, Razaq et al. [17] observed another modification with only three-segmented labium in *Aphis citricola* van der Goot (Aphidinae). Labium exhibits variation in length, and in most of the species, it reaches the coxa of the third pair of legs. However, it can be exceptionally long (as long as the body) in spe-cies that feed on the trunk, branches, and roots of trees as in members of families Lachninae and Eriosomatinae.

## 4. Compatible aphid-plant interactions

Aphids are specialized phloem sap feeders which insert their needle like stylets in the plant tissue avoiding/counteracting the different plant defenses and withdrawing large quantities of phloem sap while keeping the phloem cells alive. In contrast to the insects with biting and chewing mouthparts which tear the host tissues, aphids penetrate their stylets between epidermal and parenchymal cells to finally reach sieve tubes with slight physical damage to the plants, which is hardly perceived by the host plant [6]. The long and flexible stylets mainly move intercellular in the cell wall apoplasm [18], although stylets also make intracellular punctures to probe the internal chemistry of a cell. The high pressure within sieve tubes helps in passive feeding [6]. During the stylet penetration and feeding, aphids produce two types of saliva. The first type is dense and proteinaceous (including phenol oxidases, peroxidases, pectinases, β-glucosidases) that forms an intercellular-tunneled path around the stylet in the form of sheath [19]. In addition to proteins, this gelling saliva also contains phospholipids and conjugated carbohydrates [20–22]. This stylet sheath forms a physical barrier and protects the feeding site from plant's immune response. When the stylet comes in contact with active flow of phloem sap, the feeding aphid releases digestive enzymes in the vascular tissue in the form of second type of "watery" saliva. The injection of watery saliva (E1) prevents the coagula-tion of proteins in plant sieve tubes, and during feeding the watery (E2) saliva gets mixed with the ingested sap which prevents clogging of proteins inside the capillary food canal in the insect stylets [6]. Though the actual biochemical mode of action of inhibition of protein coagulation is unknown, the calcium-binding proteins of aphid saliva are reported to interact with the calcium of plant tissues resulting in suppression of calcium-dependent occlusion of sieve tubes and subsequent delayed plant response [23, 24]. This mechanism of feeding is more specialized and precise which avoids different allelochemicals and indigestible compounds abundant in other plant tissues [25]. In addition to this, aphid saliva also contains nonenzymatic-reducing compounds which in the presence of oxidizing enzymes inactivate different defense-related compounds produced by plants after insect attack [21].

The early response of plants to feeding by insects or infection by patho-gens shares some common events such as protein phosphorylation, mem-brane depolarization, calcium influx, and release of reactive oxygen species (ROS, such as hydrogen peroxide) [26], which leads to the activation of phytohormone-dependent pathways. In response to infestation/infection, different phytohormone-dependent pathways are activated. The ethylene (ET) and jasmonate (JA) pathways are activated by different necrotrophic pathogens [27] and grazing insects [28], while salicylate (SA)-dependent responses are activated by biotrophic pathogens [27]. These responses lead to the production of various defense-related proteins and secondary metabolites with antixenotic or antibiotic properties. In the case of infestation by aphids, a SA-dependent response appears to be activated, while the expression of JA-dependent genes is repressed [29–32]. All these responses lead to the manipulation of the plant metabolism to ensure compatible aphid-plant interactions.

## 5. Aphid endosymbionts

The plant phloem sap is a highly unbalanced diet composed principally of sugars and amino acids with high C:N content. To cope with excess of sugars in their diet, aphids have evolved modification in their intestinal tract and filter out excess of sugars and water in the form of honeydew [33]. The most of amino acids are present at very low concentrations. Despite their nutritionally poor diet, aphids exhibit high growth and reproduction rates. Since aphids directly feed on the sugars and amino acids, they need not spend extra energy to digest complex nutrients such as proteins which remarkably increases their assimilation efficiency. In addition to this, the essential amino acids required by their growth and development are synthesized by symbiotic bacteria present in their body. Generally two types of symbiotic bacteria are known to be present in aphids: the primary (obligate) symbionts and second-ary (facultative) symbionts. *Buchnera aphidicola* (γ3-proteobacteria: *Escherichia coli* is also a member of this group) is the most common vertically transmitted primary symbiont present in most aphid species [34]. Some species of aphids also bear other bacteria, i.e., "secondary symbionts." These include several species of γ-proteobacteria such as *Serratia symbiotica*, *Regiella insecticola*, and *Hamiltonella defensa* [35–43]. *B. aphidicola* is a coccoid hosted in the cytoplasm of specialized cells called mycetocytes/bacteriocytes in the hemocoel of insect. These endosym-bionts upgrade the aphid diet by converting nonessential amino acids to essential amino acids. The evolution of symbiotic relationship with endosymbionts has enabled aphids to exploit new ecological niches, i.e., to feed on the plant phloem sap which is otherwise the nutritionally poor diet.

## 6. Response of aphids to plant characters

The decision for suitability of the plant as a host is made in the very first phase of the host selection. *Alate* aphids use both visual [44] and chemical cues [45] to decide landing on a plant. Upon landing aphids encounter trichomes as the first line of defense. Trichomes can be either glandular or nonglandular. Regardless of their structure, trichome density has significant influence on aphid feeding [46]. Many crop wild relatives (CWRs) of cultivated plants and resistant varieties are resistant to aphid attack due to the presence of trichomes that affect aphid movement and stylet insertion [47]. For example, the presence of high density of trichomes (both simple and glandular) in wild tomato, *Lycopersicon pennellii* (Corr.) D'Arcy, imparts high level of resistance (R) to aphid attack. In addition, the glandular trichomes produce toxic exudates that trap aphids and kill them.

In addition to trichomes, plants possess other constitutive defenses such as thorns and thick cell walls that provide direct resistance to plants against aphid feeding. Though these mechanical barriers are constitutive defenses, they can also be produced in response to aphid feeding (directly induced defenses).

In addition to these structural defenses, constitutive defenses can also be chemical. For example, glandular trichomes of *Solanum berthaultii* Hawkes produce (E)-farnesene—aphid alarm pheromone that triggers aphid dispersal and prevents colonization [48]. Such antixenotic defenses are of great significance and particularly effective against aphid species that act as vectors of plant pathogenic viruses. However, successful virus transmission can occur even on nonhost plants as stylet insertion is sufficient for some successful infection by quickly acquired viruses. Aphid salivation occurs on even resistant plants even if they do not feed on such resistant plants [23].

The depth of the sieve elements is an important factor determining successful feeding. The length of the aphid stylets must be compatible with the depth of sieve elements. In addition, thickness at the tip of stylets is also crucial for successful feeding [49]. The movement of stylets through plant tissue is mostly intercel-lular, and aphids probe all the cells that they encounter during probing. Sensorial structures located at the back of the mouth characterize the plant sap, and aphids recognize the substrate as host or nonhost. On nonhost plants, aphids retract the stylets and leaves in search of suitable host unless the plant produces toxins [50]. Many plant species possess toxic compounds that can be either constitutive or induced that have detrimental effect on insects. The well-known examples include plants in the family Brassicaceae and Solanaceae.

Brassica plants possess a well-studied class of sulfur-containing secondary metabolites—glucosinolates—that defend them from insects. However, during the course of evolution, some (though only a few) insects have been specialized to feed even on these plants. The examples include the turnip aphid, *Lipaphis erysimi* (Kaltenbach); cabbage aphid, *Brevicoryne brassicae* (L.); and cabbage white butterflies, *Pieris brassicae* and *P. rapae* [51]. These insects have evolved to use otherwise toxic compounds to their advantage—as cues for the identification of host plants and for development.

Similarly, members of family Solanaceae, e.g., potato and tomato, possess glyco-sidic alkaloids (tomatine, solanine) that defend them from not only insect pests but bacteria and fungi as well. However, some of the species have evolved to overcome this defense, for example, *Macrosiphum euphorbiae* (Thomas) and *Myzus persicae* (Sulzer). The well-known insecticidal compound, nicotine, found in *Nicotiana* spp. provides protection against feeding aphids. However, continuous selection pressure exerted by these compounds leads to the development of resistance in aphid popu-lations to these compounds. The presence of both sexual (that includes a genetic variability) and asexual modes of reproduction (that leads to faster multiplication) aid in faster resistance development [52].

The resistance gene present in resistant plant provides protection against avirulent strains of insects. To date, one R gene (*Mi-1.2*) has been characterized at molecular level. Plants that possess *Mi-1.2* gene are resistant to potato aphid, two whitefly biotypes (silverleaf whitefly and biotype Q), syllid, and three nematode species [53–55]. Due to the high selection pressure on insect population, there are chances of resistance breakdown in plants due to the development of counter resistance to the *Mi-1.2* [56]. The other genes associated with aphid resistance include virus aphid transmission (*Vat*) resistance gene in melon that confers antixenotic resistance to melon aphid, *Aphis gossypii* Glover, and to virus transmission associated with this species [57] and recombination-activating gene (*Rag1*) in soybean that provides resistance to soybean aphid, *Aphis glycines* Matsumura [58].

The defense-signaling mechanism in plants after aphid attack is similar to incompatible responses in plant-pathogen interactions. Aphid feeding triggers SA-dependent response similar to that triggered by biotrophic pathogens and/or *PR* gene RNAs in resistant than in susceptible plants [59–61], while there is downregulation of jasmonic acid-dependent genes [62]. From the very first stylet insertion in epidermal tissues to sustained feeding on sieve elements, aphids continuously inject saliva in the plant tissue which continuously interacts with plant cells to determine compatible/incompatible aphid-plant interactions. However, such interactions have been partially understood. Aphid saliva plays an important role in countering plant defense response and modifying the incompatible interaction to compatible one by modifying the plant metabolism. Aphid feeding may lead to alterations in host plants, including morphological changes, alteration in resource allocation and production of local, and systemic symptoms [32].

## 7. Response of host plants to aphid infestation

Plants respond in a variety of ways to attack by aphid herbivores. Simple feeding by aphids leads to withdrawal of large quantities of plant sap leading to local chlorosis, weakening of the plant, and increase in susceptibility to other insects or pathogens. The well-known examples include infestation of Brassica plants by *Lipaphis erysimi* and *Brevicoryne brassicae* [63] and of beans by *Aphis fabae* [64]. On the contrary, large aphid populations can also develop on host plant without manifestation of symptoms such as infestation of tomato plants by *Macrosiphum euphorbiae* [65]. The visible symptoms after aphid attack can vary from localized chlorosis at the feeding site or along the stylet path due to damage to the chloroplast [64]; local-ized tissue damage, e.g., *Dysaphis plantaginea* (Passerini) on apple fruits; curling of leaves, flower buds, and pods of mustard plants by *L. erysimi* [66]; leaf curling to cigar shape in peach by *Myzus varians* Davidson; growth distortions on citrus by *Aphis spiraecola* Patch; to systemic effects caused by feeding of *Acyrthosiphon pisum* (Harris) and *Therioaphis trifolii* (Monell on alfalfa) [52]. All the manifestations are in part due to the toxic effect of saliva on host plant. Further, saliva may also have effect on the hormonal balance of plants leading to changes in normal cell division (hypertrophy) that can result in gall formation on host plant. The actual mecha-nism of gall formation is still not fully understood. Detailed studies on aphid saliva have found no evidence of any cecidogenic compound that can result in gall forma-tion on host plant [67]. However, it has been postulated that galls contain higher concentration of nutrients than the uninfested plant part which may be of adaptive advantage to the insect that develops inside the gall. Koyama et al. [68] analyzed the concentration of amino acids in galled leaves of *Sorbus commixta* Hedl induced by *Rhopalosiphum insertum* (Walker) and found it to be five times higher than that in ungalled leaves without any difference in the composition. In addition to providing better nutrition, galls also provide conducive microclimate to the aphid species that develops within and protects it from its natural enemies as well as insecticides [69].

Unlike other herbivores that only cause direct feeding damage, aphids also cause indirect damage to plants. The honeydew drops deposited on the leaves act as magnifying lenses that may burn the leaf tissue beneath on sunny days. In addition, black sooty mold develops on the honeydew that interferes with normal photosynthetic activity and blocks the stomata which interferes with gas exchange leading to leaf fall. Some of the aphid species also act as vectors of phytopathogenic viruses, and the association is of advantage to both the aphid vector and the phytopathogenic virus. Aphids serve as an important mean of dispersion, and some species of viruses (replicative) even use aphids as favorable host for replication. Once inside the aphid body, both replicative and circulative viruses make aphids infective for the rest of its life. When aphid density increases on a virus-infected plant due to it being more nutritious than healthy plant, they produce *alate* forms that disperse to new uninfected plants, which further aids in their dispersal [52]. In addition to being adaptive advantage to virus, this association is beneficial for aphids as well. The virus-infected plants become more nutritious to aphids than uninfected plants [52]. For example, the concentration of free amino acids is more in virus-infected plants. Virus infection also leads to downregulation of plant defenses, thus making the plant more suitable host for aphids. Further, virus-infected plants assume yellowish coloration making them more attractive to aphids.

## 8. Aphid-plant-natural enemy tritrophic interaction: the "cry or call for help"

In response to aphid feeding, plants release a number of volatile compounds which are perceived by aphid natural enemies. Since plants employ these natural enemies to defend themselves, the release of volatile compounds is analogous to "cry or call for help" by plants. This type of defense is referred to as indirect defense. A number of insects are associated with natural suppression of aphid population which includes predators such as ladybird beetles (e.g., *Coccinella* spp., Brumus sp., *Adalia bipunctata* L., *Menochilus* sp., etc.), green lacewing (*Chrysoperla carnea* Stephens), syrphids (*Episyrphus balteatus* De Geer), mirid bugs, and parasitoids (*Aphidius* spp., *Diaeretiella rapae* M'Intosh, *Praon* spp., etc.). However, these natural control agents are not efficient in suppressing aphid population, and there is a lack of synchrony in the peak activity of aphids and their natural enemies [63]. Aphid populations generally develop early in the season (mostly in spring) with delayed action of natural control agents. But once their action has started, there is sudden decline in aphid population as observed in oilseed Brassica [66] and organic crops [70].

The feeding by aphids triggers the release of volatile compounds from infested plants making them more attractive to parasitoids. For example, *Acyrthosiphon pisum*-infested broad bean plants are six times more attractive to *Aphidius ervi* Haliday than uninfested plants [71]. Similarly, *Brassica rapa* L. var. *rapifera* plants infested either by *L. erysimi* or *M. persicae* become more attractive to *D. rapae*. This increase in attractiveness has potential implications in aphid control, and researchers are working to find possible ways to elicit this attractiveness in uninfested plants. For example, exogenous application of (Z)-jasmone, a compound derived from jasmonic acid, results in increased attractiveness of uninfested broad bean plants to *A. ervi* similar to those infested by *A. pisum* [72].

## 9. Potential applications for aphid management

The current understanding of these interactions can help find ways to improve plant resistance to aphids. Since aphids cause serious damage to many agricultural crops, there is a need to find sustainable solution for the management as an effective alternative strategy to synthetic insecticides. There are accelerated global research efforts to search for source(s) of aphid resistance especially in crop wild relatives (CWRs) [4, 73–75]. There is a growing body of literature that suggests that almost all the variations necessary for crop improvement can be found in their CWRs that were lost over the course of domestication [76–80]. The use of CWRs is continuously increasing over the years for a range of beneficial traits including pest and disease resistance [81–83]. In a comprehensive survey by Hajjar and Hodgkin [83] about the use of CWRs in crop improvement for the period 1986–2005, over 80% of the beneficial traits involved pest and disease resistance. The present knowledge of genomics and availability of tools of biotechnology have erased the boundaries of crossing the species from different gene pools, and there has been a significant increase in the number of wild species in gene banks. Despite this, the use of CWRs in their contributions in providing useful genes for improvement of crop plants has been less than expected. In addition to this, the external application of analogues of jasmonic acid and salicylic acid can also be used to further enhance the level of resistance in crop plants [84].

In recent years, there has been an increase in the knowledge on resistance genes, but only a few *R* genes that confer resistance against hemipteran insects have been identified. Some of them include *Vat* that confers resistance to *Aphis gossypii* in melon [85], *Bph 14* and *Bph 26* genes in rice that confer resistance to *Nilaparvata lugens*, and *Mi-12.* gene in tomato that confers resistance to *Macrosiphum euphorbiae* [32]. The *Vat* gene in melon enhances SE wound healing and thus confers resistance to *A. gossypii* [86]. The cloning of *Mi-1.2* gene has been a milestone in plant resistance to aphids [54, 55, 86–88], and it has distinct resistance mechanisms against different pests. Against root-knot nematode, *M. incognita*, plants exhibit hypersensitive response, and this response is not manifested upon aphid infestation. The resistance to aphids is antibiotic and phloem based, while it is antixenotic to psyllids. On the other hand, *Mi-1.2*-mediated resistance to whiteflies deters insect settling. However, if the insect establishes a feeding site, it can develop even on the *Mi-1.2* plants. The resistant plants exhibit distinct mechanism of resistance against members of four different animal taxa; however, the biochemical basis of such resistance is not yet known.

The attractiveness of the crop plants to aphids and subsequently to their parasitoids can also be augmented to increase effectiveness of parasitoids/natural enemies provided aphids do not act as vector of the phytopathogenic virus. This strategy is especially important as it does not exert any ecological pressure on the aphids. Germplasm screening can be targeted for genotypes that are good at defending themselves from aphid attack and simultaneously attractive to aphid natural enemies. For example, *Eruca sativa* genotypes are particularly attractive to coccinellid beetles in *Brassica* systems compared to *B. juncea*, *B. napus*, *B. carinata*, or *B. rapa*.

Another area of potential application in aphid control is the development of transgenic plants expressing resistance against aphids. Modern breeding techniques can be of great help in transferring target trait to the cultivated plant compared to traditional breeding methods. The commercial insect-resistant GM crops that express *Bt* toxins are particularly effective against Lepidoptera and Coleoptera [89] with no efficacy against phloem feeders including aphids [90]. This accelerated the work on finding alternate strategies such as protease inhibitors, RNAi, anti-microbial peptides (AMPs), etc. Protease inhibitors which may be small peptides or protein molecules inhibit the activity of proteases, thus disrupting the normal protein digestion and consequent amino acid assimilation vital for insect growth. These are already present in plant storage organs and are induced upon insect feed-ing. Significantly high activity of PI was reported in barley infested with *Schizaphis graminum* with minor effect on its survival, while survival of *Rhopalosiphum padi* was significantly affected [91]. Oryzacystatin-I in transgenic rapeseed [92] and egg plants [93] and cysteine in *Arabidopsis thaliana* [94] from barley are known to provide protection against aphid infestation with their effect on aphid survival, growth, and reproduction. Thus, the use of PIs in aphid management has a good promise as an alternate control strategy [92–94].

Another potential area in aphid management is the exploitation of RNAi technology, which is posttranslational RNA-mediated gene silencing. Plants can be genetically engineered to produce dsRNA to provide protection against a target pest. Transgenic maize plants that produce dsRNA significantly reduced feeding damage by Western corn rootworm, *Diabrotica virgifera* larvae [95]. In the case of aphids, different workers have achieved RNAi-mediated gene silencing either by injecting the siRNA (short-interfering RNA) [96, 97] or dsRNA into insect hemolymph or feeding the insect with dsRNA [98, 99]. A temporary mRNA inhibition of about 30–40% in aphids was observed by single dose of dsRNA [96]. Similarly, 50% reduction in salivary gland protein expression was observed by Mutti et al. [97].

All the organisms synthesize small 12–50 amino acid long peptides which have antibiotic activity and are termed antimicrobial peptides. They are generally synthesized ribosomally but are also produced enzymatically in fungi and bacteria. They are known to possess antibiotic activity against both gram-positive and gram-negative bacteria and provide immunity against microbial infection. Many insect species are known to produce AMPs [100, 101]. On the contrary aphids do not produce AMPs [95] as they have mutual relationship with endosymbiotic bacteria such as *Buchnera aphidicola*, *Hamiltonella*, *Serratia*, *Rickettsia*, and *Regiella* spp. [102] which play an important role of converting nonessential amino acids in phloem sap to essential ones [103]. Thus, aphid bacterial endosymbionts can be a useful target for AMPs. Any adverse effect on aphid endosymbionts can adversely affect aphid fecundity and can prolong development period [104, 105]. So far, there is only one report on the effect of AMP (indolicidin) on aphids, ingestion of which reduces the number of bacteriocytes and number of bacteria in *M. persicae*, which have significant negative effect on aphid survival, development, and fecundity [106]. This suggests that AMPs expressed in GM plants offer a promising approach for aphid control.

Production of volatile compounds by plants is another area that can be explored. Aphids respond to plant volatiles and use them for long-range orientation as recorded in *Aphis fabae*, *A. pisum*, *Brevicoryne brassicae*, and *M. persicae* [107–110]. Many plants synthesize E-ß-farnesene (Eßf), a well-known alarm pheromone of aphids, as aphid repellent such as wild potato species [48]. Choice experiments by these authors indicated that aphids remain at a distance of 1–3 mm from leaf surface. Apart from general avoidance, aphids also responded to Eßf by produc-ing higher proportion of *alate* (migratory) individuals on treated plants under controlled conditions [111] as well in the field [112]. Thus, plants are exposed to reduced number of apterous (feeding) forms and high proportion of *alates* (migratory forms) that have greater tendency to leave the plant [111]. Besides a repellent effect on aphids, Eßf is also known to attract natural enemies of aphids such as ladybirds *Coccinella septempunctata* and *Harmonia axyridis*, parasitoids *Aphidius uzbekistanicus* and *A. ervi*, and syrphid fly *Episyrphus balteatus* [113–117]. Thus, production of transgenic plants expressing Eßf can have dual effect on aphids and can increase the benefits of Eßf production.

## 10. Conclusion

The aphid-plant coevolution is a continuous arms race that helps to improve defense strategies employed by plants to ward off aphids and counter defense mechanisms employed by aphid herbivores. For a compatible aphid-plant interaction, aphids not only need to alter local and systemic events but also need to modify resource allocation to suit phloem sap to their requirements. Generally, the JA-mediated defenses are employed by plants to control aphids. But aphids through the use of specific effectors are able to modify the JA-mediated defense response of plant and are able to establish successful feeding. Plants, on the other hand, have evolved to use aphid salivary components as elicitors of defense response. The phloem sealing mechanism is one such response observed in resistant plants. In addition, plants have also evolved a plethora of plant secondary metabolites (PSMs) that have defensive functions. But some specialist aphids have learned to use these compounds to their own advantage and use them as cues for feeding and coloniza-tion and even sequester them for their advantage.

The current knowledge on aphid-plant interactions is still in its infancy. But the recent studies have provided insights into such interactions which will have far-reaching implications at different levels including development of novel aphid management strategies.

## Author details

Sarwan Kumar
Department of Plant Breeding and Genetics, Punjab Agricultural University, Ludhiana, India

*Address all correspondence to: sarwanent@pau.edu

## References

[1] Remaudiere G, Remaudiere M. Catalogue des Aphididae du Monde. Paris: INRA; 1997. 473 p

[2] Dedryver CA, Le Ralec A, Fabre F. The conflicting relationship between aphids and men: A review of aphid damages and of their control strategies. Comptes Rendus Biologies. 2010;**333**:539-553

[3] Blackman RL, Eastop VF. Taxonomic issues. In: van Emden HF, Harrington R, editors. Aphids as Crop Pests. 2nd ed. UK: CAB International; 2017. pp. 1-36

[4] Goggin FL. Plant-aphid interactions: Molecular and ecological perspectives. Current Opinion in Plant Biology. 2007;**10**:399-408

[5] Peccoud J, Simon JC, von Dohlen C, Coeur d'acier A, Plantegenest M, Vanlerberghe-Masutti F, et al. Evolutionary history of aphid-plant associations and their role in aphid diversification. Comptes Rendus Biologies. 2010;**333**:474-487. DOI: 10.1016/j.crvi.2010.03.004

[6] Bhatia V, Uniyal PL, Bhattacharya R. Aphid resistance in Brassica crops: Challenges, biotechnological progress and emerging possibilities. Biotechnology Advances. 2011;**29**:879-888

[7] Williams IS, Dixon AFG. Life cycles and polymorphism. In: van Emden HF, Harrington R, editors. Aphids as Crop Pests. 1st ed. UK: CAB International; 2007. pp. 69-85

[8] Capinera JL. Green peach aphid, *Myzus persicae* (Sulzer) (Insecta: Hemiptera: Aphididae). In: Capinera JL, editor. Encyclopedia of Entomology. Dordrecht, The Netherlands: Springer; 2008. pp. 1727-1730

[9] Forbes AR. The mouthparts and feeding mechanism of aphids. In: Harris K, Maramorosch K, editors. Aphids as Virus Vectors. New York: Academic Press; 1977. pp. 83-103

[10] Powell G, Tosh CR, Hardie J. Host plant selection by aphids: Behavioral, evolutionary, and applied perspectives. Annual Review of Entomology. 2006;**51**:309-330

[11] Uzest M, Gargani D, Dombrovsky A, Cazevieille C, Cot D, Blanc S. The "acrostyle": A newly described anatomical structure in aphid stylets. Arthropod Structure and Function. 2010;**39**:221-229

[12] Guyton TL. A taxonomy, ecologic and economic study of Ohio Aphididae. The Ohio Journal of Science. 1924;**26**:1-26

[13] Skelett WH. Muskulatur und Darm der Schwarzen Blattlaus *Aphis fabae* SCOP. Zoologica. 1928;**76**:1-120

[14] Forbes AR. The stylets of the green peach aphid, *Myzus persicae* (Homoptera: Aphididae). The Canadian Entomologist. 1969;**101**:31-41

[15] Saxena PX, Chada HL. The greenbug, *Schizaphis graminum*. Mouthparts and feeding habits. Annals of the Entomological Society of America. 1971;**64**:897-904. DOI: 10.1093/aesa/64.4.897

[16] Wojciechowski W. Studies on the systematic system of aphids (Homoptera: Aphidinea). Katowice: Uniwersytet Slaski; 1992

[17] Razaq A, Toshio K, Pear M, Masaya S. SEM observations on the citrus green aphid, *Aphis citricola* van der Goot (Homoptera: Aphididae). Pakistan Journal of Biological Sciences. 2000;**3**:949-952. DOI: 10.3923/pjbs.2000.949.952

[18] Giordanengo P, Brunissen L, Rusterucci C, Vincent C, van Bel A, Dinant S, et al. Compatible plant-aphid interactions: How aphids manipulate plant response. Comptes Rendus Biologies. 2010;**333**:516-523. DOI: 10.1016/j.crvi.2010.03.007

[19] Felton GW, Eichenseer H. Herbivore saliva and induction of resistance to herbivores and pathogens. In: Agrawal AA, Tuzun S, Bent E, editors. Induced Plant Defenses Against Pathogens and Herbivores: Biochemistry, Ecology, and Agriculture. St. Paul, MN: APS Press; 1999. pp. 19-36

[20] Urbanska A, Tjallingii WF, Dixon AFG, Leszczynski B. Phenol oxidizing enzymes in the grain aphid's saliva. Entomologia Experimentalis et Applicata. 1998;**86**:197-203

[21] Miles PW. Aphid saliva. Biological Reviews. 1999;**74**:41-85

[22] Cherqui A, Tjallingii WF. Salivary proteins of aphids, a pilot study on identification, separation and immunolocalisation. Journal of Insect Physiology. 2000;**46**:1177-1186

[23] Will T, Tjallingii WF, Thonnessen A, van Bel AJE. Molecular sabotage of plant defense by aphid saliva. Proceedings of the National Academy of Sciences of the United States of America. 2007;**104**:10536-10541

[24] Will T, Kornemann SR, Furch ACU, Tjallingii WF, van Bel AJE. Aphid watery saliva counteracts sieve-tube occlusion: A universal phenomenon? Journal of Experimental Biology. 2009;**212**:3305-3312

[25] Schoonhoven LM, van Loon JJA, Dicke M. Insect-Plant Biology. Oxford: Oxford University, Press; 2005. 421 p

[26] Garcia-Brugger A, Lamotte O, Vandelle E, Bourque S, Lecourieux D, Poinssot B, et al. Early signaling events induced by elicitors of plant defenses. Molecular Plant-Microbe Interactions. 2006;**19**:711-724

[27] Thomma BPHJ, Penninckx IAMA, Cammue BPA, Broekaert WF. The complexity of disease signaling in Arabidopsis. Current Opinion in Immunology. 2001;**13**:63-68

[28] Maffei ME, Mithofer A, Boland W. Before gene expression: Early events in plant-insect interaction. Trends in Plant Science. 2007;**12**:310-316

[29] Zhu-Salzman K, Salzman RA, Ahn JE, Koiwa H. Transcriptional regulation of sorghum defense determinants against a phloem-feeding aphid. Plant Physiology. 2004;**134**:420-431

[30] Thompson GA, Goggin FL. Transcriptomics and functional genomics of plant defence induction by phloem-feeding insects. Journal of Experimental Botany. 2006;**57**:755-766

[31] Gao LL, Anderson JP, Klingler JP, Nair RM, Edwards OR, Singh KB. Involvement of the octadecanoid pathway in bluegreen aphid resistance in *Medicago truncatula*. Molecular Plant-Microbe Interactions. 2007;**20**:82-93

[32] Walling LL. Avoiding effective defenses: Strategies employed by phloem-feeding insects. Plant Physiology. 2008;**146**:859-866

[33] Dixon AFG. Aphid Ecology: An Optimization Approach. 2nd ed. London: Chapman and Hall; 1998. 300 p

[34] Munson MA, Baumann P, Kinsey MG. *Buchnera* gen. nov. and *Buchnera aphidicola* sp. nov., a taxon consisting of the mycetocyte-associated, primary endosymbionts of aphids. International Journal of Systematic Bacteriology. 1991;**41**:566-568

[35] Loudit SMB, Bauwens J, Francis F. Cowpea aphid-plant interactions:

Endosymbionts and related salivary protein patterns. Entomological Experimentalis et Applicata. 2018;**166**:460-473. DOI: 10.1111/eea.12687

[36] Chen DQ, Purcell AH. Occurrence and transmission of facultative endosymbionts in aphids. Current Microbiology. 1997;**34**:220-225

[37] Fukatsu T, Nikoh N, Kawai R, Koga R. The secondary endosymbiotic bacterium of the pea aphid *Acyrthosiphon pisum* (Insecta: Homoptera). Applied and Environmental Microbiology. 2000;**66**:2748-2758

[38] Fukatsu T, Tsuchida T, Nikoh N, Koga R. *Spiroplasma* symbiont of the pea aphid *Acyrthosiphon pisum* (Insecta: Homoptera). Applied and Environmental Microbiology. 2001;**67**:1284-1291

[39] Darby AC, Birkle LM, Turner SL, Douglas AE. An aphid-borne bacterium allied to the secondary symbionts of whitefly. FEMS Microbiology Ecology. 2001;**36**:43-50

[40] Sandstrom JP, Russell JA, White JP, Moran NA. Independent origins and horizontal transfer of bacterial symbionts of aphids. Molecular Ecology. 2001;**10**:217-228

[41] Haynes S, Darby AC, Daniell TJ, Webster G, van Veen FJF, Godfray HCJ, et al. Diversity of bacteria associated with natural aphid populations. 2003;**69**:7216-7223

[42] Russell JA, Latorre A, Sabater-Munoz B, Moya A, Moran NA. Independent origins and horizontal transfer of bacterial symbionts of aphids. Molecular Ecology. 2003;**12**:1061-1075

[43] Moran NA, Russell JA, Koga R, Fukatsu T. Evolutionary relationships of three new species of enterobacteriaceae living as symbionts of aphids and other insects. Applied and Environmental Microbiology. 2005;**71**:3302-3310

[44] Doring TF, Chittka L. Visual ecology of aphids: A critical review on the role of colours in host finding. Arthropod-Plant Interactions. 2007;**1**:3-16

[45] Pickett JA, Birkett MA, Bruce TJA, Chamberlain K, Gordon-Weeks R, Matthes MC, et al. Developments in aspects of ecological phytochemistry: The role of cis-jasmone in inducible defence systems in plants. Phytochemistry. 2007;**68**:2937-2945

[46] Musetti L, Neal JJ. Resistance to the pink potato aphid, *Macrosiphum euphorbiae*, in two accessions of *Lycopersicon hirsutum* f. *glabratum*. Entomologia Experimentalis et Applicata. 1997;**84**:137-146

[47] Bin F. Influenza dei peli glandolari sugli insetti in *Lycopersicon* spp. Frust Entomology. 1979;**15**:271-283

[48] Gibson RV, Pickett JA. Wild potato repels aphids by release of aphid alarm pheromone. Nature. 1983;**302**:608-609

[49] Will T, van Bel AJE. Physical and chemical interactions between aphids and plants. Journal of Experimental Botany. 2006;**57**:729-737

[50] Martinez CE, Leybourne DJ, Bos JIB. Non-host and poor host resistance against aphids may reside in different plant cell layers depending on the plant species-aphid species interaction. BioRXiv. 2018. DOI: 10.1101/372839

[51] Hopkins RJ, van Dam NM, van Loon JJA. Role of glucosinolates in insect plant relationships and multitrophic interactions. Annual Review of Entomology. 2009;**54**:57-83

[52] Guerrieri E, Digilio MC. Aphid-plant interactions: A review. Journal of Plant Interactions. 2008;**3**:223-232

[53] Nombela G, Williamson VM, Muniz M. The rootknot nematode resistance gene Mi-1.2 of tomato is responsible for resistance against the whitefly *Bemisia tabaci*. Molecular Plant-Microbe Interactions. 2003;**16**:645-649

[54] Kaloshian I, Walling LL. Hemipterans as plant pathogens. Annual Review of Phytopathology. 2005;**3**:491-521

[55] Casteel C, Walling LL, Paine T. Behavior and biology of the tomato psyllid, *Bactericera cockerelli*, in response to the Mi-1.2 gene. Entomologia Experimentalis et Applicata. 2006;**121**:67-72

[56] Goggin FL, Williamson VM, Ullman DE. Variability in the response of *Macrosiphum euphorbiae* and *Myzus persicae* (Hemiptera: Aphididae) to the tomato resistance gene Mi. Environmental Entomology. 2001;**30**:101-106

[57] Chen JQ , Rahbe' Y, Delobel B, Sauvion N, Guillaud J, Febvay G. Melon resistance to the aphid *Aphis gossypii*: Behavioral analysis and chemical correlations with nitrogenous compounds. Entomologia Experimentalis et Applicata. 1997;**85**:33-44

[58] Li Y, Hill C, Carlson S, Diers B, Hartman G. Soybean aphid resistance genes in the soybean cultivars Dowling and Jackson map to linkage group M. Molecular Breeding. 2007;**19**:25-34

[59] Forslund K, Pettersson J, Bryngelsson T, Jonsson L. Aphid infestation induces PR proteins differently in barley susceptible or resistant to the bird cherry-oat aphid (*Rhopalosiphum padi*). Physiologia Plantarum. 2000;**110**:496-502

[60] Mohase L, van der Westhuizen AJ. Salicylic acid is involved in resistance responses in the Russian wheat aphid-wheat interaction. Journal of Plant Physiology. 2002;**159**:585-590

[61] Martinez de Ilarduya O, Xie QG Kaloshian I. Aphid-induced defense responses in *Mi-1*-mediated compatible and incompatible tomato interactions. Molecular Plant-Microbe Interactions. 2003;**16**:699-708

[62] Zarate SI, Kempema LA, Walling LL. Silverleaf whitefly induces salicylic acid defenses and suppresses effectual jasmonic acid defenses. Plant Physiology. 2007;**143**:866-875

[63] Kumar S, Singh YP. Insect pests. In: Kumar A, Banga SS, Meena PD, Kumar PR, editors. Brassica Oilseeds: Breeding and Management. Wallingford, UK: CABI Publishing; 2015. pp. 193-232

[64] Miles PW. Specific responses and damage caused by Aphidoidea: Principles. In: Minks AK, Harrewijn P, editors. Aphids: Their Biology, Natural Enemies and Control. New York: Elsevier; 1989. pp. 23-47

[65] Guerrieri E. Afidone verde del pomodoro e della patata Macrosiphum euphorbiae. Lavoro pubblicato dalla Regione Campania nell'ambito dei finanziamenti UE Obiettivo 1 _ Quadro comunitario di sostegno; 2001. pp. 94-99. Regg. Ce 2052/88, 2081/93 _ POM

[66] Kumar S. Relative abundance of turnip aphid and the associated natural enemies on oilseed *brassica* genotypes. Journal of Agricultural Science and Technology. 2015;**17**:1209-1222

[67] Otha S, Kajino N, Hashimoto H, Hirata T. Isolation and identification of cell hypertrophy-inducing substances in the gall forming aphid *Colopha moriokaensis*. Insect Biochemistry and Molecular Biology. 2000;**30**:947-952

[68] Koyama Y, Yao I, Akimoto SI. Aphid galls accumulate high concentrations of amino acids: A support for the

nutrition hypothesis for gall formation. Entomologia Experimentalis et Applicata. 2004;**113**:135-144

[69] Wool D. Galling aphids: Specialization, biological complexity and variation. Annual Review of Entomology. 2004;**49**:75-192

[70] Trembley E. Possibilities for utilization of *Aphidius matricariae* Hal. (*Hymenoptera Ichneumonoidae*) against *Myzus persicae* (Sulz.) (*Homoptera Aphidoidea*) in small glasshouses. Journal of Plant Diseases and Protection. 1974;**81**:612-619

[71] Guerrieri E, Pennacchio F, Tremblay E. Flight behaviour of the aphid parasitoid *Aphidius ervi* Haliday (Hymenoptera: Braconidae) in response to plant and host volatiles. European Journal of Entomology. 1993;**90**:415-421

[72] Birkett MA, Campbell CAM, Chamberlain K, Guerrieri E, Hick AJ, Martin JL, et al. New roles for cis-jasmone as an insect semiochemical and in plant defence. Proceedings of the National Academy of Sciences of the United States of America. 2000;**97**:9329-9334

[73] Dosdall LM, Kott L. Introgression of resistance to cabbage seed pod weevil to canola from yellow mustard. Crop Science. 2006;**46**:2437-2445. DOI: 10.2135/cropsci2006.02.0132

[74] Gos R, Wagenaar R, Bukovinszky T, van Dam NM, Dicke M, Bullock JM, et al. Genetic variation in defense chemistry in wild cabbages affects herbivores and their endoparasitoids. Ecology. 2008;**89**:1616-1626

[75] Edwards D, Henry RJ, Edwards KJ. Advances in DNA sequencing accelerating plant biotechnology. Plant Biotechnology Journal. 2012;**10**:621-622. DOI: 10.1111/j.1467-7652.2012.00724.x

[76] Tanksley SD, McCouch SR. Seed banks and molecular banks: Unlocking genetic potential from the wild. Science. 1997;**277**:1063-1066. DOI: 10.1126/science.277.5329.1063

[77] Fernie AR, Tadmor Y, Zamir D. Natural genetic variation for improving crop quality. Current Opinion in Plant Biology. 2006;**9**:196-202. DOI: 10.1016/j. pbi.2006.01.010

[78] Vaughan DA, Balazs E, Heslop-Harrison JS. From crop domestication to super domestication. Annals of Botany. 2007;**100**:893-901. DOI: 10.1093/aob/mcm224

[79] Burger JC, Chapman MA, Burke JM. Molecular insights into the evolution of crop plants. American Journal of Botany. 2008;**95**:113-122

[80] Pelgrom K, Broekgaarden C, Voorrips RE, Vosman BJ. Successful use of crop wild relatives in breeding: Easier said than done. In: International Conference on Enhanced Gene Pool Utilization-Capturing Wild Relative and Landrace Diversity for Crop Improvement. Vol. 10. University of Birmingham; 2015. 15 p

[81] Prescott-Allen C, Prescott-Allen R. The First Resource: Wild Species in the North American Economy. New Haven: Yale University; 1986

[82] Prescott-Allen C, Prescott-Allen R. Genes from the Wild: Using Wild Genetic Resources for Food and Raw materials. London: International Institute for Environment and Development; 1988

[83] Hajjar R, Hodgkin T. The use of wild relatives in crop improvement: A survey of developments over the last 20 years. Euphytica. 2007;**156**:1-13. DOI: 10.1007/s10681-007-9363-0

[84] Cooper WC, Jia L, Goggin FL. Acquired and R-gene-mediated

resistance against the potato aphid in tomato. Journal of Chemical Ecology. 2004;**30**:2527-2542

[85] Martin B, Rahbe Y, Fereres A. Blockage of stylet tips as the mechanism of resistance to virus transmission by Aphis gossypii in melon lines bearing the Vat gene. The Annals of Applied Biology. 2003;**142**:245-250

[86] Kaloshian I, Kinsey DE, Ullman DE, Williamson VM. The impact of *Meul*-mediated resistance in tomato on longevity, fecundity, and behavior of the potato aphid, *Macrosiphum euphorbiae*. Entomologia Experimentalis et Applicata. 1997;**83**:181-187

[87] Rossi M, Goggin FL, Milligan SB, Kaloshian I, Ullman DE, Williamson VM. The nematode resistance gene *Mi* of tomato confers resistance against the potato aphid. Proceedings of the National Academy of Sciences of the United States of America. 1998;**95**:9750-9754

[88] Vos P, Simons G, Jesse T, Wijbrandi J, Heinen L, Hogers R, et al. The tomato *Mi-1* gene confers resistance to both root-knot nematodes and potato aphids. Nature Biotechnology. 1998;**16**:1365-1369

[89] Kumar S, Chandra A, Pandey KC. *Bacillus thuringiensis* (Bt) transgenic crop: An environment friendly insect-pest management strategy. Journal of Environmental Biology. 2008;**29**:641-653

[90] Raps A, Kehr J, Gugerli P, Moar WJ, Bigler F, Hilbeck A. Immunological analysis of phloem sap of *Bacillus thuringiensis* corn and of the non-target herbivore *Rhopalosiphum padi* (Homoptera: Aphididae) for the presence of Cry1Ab. Molecular Ecology. 2001;**10**:525-533

[91] Ryan JD, Morgham AT, Richardson PE, Johnson RC, Mort AJ, Eikenbary R. Greenbugs and wheat: A model system for the study of phytotoxic Homoptera. In: Campbell RK, Eikenbary RD, editors. Aphid-Plant Genotype Interactions. Amsterdam: Elsevier; 1990

[92] Rhabé Y, Deraison C, Bonadé-Bottino M, Girard C, Nardon C, Jouanin L. Effects of the cysteine protease inhibitor oryzacystatin (OC-I) of different aphids and reduced performance of *Myzus persicae* on OC-I expressing transgenic oilseed rape. Plant Science. 2003;**164**:441-450

[93] Ribeiro APO, Pereira EJG, Galvan TL, Picanco MC, Picoli EAT, da Silva DJH, et al. Effect of eggplant transformed with oryzacystatin gene on *Myzus persicae* and *Macrosiphum euphorbiae*. Journal of Applied Entomology. 2006;**130**:84-90

[94] Carrillo L, Martinez M, Álvarez-Alfageme F, Castanera P, Smagghe G, Diaz I, et al. A barley cysteine-proteinase inhibitor reduces the performance of two aphid species in artificial diets and transgenic Arabidopsis plants. Transgenic Research. 2011;**20**:305-319

[95] Baum JA, Bogaert T, Clinton W, Heck GR, Feldmann P, Ilagan O, et al. Control of coleopteran insect pests through RNA interference. Nature Biotechnology. 2007;**25**:1322-1326

[96] Jaubert-Possamai S, Trionnair GL, Bonhomme J, Christophides GK, Rispe C, Tagu D. Gene knockdown by RNAi in the pea aphid *Acyrthosiphon pisum*. BMC Biotechnology. 2007;**7**:63

[97] Mutti NS, Park Y, Reese JC, Reek GR. RNAi knockdown of a salivary transcript leading to lethality in the pea aphid *Acyrtosiphon pisum*. Journal of Insect Science. 2006;**6**:38

[98] Whyard S, Singh AD, Wong S. Ingested double-stranded RNAs can act

as species specific insecticides. Insect Biochemistry and Molecular Biology. 2009;**39**:824-832

[99] Shakesby AJ, Wallace LS, Isaacs HV, Pritchard J, Roberts DM, Douglas AE. A water-specific aquaporin involved in aphid osmoregulation. Insect Biochemistry and Molecular Biology. 2009;**39**:1-10

[100] Bulet P, Stöcklin R. Insect antimicrobial peptides: Structures, properties and gene regulation. Protein and Peptide Letters. 2005;**12**:3-11

[101] Vilcinskas A. Evolutionary plasticity of insect immunity. Journal of Insect Physiology. 2013;**59**:123-129

[102] Moran NA, Russell JA, Koga R, Fukatsu T. Evolutionary relationship of three new species of enterobacteriaceae living as symbionts of aphids and other insects. Applied Environmental Biology. 2005;**71**:3302-3310

[103] The International Aphid Genomics Consortium. Genome sequence of the pea aphid *Acyrthosiphon pisum*. PLoS Biology. 2010;**8**(2):e1000313. DOI: 10.1371/journal.pbio.1000313

[104] Keymanesh K, Soltani S, Sardari S. Application of antimicrobial peptides in agriculture and food industry. World Journal of Biotechnology and Biotechnology. 2009;**25**:933-944

[105] Koga R, Tsuchida T, Sakurai M, Fikatsu T. Selective elimination of aphid endosymbionts: Effects of antibiotic dose and host genotype, and fitness consequences. FEMS Microbiology Ecology. 2007;**60**:229-239

[106] Le-Feuvre RR, Ramirez CC, Olea N, Meza-Basso L. Effect of the antimicrobial peptide indolicidin on the green peach aphid *Myzus persicae* (Sulzer). Journal of Applied Entomology. 2007;**131**:71-75

[107] Hardie J, Visser JH, Piron PGM. Peripheral odour perception by adult aphid forms with the same genotype but different host-plant preferences. Journal of Insect Physiology. 1995;**41**:91-97

[108] van Giessen WA, Fescemyer HW, Burrows PM, Peterson JK, Barnett OW. Quantification of electroantennogram responses of the primary rhinaria of *Acyrthosiphon pisum* (Harris) to C4–C8 primary alcohols and aldehydes. Journal of Chemical Ecology. 1994;**20**:909-927

[109] Visser JH, Piron PGM, Hardie J. The aphid's peripheral perception of plant volatiles. Entomologia Experimentalis et Applicata. 1996;**80**:35-38

[110] Wang Q, Zhou JJ, Liu JT, Huang GZ, Xu WY, Zhang Q, et al. Integrative transcriptomic and genomic analysis of odorant binding proteins and chemosensory proteins in aphids. Insect Molecular Biology. 2019;**28**:1-22. DOI: 10.1111/imb.12513

[111] Kunert G, Otto S, Röse USR, Gershenzon J, Weisser WW. Alarm pheromone mediates production of winged dispersal morphs in aphids. Ecology Letters. 2005;**8**:596-603

[112] Hatano E, Kunert G, Weisser WW. Aphid wing induction and ecological costs of alarm pheromone emission under field conditions. PLoS One. 2010;**5**(6):e11188. DOI: 10.1371/journal.pone.0011188

[113] Abassi A, Birkett S, Petterson MA, Pickett JA, Wadhams LJ, Woodcock CM. Response of the seven-spot ladybird to an aphid alarm pheromone and an alarm pheromone inhibitor is mediated by paired olfactory cells. Journal of Chemical Ecology. 2000;**26**:1765-1771

[114] Zhu JW, Cosse AA, Obrychi JJ, Boo KS, Baker TC. Olfactory reactions of the

twelve-spotted lady beetle, *Coleomegilla maculata* and the green lacewing, *Chrysoperla carnea* to semiochemicals released from their prey and host plant: Electroantennogram and behavioral responses. Journal of Chemical Ecology. 1999;**25**:1163-1177

[115] Micha SG, Wyss U. Aphid alarm pheromone (E)-ß-farnesene: A host finding kairomone for the aphid primary parasitoid *Aphidius uzbekistanicus* (Hymenoptera: Aphidiinae). Chemoecology. 1996;7:132-139

[116] Du YJ, Poppy GM, Powell W, Pickett JA, Wadhams LJ, Woodcock CM. Identification of semiochemicals released during aphid feeding that attract parasitoid *Aphidius ervi*. Journal of Chemical Ecology. 1998;**24**:1355-1368

[117] Verheggen FJ, Haubruge E, De Moraes CM, Mescher MC. Social environment influences aphid production of alarm pheromone. Behavioral Ecology. 2009;**20**:283-288

# 3

# Responses of Community Structure, Productivity and Turnover Traits to Long-Term Grazing Exclusion in a Semiarid Grassland on the Loess Plateau of Northern China

*Jimin Cheng, Wei Li, Jishuai Su, Liang Guo, Jingwei Jin and Chengcheng Gang*

**Abstract**

Grazing exclusion has been widely used for restoration of degraded grassland all over the world. Based on over a 30-year (from 1982 to 2011) vegetation survey and a 2-year (from 2013 to 2014) field decomposition experiment in Yunwu Mountain Grassland Nature Reserve on the Loess Plateau of China, responses of community structure and productivity and decomposition traits of dominant Stipa species (Stipa bungeana, Stipa grandis and Stipa przewalskyi) litters were determined to reveal the ecosystem cyclic process. Results showed that grassland coverage, plant density, Shannon-Wiener index and aboveground productivity changed in a hump pattern with peaks in 2002. Productivity was significantly positively correlated with mean annual temperature. The direction and magnitude about effects of climatic changes on productivity depended on phonological stages of plant community. Warming in early stage of growing season (April–May) contributed the increase of productivity, while temperature rise after the growing season (September–March in the next following year) was negatively correlated with productivity in the following year. Leaf litters of three Stipa species (S. bungeana, S. grandis and S. przewalskyi) had higher decomposition rates in the growing season than that in the nongrowing season. Nutrient-releasing pattern in litters of three Stipa species followed a different pattern: S. bungeana > S.grandis>S. przewalskyi. Considering productivity and decomposition traits, grazing exclusion promotes carbon seques-tration of semiarid grassland, while adjustments in nutrient cycling might explain fluctuations of community structure.

**Keywords:** yunwu mountain, loess plateau, grazing exclusion, climate variation, decomposition, *Stipa*

## 1. Introduction

As one of the most important and largest terrestrial ecosystems in the world, grasslands cover 30% of the land surface and are mainly distributed in arid and semiarid regions [1]. Due to global climate change and human activity, such as heavy grazing, grasslands in this area have undergone desertification and even virtually disappeared in recent decades [2, 3], making restoration process urgent for degraded grasslands [4, 5]. Current studies about grassland restoration mainly focus on several key components: community composition and structure, species diversity, soil properties and vegetation succession process [6–10]. Grassland is considered very sensitive to climate changes [11–14] and also is influenced by soil resource availability [15, 16].

Compared with forest ecosystem and cropland ecosystem, aboveground net primary productivity (ANPP) of grasslands is highly temporally variable [16, 17]. Specifically, climate-driven variability in grassland productivity has important effects on the global carbon balance, ecosystem service delivery, profitability of pastoral livelihoods and the sustainability of grassland resources [11, 18, 19]. Many ecologists have analysed the impacts of annual precipitation and temperature on ANPP at regional and continental scales [17, 20–23], while numerous site-specific reports have indicated that interannual variability in ANPP is poorly or even not at all correlated with annual climate conditions [19, 24, 25]. Changes in precipita-tion or temperature during certain parts of the year have been proven to be more relevant drivers of ANPP than annual changes [26–29], and the impacts on vegeta-tion production varied with seasons [13, 28, 30, 31]. For instance, warming in early spring increased grassland productivity by ameliorating cold temperature constraints on plant growth in northern mid- and high latitudes [32, 33] and advancing spring greening phenology [34–36]. Temperature increases in summer; however, it can depress productivity by reducing soil moisture and intensifying physiological stress [13].

The Loess Plateau of China has a total area of about 52 million hectares and is widely known for its fragile ecological environment, frequent severe droughts and problems with water runoff and soil erosion [37]. In recent years, the complicated landscape, frequent droughts and severe soil erosion have attracted worldwide attention and caused sustained deterioration of the ecosystem of this region. In contrast to numerous studies in the temperate grasslands of Inner Mongolia and the alpine grasslands of the Tibetan Plateau, very few reports are available on responses of grassland productivity to climate variability on the more arid Loess Plateau in China [3], especially with respect to responses to seasonal climatic variability. Restoration of the natural vegetation is regarded as the most effective method for changing the ecological environment of the Loess Plateau [7, 8, 38].

As a major determinant of nutrient cycling, litter decomposition is a fundamen-tal process of grassland ecosystem functioning [39]. Decomposition traits of plant litters are affected by a number of factors, including litter quality, abiotic environ-ment and soil organisms [40]. In general, plant litters with high C:N ratio and lignin concentration are supposed to have slow decomposition and nutrient immobilisa-tion processes, whereas low C:N ratio and low lignin concentration contribute to fast decomposition and nutrient mineralisation processes. Decomposition traits of plant materials may vary with succession stages. For example, late-seral domi-nant grasses normally had high tissue N concentrations, low C:N ratios and lignin concentrations, which result into fast decomposition rate and enhanced nutrient mineralisation.

Most previous studies have focused on plant species richness and diversity in aban-doned croplands following short-term grazing exclusion in China [8, 41, 42]. Few studies have reported on the restoration succession of typical natural steppe under long-term grazing exclusion [19, 43]. In the present study, *Stipa* steppe has been fenced from 1982 to the present at Yunwu Mountain National Nature Reserve, and long-term grassland ecological characteristics, productivity and weather records have been collected.

The community in the study area consists of 313 plant species, covering 56 families and 165 genera, with five main families being Compositae, Gramineae, Leguminosae, Rosaceae and Labiatae [44]. The dominant *Stipa* plants include *S. bungeana*, *S. grandis* and *S. przewalskyi*, and main forbs are *Thymus mongolicus*, *Artemisia sacrorum* and *Potentilla acaulis* [45]. Genus-specific morphological and functional traits contrib-uted the dominance of *Stipa* plants in temperate, subtropical and tropical steppe in semiarid areas worldwide [46]. Meanwhile, *Stipa* species showed various adapt-abilities to environmental changes, presenting an ecological distribution pattern along the climate gradients [47]. There are 32 species, 1 subspecies and 3 variations in genus *Stipa* plants in China, mainly distributed in western and northeastern area, and 5 *Stipa* species are found in our study area. As the constructive species, *S. bungeana* mainly distributed on the Loess Plateau [48]. Noticeably, replacement of dominant *Stipa* species occurred during the long-term restoration process, with *Stipa bungeana* being replaced by *S. grandis* and *S. przewalskyi* [49]. The three *Stipa* species differentiated in their phenotypic traits. In detail, *S. grandis* owns higher plant height, and *S. przewalskyi* possesses more tillers. Besides, *S. bungeana* and *S. przewalskyi* consistently flower and produce seeds earlier than *S. grandis* [49].

The temperature and precipitation variability during 1982–2011 were assessed in this study; the ecological characteristics during long-term grazing exclusion were examined; the relationship between grassland productivity and variation in climate variables were explored; and the variations in decomposition traits of three *Stipa* dominant species (*S. bungeana*, *S. grandis* and *S. przewalskyi*) were determined.

## 2. Material and method

### 2.1 Study site

This study was conducted in Yunwu Mountain National Nature Reserve on the Loess Plateau (106°24′–106°28′ E, 36°13′–36°19′ N) (**Figure 1**) [45, 50].

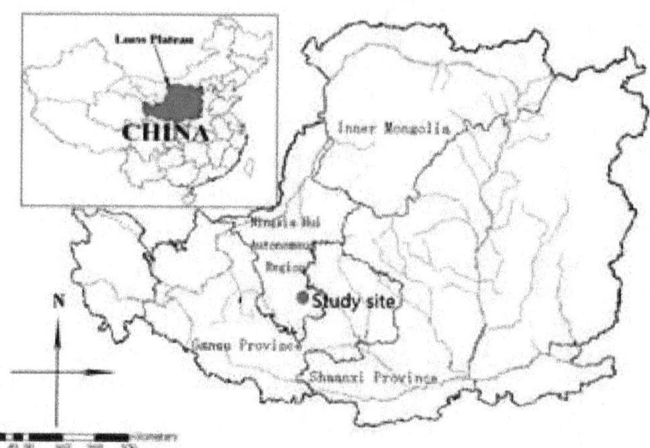


**Figure 1.**
*Location of experimental site.*

Grassland in this area was restored from grazing as a long-term ecological monitoring station since 1982. The elevation of this study area is 1800–2180 m and has a total area of 6660 $hm^2$. The mean annual temperature is 7.01°C, and there are on average 137 frost-free days per year [49]. The mean annual precipitation is 425 mm, with 60–75% of rainfall falling during July–September. The mean annual evaporation is 1017–1739 mm. Snow cover depth in winters averaged 1.2 cm during the dormancy period. The vegetation type is typical steppe. *Gentianaceae*, *Stipa* and *Potentilla* are important plant components, and the main dominant species include *S. bungeana*, *Stipa grandis*, *S. przewalskyi*, *Thymus mongolicus*, *Artemisia sacrorum*, *Potentilla acaulis* and *Androsace erecta* [45]. Soil type is montane grey-cinnamon soil [45].

## 2.2 Experimental design and sampling

### *2.2.1 Grassland ecological survey*

The grassland sites have been restored from grazing exclusion since 1982, and consequently goat grazing was excluded [45, 49, 50]. Three equal-sized transect of 300 × 100 m was established at the top, middle and down positions of the same slope, respectively. And, 15 quadrats (1 × 1 m) were established within each transect. The vegetation survey was carried out in mid- or late August each year during 1982–2011. Plant coverage, height, species abundance and plant density in each quadrat were measured. Aboveground parts of grassland plants were clipped and dried at 65°C for 48 h to determine aboveground biomass [43]. Plant roots of 0–120 cm soil layers were collected with a soil auger of 9 cm diameter, then were washed and dried to determine belowground biomass.

Important value (IV) was used to describe the importance of species in grassland community during the restoration process. Shannon-Wiener index was used to indicate diversity and evenness of plant community [50]. All indices were calculated according to 8 and 43.

Important value (IV)

$$IV = \frac{RH + RC + RA + RF}{4} \tag{1}$$

where IV is the important value, RH is the relative height, RC is the relative coverage, RA is the relative abundance and RF is the relative frequency.

Diversity index (H), using Shannon-Wiener index

$$H = -\sum_{i=1}^{S} P_i \ln P_i \tag{2}$$

where S is the total species number of a quadrat and $P_i$ is the relative importance value of species i.

### *2.2.2 Litter decomposition experiment*

Considering the difficulty of gathering sufficient senesced leaves, leaves of three *Stipa* species (*S. bungeana*, *S. grandis* and *S. przewalskyi*) were collected in August of 2013 and then dried at 40°C as decomposition materials, according to other decomposition studies [51–53]. Leaf litters were cut into pieces of 10 cm in length and enclosed in nylon bag (15 g $bag^{-1}$, 15 × 10 cm, 0.15 mm mesh).

In early October of 2013, the leaf litterbags of three *Stipa* species were transferred to grassland site restored for 23 years. Four plots of 10 × 10 m were established, and seven leaf litterbags of each *Stipa* species were placed on the soil surface and secured in place with iron nails on each of four plots. Four leaf litterbags of each *Stipa* species were harvested after 1, 3, 6, 9, 12, 18 and 24 months of incubation.

In the laboratory, leaf litters were removed from bags, cleaned to remove any extraneous material and weighed after drying at 65°C for 48 h. Leaf litters were analysed for carbon (C), nitrogen (N) and phosphorus (P). C was determined by oxidation with potassium dichromate in a heated oil bath. N was determined by the semimicro Kjeldahl method. P was determined by Olsen method [54].

According to [55], decomposition rate (k) of leaf litters was estimated by the negative exponential decay function:

$$\frac{X}{X_0} = e^{-kt} \tag{3}$$

where X is the remaining mass, $X_0$ is the initial mass and t is the decaying time (year).

Based on the nutrient concentration and remaining mass, we further calculated nutrient accumulation index (NAI) for C, N and P of leaf litters during decomposition process [56, 57]:

$$NAI = \frac{X_t \times C_t}{X_0 \times C_0} \times 100\% \tag{4}$$

where $X_0$ and $C_0$ indicate initial leaf litter mass and chemical element concentration, respectively. $X_t$ and $C_t$ indicate remaining leaf litter mass and chemical element concentration after a period of time t (year), respectively.

### 2.3 Data analyses

All data in the paper are presented as mean ± standard error. A two-way analysis of variance was conducted to determine the effects of decomposition time, species and their interaction on decomposition rate, nutrient concentration and NAI of leaf litters. A linear mixed model was used to examine correlations of vegetative indices with restoration time, productivity with climate variables and remaining mass with decomposition time. Significant differences of all statistical tests were estimated at a significance level of $P < 0.05$. All statistical analyses were performed using SPSS 18.0 (SPSS Inc., Chicago, IL, USA).

Partial least squares (PLS) regression was used to analyse the responses of grassland productivity to variation in daily temperature and precipitation dur-ing all 365 days of the year based on data for 1992–2011 [58, 59]. The two major outputs of PLS analysis are the variable importance in the projection (VIP) and standardised model coefficients. The VIP threshold for considering variables as important is often set to 0.8 [60]. The standardised model coefficients indicate the strength and direction of the impacts of each variable in the PLS model. The root-mean-square errors (RMSE) of the regression analyses were calculated to determine the accuracy of the PLS model. In the PLS analyses, periods with VIP greater than 0.8 and high absolute values of model coefficients represent the relevant phases influencing grassland productivity. Positive model coefficients indicate that increasing temperature or precipitation during the respective period should increase ANPP, while negative model coefficients imply negative impacts on productivity.

# 3. Results

## 3.1 Temperature and precipitation changes

The annual mean air temperature had an increasing trend and increased by 1.17°C from 1982 to 2011(**Figure 2A**). In contrast with mean annual temperature, mean annual precipitation showed a decreasing trend and larger intra- and interan-nual variations in our study, indicating the warmer and drier climate. The mean annual precipitation from 1982 to 2011 was 425.42 mm, with markedly lower values in 1986, 1991 and 1999 and with higher values in 2003 (**Figure 2B**).

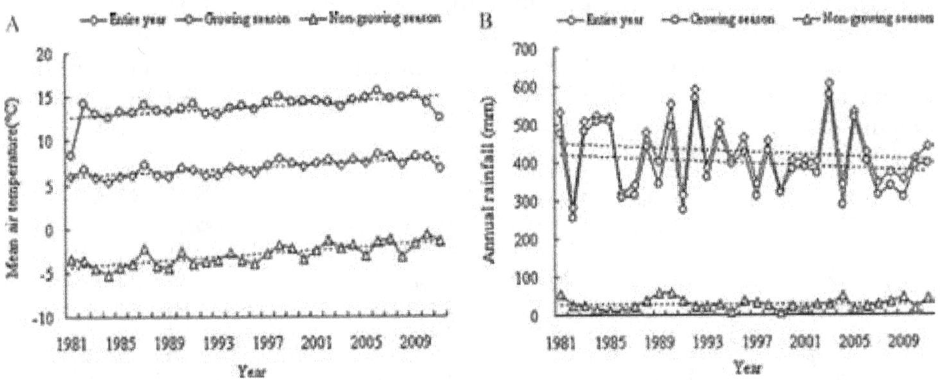


**Figure 2.**
*Mean annual air temperature (a) and mean annual rainfall (B) of growing season, nongrowing season and entire year at Yunwu Mountain during 1982–2011.*

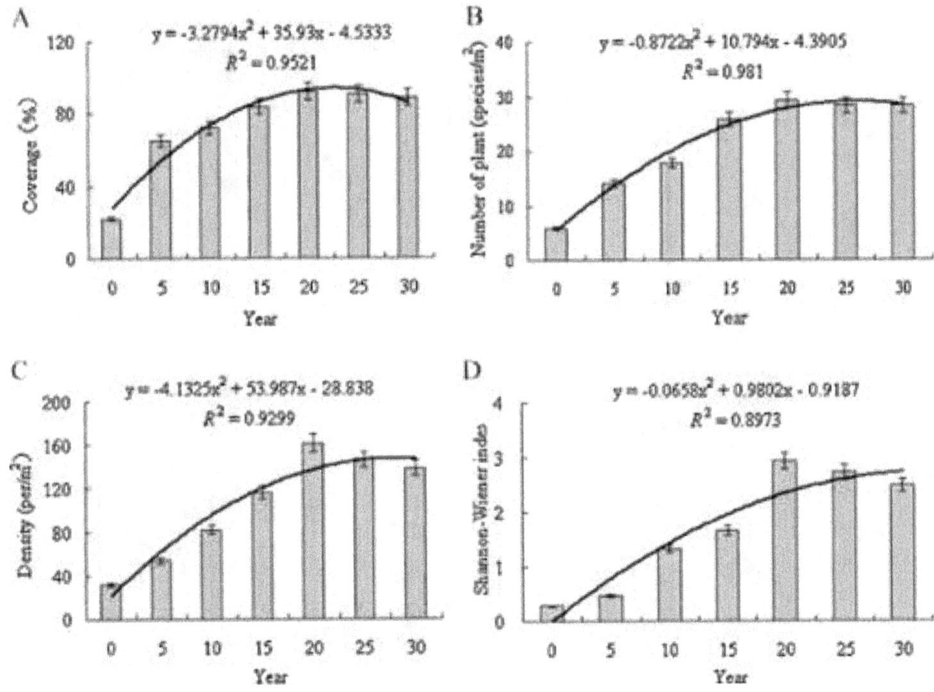


**Figure 3.**
*Changes of coverage (a), number of plants (B), plant density (C) and Shannon-wiener index (D) of grassland with grazing exclusion time at 5, 10, 15, 20, 25 and 30 years.*

### 3.2 Vegetative ecological characteristics after grazing exclusion

Grassland coverage, plant species richness (number of plant species), plant den-sity (number of plant individuals) and Shannon-Wiener index had similar variation tendencies during the three-decade restoration process (**Figure 3**). Initially, the coverage, plant richness, plant density and Shannon-Wiener index significantly increased. After 20 years' restoration, they reached peak values of 92.47%, 29.33 species $m^{-2}$, 161.8 individuals $m^{-2}$ and 2.93, respectively. With grazing exclusion process continuing, the four indices' values decreased to 88.73%, 28.2 species $m^{-2}$, 138.7 individuals $m^{-2}$ and 2.47, respectively (**Figure 3**).

### 3.3 Biomass changes in grassland community after grazing exclusion

There were significant differences in aboveground biomass between four groups and between total aboveground biomass and total belowground biomass (**Figure 4**).

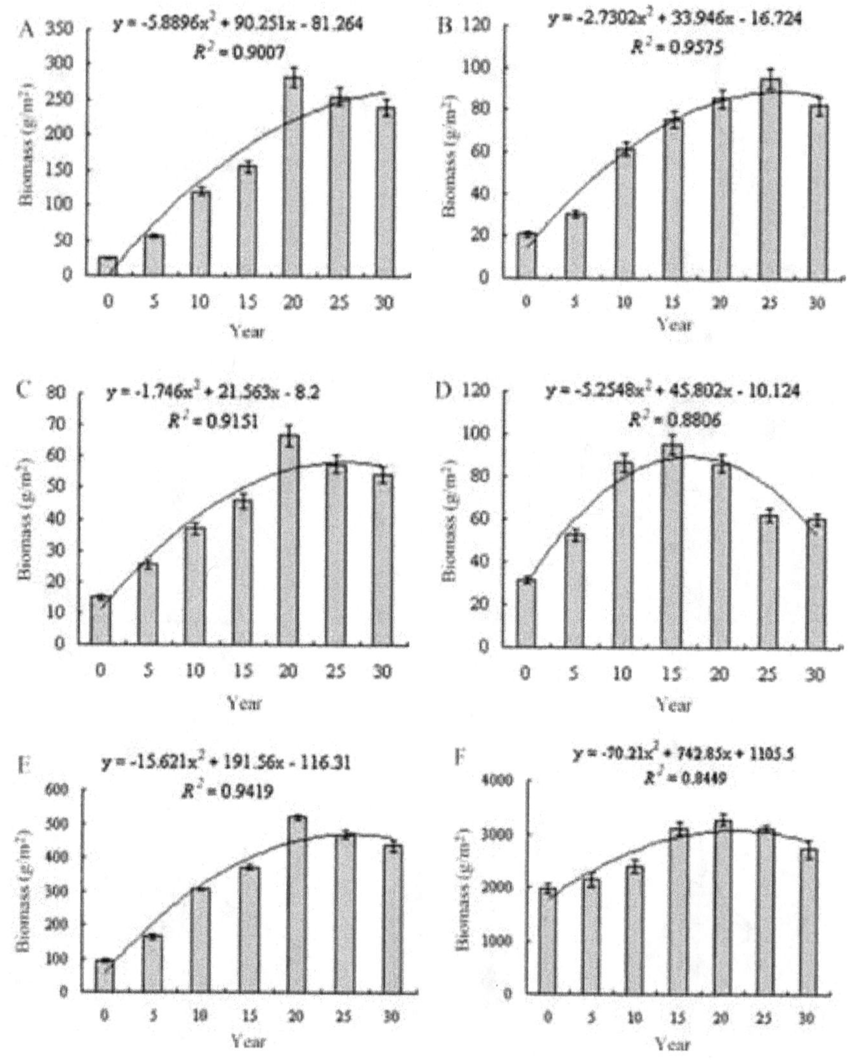


**Figure 4.**
*Biomass changes of Gramineae (a), Leguminosae (B), Compositae (C), weeds (D), aboveground community (E) and belowground community (F) of grassland with grazing exclusion time at 5, 10, 15, 20, 25 and 30 years.*

Aboveground biomasses of four plant groups increased with restoration time after grazing exclusion. Aboveground biomass of Gramineae and Compositae peaked at the 20th year, while that of Leguminosae peaked at the 25th year, and that of Weeds families peaked at the 15th year during restoration process after grazing exclusion. Considering the reduced biomasses of weed families, long-term grazing exclusion improved forage quality of grassland. Meanwhile, aboveground and belowground community biomasses were both increased by grazing exclusion. Since grassland mainly consisted of plants belonging to Gramineae and Compositae, peaks of the total above- and belowground community biomass both occurred at the 20th year, with aboveground community biomass of 520.5 g $m^{-2}$ and belowground community biomass of 3240.2 g $m^{-2}$ (**Figure 4**).

### 3.4 Responses of aboveground productivity to climate variation

Regression analysis showed that ANPP was significantly correlated with MAT (**Figure 5b**) but was little influenced by AP variations (**Figure 5a**).

The VIP and standardised model coefficients of the PLS analysis showed that impacts of warming on grassland productivity varied with season periods (**Figure 6a**). Different with the clear-cut impacts of temperature on ANPP, precipitation showed more complex impacts (**Figure 6b**).

### 3.5 Decomposition traits of leaf litters of three dominant *Stipa* species

The remaining mass of leaf litters decreased with decomposition time and showed significant differences among three *Stipa* species (**Figure 7**). At the end of decomposition experiment, the remaining masses of leaf litters of *S. bungeana*, *S. grandis* and *S. przewalskyi* were 64.47%, 61.53% and 65.78%, respectively (**Table 1**).

Different lowercase letters in the same column indicate significant differences ($P < 0.05$).

During 2 years' decomposition process, variations of nutrient concentration were affected by the nutrient type (**Figure 8**). In detail, concentrations of carbon and nitrogen showed species-specific fluctuations with decreasing tendency among three *Stipa* species. In contrast, phosphorus concentrations in leaf litters were averaged doubled. There were significant differences in C:N ratio and nutrient accumulation index (NAI) of leaf litters among three *Stipa* species (**Table 2**).

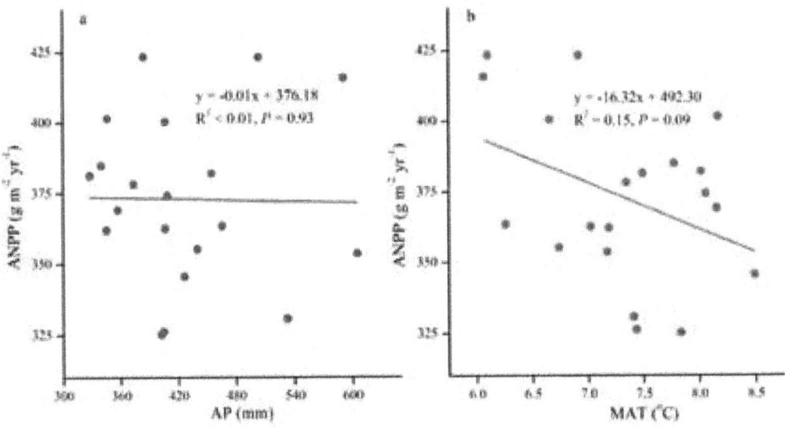


**Figure 5.**
*Correlations between ANPP and annual precipitation (a) and mean annual temperature (b) during 1992–2011 at Yunwushan. AP means annual precipitation and MAT represents mean annual temperature.*

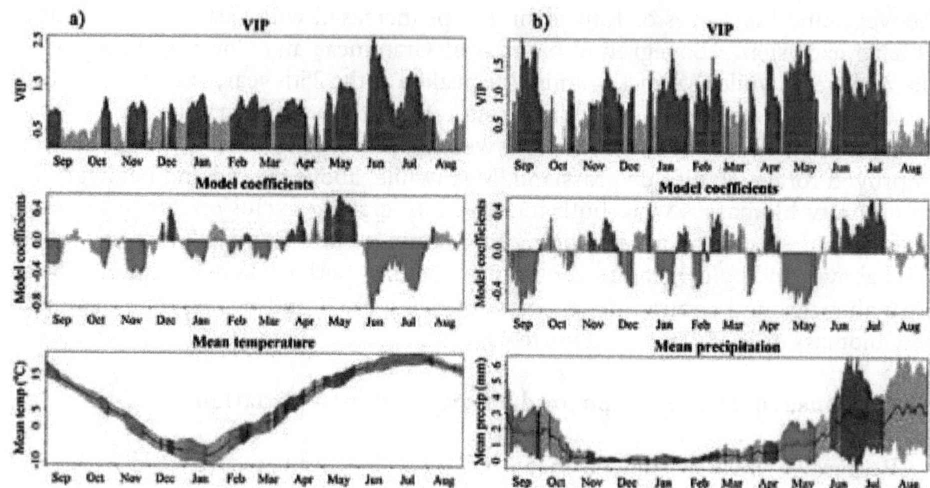


**Figure 6.**
*Results of partial least squares (PLS) regression correlating grassland productivity at Yunwu Mountain during 1992–2011 with 15-day running means of (a) daily mean temperature and (b) daily precipitation previously from September to august. Blue bars in the top row indicate that VIP values are greater than 0.8, the threshold for variable importance. In the middle row, red colour means model coefficients are negative and important, while green colour indicates important positive relationships between grassland productivity and climate variables. The black lines in the bottom panel stand for daily mean temperature and precipitation, while grey, green and red areas represent the standard deviation of daily climate variables.*

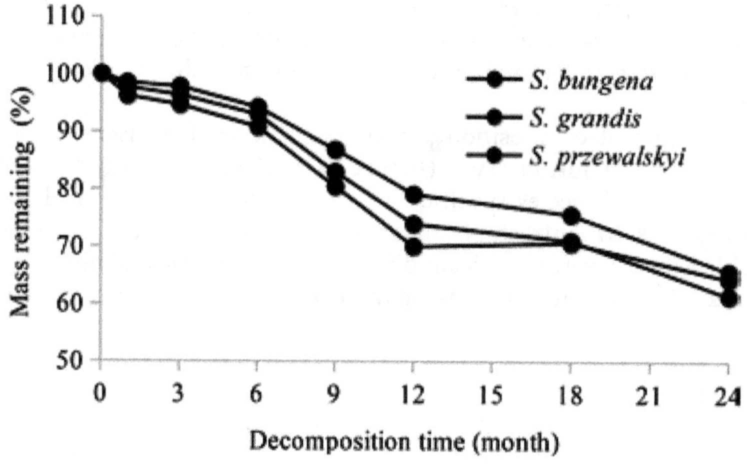


**Figure 7.**
*The remaining mass dynamics of leaf litters of three Stipa species during 2 years' field decomposition process.*

| Species | Remaining mass | | k-Value | |
|---|---|---|---|---|
| | First year | Second year | First year | Second year |
| *S. bungeana* | 70.05 ± 3.91 b | 64.47 ± 3.66 ab | 0.360 | 0.236 |
| *S. grandis* | 73.97 ± 1.81 ab | 61.53 ± 5.24 b | 0.320 | 0.242 |
| *S. przewalskyi* | 79.18 ± 1.49 a | 65.77 ± 1.80 a | 0.237 | 0.225 |

**Table 1.**
*Comparisons of litter decomposition traits after 1 and 2 years' decomposition between three Stipa species.*

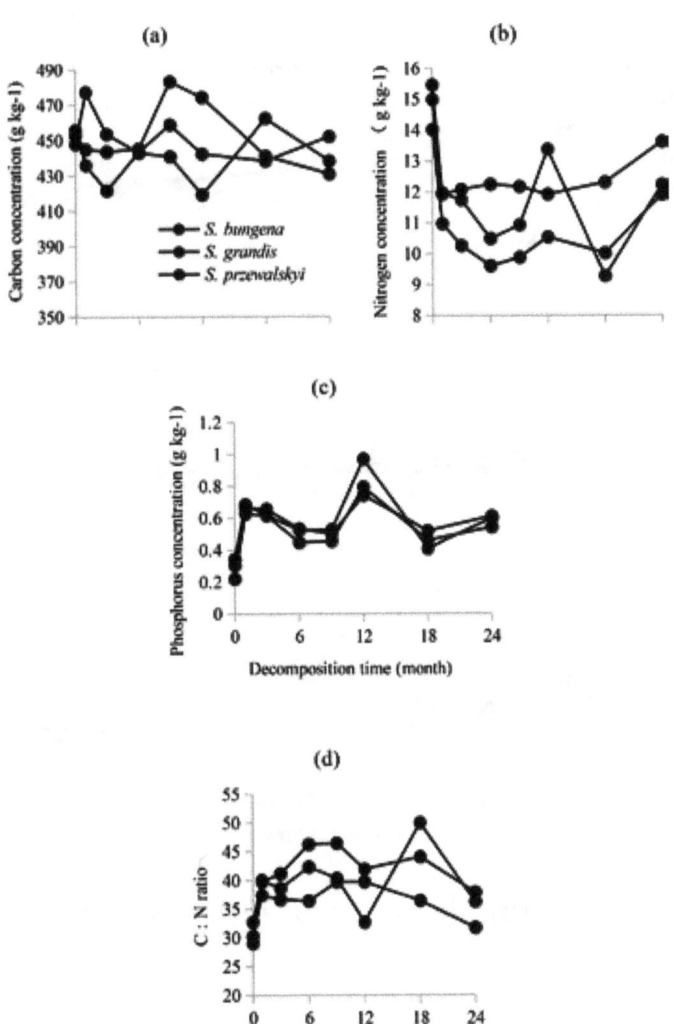


**Figure 8.**
*Dynamic of carbon (a), nitrogen (b), phosphorus (c), concentrations and C:N ratio (d) of leaf litters of three Stipa species during 2 years' field decomposition process.*

| Variables | df | Concentration ($g{\cdot}kg^{-1}$) | | | C/N | NAI | | |
|---|---|---|---|---|---|---|---|---|
| | | C | N | P | | C | N | P |
| Time | 6 | 0.575 ns | 4.701 ** | 39.564 *** | 3.877** | 49.738 *** | 23.944 *** | 53.070 *** |
| Species | 2 | 0.613 ns | 18.860 *** | 2.991 ns | 9.074** | 0.560 ns | 11.026 *** | 50.008 *** |
| Time× Species | 12 | 1.163 ns | 1.843 ns | 1.224 ns | 1.889 ns | 1.663 ns | 1.014 ns | 1.185 ns |

**Table 2.**
*Analysis of variance of decomposition time, species for nutrient concentration, C:N ratio and NAI.*

NAI, nutrient accumulation index; ns indicates no significant effects ($P > 0.05$). ** and *** indicate significant effects at $P < 0.01$ and $P < 0.001$ level, respectively.

Different with nutrient concentrations, nutrient accumulation indices in **Figure 9** indicated that C, N and P were all mineralised into soils during the decomposition process. There was no significant difference between species for carbon-releasing pattern (**Figure 9**).

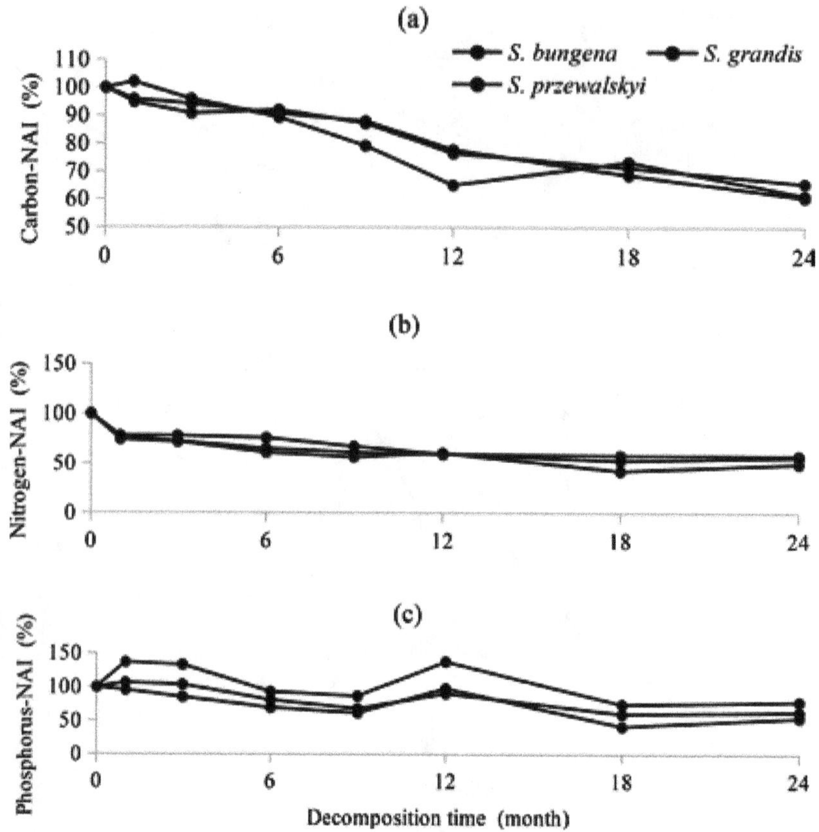


**Figure 9.**
*NAI dynamics for carbon (a), nitrogen (b), phosphorus (c) of leaf litters of three Stipa species during 2 years' field decomposition process.*

## 4. Discussion

Anthropogenic activities and climate changes have made important impacts on terrestrial ecosystem structure and functions in the last century [30]. Global surface temperatures during the twentieth century was increased by 0.56–0.92°C, while temperatures are predicted to have an increment of 2.0–4.5°C in the twenty-first century [61]. Annual mean air temperature was increased by 1.17°C from 1982 to 2011 in this study, having similar temperature changing trends with study in Xilingol steppe of Inner Mongolia [61]. In detail, temperature rises differenti-ated with seasons, with temperature rises of 1.01°C and 1.68°C in growing season and nongrowing season, respectively. Thus, the nongrowing season experienced a higher temperature rise than the growing season. In contrast with mean annual temperature, mean annual precipitation showed a decreasing trend and larger intra- and interannual variations in our study, indicating the warmer and drier climate. Previous researches have shown that vegetation characteristics could be improved using grazing exclusion in the degraded sandy grasslands, alpine meadow and wetlands in China [5, 62]. However, many of these restoration studies were based on a relatively short-term scale and the research strategy focusing on the spatial series substitute for temporal series methods [5, 63]. In this study, community coverage, plant species richness, plant density and Shannon-Wiener index had similar variation tendencies during the three-decade restoration process. After 20 years' restoration, they reached peak values, but these four index values decreased in the following years. These decreases mainly resulted from accumulation of litter, which reduced the access to light for plant seedlings [64, 65]. Overall, 30 years' restoration made plant species richness increase from 9.5 species $m^{-2}$ to 28 species $m^{-2}$ and make grassland coverage increase from 25 to 85%. In addition, plants were categorised into four groups: Gramineae, Leguminosae, Compositae and weeds. Considering the reduced biomasses of weeds, long-term grazing exclusion improved forage quality of grassland. Meanwhile, aboveground and belowground community biomasses were both increased by grazing exclusion.

The rapid recovery due to grazing exclusion played a more important role than climatic variations in regulating grassland ecosystem. Therefore, datasets of aboveground grassland biomass and climate variables during 1992–2011 were used to examine the impacts of climate variations on aboveground net primary productivity (ANPP). Regression analysis showed that ANPP was significantly correlated with MAT and was little influenced by AP variations, while precipitation is regarded as the most important determinant of grassland productivity in arid and semiarid regions [19, 21, 66]. Considering the neglected temporal variation of annual climate variables, more attentions should be paid to studies at higher temporal resolution attributing impacts of climate variation on grassland produc-tivity to seasonal or even daily variation in climatic variables rather than to annual variation [26, 27, 28, 29, 31]. A low root-mean-square error (RMSE) of 8.13 g $m^{-2}$ indicated a good fit of the data for the resulting PLS model. The VIP and stan-dardised model coefficients of the PLS analysis showed that impacts of warming on grassland productivity varied with season periods. Since model coefficients in April and May were always positive and VIP values mostly exceeded 0.8, warming in this period had a positive impact on grassland productivity. The positive impacts of warming in spring on grassland productivity may result from increased water absorption, N mineralisation, accelerated snowmelt and advanced spring greening for plants, which may lengthen the growing season and increase photosynthesis and carbon acquisition for plants [13, 67–69].

Warming in summer (June–July) depressed productivity, forming a striking contrast with the impacts of spring warming. The results can be explained by physi-ological stress for plant growth generated by warming in summer coinciding with drought [70]. Moreover, warming in summer may reduce soil moisture by increas-ing evapotranspiration [71]. It is believed that climate variations make impacts on grassland productivity through changes of soil moisture [24, 72, 73]. Furthermore, continuous warming and drought in summer reduced productivity by limiting soil resource availability [74, 75]. And, temperature variation in August had no apparent impacts on grassland productivity.

The majority of published studies have focused on productivity responses to climate variability during the growing season. However, the importance of winter climate is getting more and more attentions [76–80]. Considering the majority of model coefficients during September–March, high temperature at that time was unfavourable for productivity of the following year. Temperature increases during September–October delay the senescence of grassland, which may increase soil nutrient and water depletion, inhibiting biomass production in the following year [36, 69, 81]. Our results were similar with warming experiments in two limestone grasslands in the UK, which showed that winter heating combined with drought reduced the biomass of both communities [11]. Besides, warmer winter can accel-erate snowmelt, resulting in declines of snow cover accompanied with increases of frequency of freezing events, which exerted negative impacts on plant growth [76, 82]. Also, warming in winter may delay the fulfilment of chilling require-ments of plants for resuming growth in the following spring or even delay onset of spring phenology [58, 59, 77–79, 83].

Interestingly, some short intervals with positive coefficients during 1 November–29 March were detected during 1992–2011, indicating a complex physi-ological and ecological process in dormancy period of grassland. Taking a broader view at model coefficients and aiming at consistency with established phenological phases, we interpreted the entire period (November–March) as another relevant period during which temperature increases appeared to reduce grassland produc-tivity. Therefore, we recommend that more scientific attention should be paid to impacts of winter warming on grassland productivity and the timing of spring phenology events.

The daily precipitation values between the previous September and August were also used as independent variables in the PLS analysis. The resulting model still proved to be a good fit for the data, with an RMSE of 6.53 g $m^{-2}$. Different with the clear-cut impacts of temperature on ANPP, precipitation showed more complex impacts. Precipitation increases in June and July had positive impacts on productivity, while increasing precipitation during the senescence period (September–October) and the early growing season (April–May) was correlated with low productivity. In contrast to studies reporting the positive impacts of precipitation during April–May on grassland productivity [29, 30], results in the present study can be explained by the site hydrology, with frequent winter snow providing suf-ficient soil water for plant growth, making sporadic precipitation during April–May (with an average of 59.5 mm during 1992–2011) which has less important direct impacts on grassland productivity. Similarly, there was also no significant relation-ship between grassland ANPP and precipitation in August. Similar results have also been reported for grasslands in Kansas, USA [13]. During the dormancy period, positive impacts of precipitation were almost offset by negative ones; thus, precipi-tation seemed to have little impacts on grassland productivity during this period.

Investigating the decomposition traits of dominant *Stipa* species' (*S. bungeana*, *S. grandis* and *S. przewalskyi*) litters can reveal the ecosystem cyclic process under grazing exclusion and climatic changes. The remaining mass of leaf litters decreased with decomposition time and showed significant differences among three *Stipa* species. At the end of decomposition experiment, the remaining mass of leaf litters of *S. bungeana*, *S. grandis* and *S. przewalskyi* were 64.47%, 61.53% and 65.78%, respectively. Therefore, *S. grandis* decomposed fast, and *S. przewalskyi* had a slow decomposition rate. Additionally, leaf litters decomposed faster in growing season (6–12 month and 18–24 month) than in nongrowing season (0–6 month and 12–18 month). The decomposition rate (k) was calculated based on the regression of negative exponential decay function, with k-values of 0.360, 0.320 and 0.237 after 1 year's decomposition for *S. bungeana*, *S. grandis* and *S. przewalskyi*, respectively. Similarly, k-values after 2 years' decomposition of *S. bungeana*, *S. grandis* and *S. przewalskyi* were 0.236, 0.242 and 0.225, respectively. Since higher k-values indicate higher decomposition rates, we concluded that litter's decaying progress became difficult as decomposition time increases, mainly due to the depletion of soluble compounds and easily decayed parts at the beginning of decomposition process, leaving hard parts such as lignin to decay slowly [39].

The variations of nutrient concentration were affected by nutrient type during 2 years' decomposition process during 2013–2014(**Figure 8** and **Table 2**). In detail, concentrations of carbon and nitrogen showed species-specific fluctuations with decreasing tendency among three *Stipa* species (**Figure 8a** and **b**). In contrast, phosphorus concentrations in leaf litters were averaged doubled (**Figure 8c**), indicating immobilisation of P in the leaf litters, possibly due to microbial immobilisation through the uptake of P from soil solution and translocation of P from fungal hyphae [84]. There were significant differences in C:N ratio of leaf litters among three *Stipa* species (**Figure 8d**). *S. przewalskyi* had higher C:N ratio than *S. bungeana*, which explained the differences of decomposition rates between them. As the dominant species in late succession stage of grassland, C:N ratio of *S. przewalskyi* litters did not show a lower value as predicted from other studies [51], possibly due to the divergences of climate and species between two regions. C:N ratio has been proven to be negatively correlated with decomposition rate. Besides, the lower k-value after 2 years' decomposition process could be explained by the increased C:N ratio of leaf litters. Compared with nutrient concentrations, nutrient accumulation indices indicated that C, N and P were all mineralised into soils during the decomposition process. There was no significant difference between species for carbon-releasing pattern. Still, NAI value for C of *S. przewalskyi* was higher than two other *Stipa* species after 2 years' decomposition (**Figure 9**). The lower NAI values for N and P of *S. bun-geana* indicated that *S. bungeana* released more N and P to soil than the two other *Stipa* species. From this perspective, replacement of *Stipa* species after long-term grazing exclusion might inhibit nutrient cycling of grassland ecosystem, due to the lower nutrient mineralisation in leaf litters of two *Stipa* species at middle and late succession stage.

## 5. Conclusion

The present study indicated that grazing exclusion induced positive effects on grassland vegetative characteristics, with peak values in the 20th year (2002), and long-term grazing exclusion led to decreased species diversity and biomass and can inhibit grassland renewing due to the litter accumulation. Besides, nutrient cycling in grassland might be slowed down through replacement of dominant species during long-term grazing exclusion. Grassland productivity was more influenced by temperature than precipitation. Results indicated that analysis of productiv-ity responses should account not only for the magnitude of climate variation but also for its timing. Climate warming might prolong/shorten growing season by advancing/delaying onset of greenness of plant community. Warmer winter further decreases ANPP, and impacts of warming in early spring should also be considered in evaluating ANPP variability. Therefore, more scientific attention should be paid to trends in spring phenology and their impacts on productivity at species and community levels.

## Acknowledgements

This work was supported by the National Key Research and Development Program of China (2016YFC0500700), National Natural Science Foundation of China (41601586, 41230852, 41701606, 31602004, 31601987, 41601053), Deployment Program of the Chinese Academy of Sciences (KJZD-EW-TZ-G10), Technical System of Modern Forage Industry of the Ministry of Agriculture (CARS- 35-40) and Key Cultivation Project of the Chinese Academy of Sciences—the promotion and management of ecosystem functions of restored vegetation on the Loess Plateau, China. We thank the staff at the management bureau of Yunwu Mountain National Nature Reserve for fieldworks and weather data collection since the early 1980s.

## Author details

Jimin Cheng[1,2], Wei Li[1,2*], Jishuai Su[3], Liang Guo[1,2], Jingwei Jin[1,2] and Chengcheng Gang[1,2]

1 Institute of Soil and Water Conservation, Northwest A&F University, Yangling, China

2 Institute of Soil and Water Conservation, Chinese Academy of Sciences and Ministry of Water Resources, Yangling, China

3 State Key Laboratory of Vegetation and Environmental Change, Institute of Botany, Chinese Academy of Sciences, Beijing, China

*Address all correspondence to: liwei2013@nwsuaf.edu.cn

## References

[1] Shi Y, Wang Y, Ma Y, Ma W, Liang C, Flynn DFB, et al. Field-based observations of regional-scale, temporal variation in net primary production in Tibetan alpine grasslands. Biogeosciences. 2014;**11**(7):2003-2016

[2] Zhang JT, Ru WM, Li B. Relationships between vegetation and climate on the loess plateau in China. Folia Geobotanica. 2006;**41**:151-163

[3] Slimani H, Aidoud A, Roze F. 30 years of protection and monitoring of a steppic rangeland undergoing desertification. Journal of Arid Environments. 2010;**74**:685-691

[4] Simmons MT, Venhaus HC, Windhager S. Exploiting the attributes of regional ecosystems for landscape design: The role of ecological restoration in ecological engineering. Ecological Engineering. 2007;**30**(3):201-205

[5] Wang XH, Yu JB, Zhou D, Dong HF, Li YZ, Lin QX, et al. Vegetative ecological characteristics of restored reed (*Phragmites australis*) wetlands in the Yellow River Delta, China. Environmental Management. 2012;**49**:325-333

[6] Walker K, Stevens P, Stevens D, Mountford J, Manchester S, Pywell R. The restoration and re-creation of species-rich lowland grassland on land formerly managed for intensive agriculture in the UK. Biological Conservation. 2004;**119**(1):1-18

[7] Zhang JT. Succession analysis of plant communities in abandoned croplands in the eastern loess plateau of China. Journal of Arid Environments. 2005;**63**(2):458-474

[8] Zhang JT, Dong YR. Factors affecting species diversity of plant communities and the restoration process in the loess area of China. Ecological Engineering. 2010;**36**(3):345-350

[9] Wei J, Cheng J, Li W, Liu W. Comparing the effect of naturally restored forest and grassland on carbon sequestration and its vertical distribution in the Chinese loess plateau. PLoS One. 2012;**7**(7):e40123

[10] Li W, Epstein HE, Wen ZM, Zhao J, Jin JW, Jing GH, et al. Community-weighted mean traits but not functional diversity determine the changes in soil properties during wetland drying on the Tibetan plateau. Solid Earth. 2017;**8**:137-147

[11] Grime JP, Brown VK, Thompson K, Masters GJ, Hillier SH, Clarke IP, et al. The response of two contrasting limestone grasslands to simulated climate change. Science. 2000; **289**(5480):762-765

[12] Knapp AK, Fay PA, Blair JM, Collins SL, Smith MD, Carlisle JD, et al. Rainfall variability, carbon cycling, and plant species diversity in a Mesic grassland. Science. 2002;**298**(5601):2202-2205

[13] Craine JM, Nippert JB, Elmore AJ, Skibbe AM, Hutchinson SL, Brunsell NA. Timing of climate variability and grassland productivity. Proceedings of the National Academy of Sciences of the United States of America. 2012;**109**(9):3401-3405

[14] Hsu JS, Powell J, Adler PB. Sensitivity of mean annual primary production to precipitation. Global Change Biology. 2012;**18**(7):2246-2255

[15] Hastings A, Byers JE, Crooks JA, Cuddington K, Jones CG, Lambrinos JG, et al. Ecosystem engineering in space and time. Ecology Letters. 2007;**10**(2):153-164

[16] Fang J, Piao S, Tang Z, Peng C, Ji W. Interannual variability in net primary production and precipitation. Science. 2001;**293**(5536):1723

[17] Knapp AK, Smith MD. Variation among biomes in temporal dynamics of aboveground primary production. Science. 2001;**291**(5503):481-484

[18] Guo Q, Hu Z, Li S, Li X, Sun X, Yu G. Spatial variations in aboveground net primary productivity along a climate gradient in Eurasian temperate grassland: Effects of mean annual precipitation and its seasonal distribution. Global Change Biology. 2012;**18**(12):3624-3631

[19] Sala OE, Gherardi LA, Reichmann L, Jobbágy E, Peters D. Legacies of precipitation fluctuations on primary production: Theory and data synthesis. Philosophical Transactions of the Royal Society B. 2012;**367**(1606):3135-3144

[20] Rosenzweig ML. Net primary productivity of terrestrial communities: Prediction from climatological data. The American Naturalist. 1968;**102**(923):67-74

[21] Lauenroth WK, Sala OE. Long-term forage production of north American shortgrass steppe. Ecological Applications. 1992;**2**(4):397-403

[22] Bai Y, Han X, Wu J, Chen Z, Li L. Ecosystem stability and compensatory effects in the Inner Mongolia grassland. Nature. 2004;**431**(7005):181-184

[23] Hu Z, Fan J, Zhong H, Yu G. Spatiotemporal dynamics of aboveground primary productivity along a precipitation gradient in Chinese temperate grassland. Science in China Series D: Earth Sciences. 2007;**50**(5):754-764

[24] Hsu JS, Adler PB. Anticipating changes in variability of grassland production due to increases in inter-annual precipitation variability. Ecosphere. 2014;**5**(5):1-15

[25] Guo L, Cheng JM, Luedeling E, Koerner SE, He JS, Xu JC, et al. Critical climate periods for grassland productivity on China's loess plateau. Agricultural and Forest Meteorology. 2017;**233**:101-109

[26] Milchunas DG, Forwood JR, Lauenroth WK. Productivity of long-term grazing treatments in response to seasonal precipitation. Journal of Range Management. 1994;**47**(2):133-139

[27] Chou WW, Silver WL, Jackson RD, Thompson AW, Allen-Diaz B. The sensitivity of annual grassland carbon cycling to the quantity and timing of rainfall. Global Change Biology. 2008;**14**(6):1382-1394

[28] La Pierre KJ, Yuan S, Chang CC, Avolio ML, Hallett LM, Schreck T, et al. Explaining temporal variation in above-ground productivity in a Mesic grassland: The role of climate and flowering. Journal of Ecology. 2011;**99**(5):1250-1262

[29] Robinson TMP, La Pierre KJ, Vadeboncoeur MA, Byrne KM, Thomey ML, Colby SE. Seasonal, not annual precipitation drives community productivity across ecosystems. Oikos. 2013;**122**(5):727-738

[30] Ma WH, Liu ZL, Wang ZH, Wang W, Liang CZ, Tang YH, et al. Climate change alters interannual variation of grassland aboveground productivity: Evidence from a 22-year measurement series in the inner Mongolian grassland. Journal of Plant Research. 2010;**123**:509-517

[31] Hovenden MJ, Newton PCD, Wills KE. Seasonal not annual rainfall determines grassland biomass response to carbon dioxide. Nature. 2014;**511**(7511):583-586

[32] Chen F, Weber KT. Assessing the impact of seasonal precipitation and temperature on vegetation in a grass-dominated rangeland. Rangeland Journal. 2014;**36**(2):185-190

[33] Chollet S, Rambal S, Fayolle A, Hubert D, Foulquié D, Garnier E. Combined effects of climate, resource availability, and plant traits on biomass produced in a Mediterranean rangeland. Ecology. 2014;**95**(3):737-748

[34] Menzel A, Estrella N, Fabian P. Spatial and temporal variability of the phenological seasons in Germany from 1951 to 1996. Global Change Biology. 2001;**7**(6):657-666

[35] Piao S, Mohammat A, Fang J, Cai Q , Feng J. NDVI-based increase in growth of temperate grasslands and its responses to climate changes in China. Global Environmental Change. 2006;**16**(4):340-348

[36] Reyes-Fox M, Steltzer H, Trlica MJ, Mcmaster GS, Andales AA, Lecain DR, et al. Elevated $CO_2$ further lengthens growing season under warming conditions. Nature. 2014;**510**(7504):259-262

[37] Xin Z, Xu J, Zheng W. Spatiotemporal variations of vegetation cover on the Chinese loess plateau (1981-2006): Impacts of climate changes and human activities. Science in China Series D: Earth Sciences. 2008;**51**(1):67-78

[38] Dang XH, Liu GB, Xue S. Models of soil and water conservation and ecological restoration in the loess hilly region of China. Transactions of the Chinese Society of Agricultural Engineering. 2010;**26**(9):72-80

[39] Moretto AS, Distel RA. Decomposition of and nutrient dynamics in leaf litter and roots of *Poa ligularis* and *Stipa gyneriodes*. Journal of Arid Environments. 2003;**55**(3):503-514

[40] Callaway RM. Positive interactions among plants. The Botanical Review. 1995;**61**:306-349

[41] Jiao JY, Tzanopoulos J, Xofis P, Bai WJ, Ma XH, Mitchley J. Can the study of natural vegetation succession assist in the control of soil erosion on abandoned croplands on the loess plateau, China? Restoration Ecology. 2007;**15**(3):391-399

[42] Jiao JY, Tzanopoulos J, Xofis P, Mitchley J. Factors affecting the distribution of vegetation types on abandoned cropland in the hilly-gullied loess plateau region of China. Pedosphere. 2008;**18**(1):24-33

[43] Cheng JM, Cheng J, Shao HB, Yang XM. Effects of mowing disturbance on the community succession of typical steppe in the loess plateau of China. African Journal of Agricultural Research. 2012;**7**(37):5224-5232

[44] Zhu RB, Cheng JM, Liu YJ, Li WJ, Wei L. Floristic study of spermatophyte in Yunwu Mountain natural reserves of China. Acta Agrestia Sinica. 2012;**20**(3):439-443

[45] Qiu LP, Wei XR, Zhang XC, Cheng JM. Ecosystem carbon and nitrogen accumulation after grazing exclusion in semiarid grassland. PLoS One. 2013;**8**(1):e55433

[46] Armas C, Kikvidze Z, Pugnaire FI. Abiotic conditions, neighbour interactions, and the distribution of *Stipa tenacissima* in a semiarid mountain range. Journal of Arid Environments. 2009;**73**:1084-1089

[47] Lv XM, Zhou GS, Wang YH, Song XL. Sensitive indicators of zonal Stipa species to changing temperature and precipitation in Inner Mongolia grassland, China. Frontiers in Plant Science. 2016;**7**:73

[48] Lu SL, Wu ZL. On geographical distribution of the genus *Stipa* L. in China. Acta Phytotaxonomica Sinica. 1996;**34**(3):242-253

[49] Su JS, Jing GH, Jin JW, Wei L, Liu J, Cheng JM. Identifying drivers of root community compositional changes in

semiarid grassland on the loess plateau after long–term grazing exclusion. Ecological Engineering. 2017;**99**:13-21

[50] Jing ZB, Cheng JM, Su JS, Bai Y, Jin JW. Changes in plant community composition and soil properties under 3-decade grazing exclusion in semiarid grassland. Ecological Engineering. 2014;**64**:171-178

[51] Moretto AS, Distel RA, Didoné NG. Decomposition and nutrient dynamic of leaf litter and roots from palatable and unpalatable grasses in a semi–arid grassland. Applied Soil Ecology. 2001;**18**(1):31-37

[52] Mathers NJ, Jalota RK, Dalal RC, Boyd SE. $^{13}$C–NMR analysis of decomposing litter and fine roots in the semi–arid Mulga lands of southern Queensland. Soil Biology and Biochemistry. 2007;**39**(5):993-1006

[53] McLaren JR, Turkington R. Plant functional group identity differentially affects leaf and root decomposition. Global Change Biology. 2010;**16**(11):3075-3084

[54] Jing ZB, Cheng JM, Chen A. Assessment of vegetative ecological characteristics and the succession process during three decades of grazing exclusion in a continental steppe grassland. Ecological Engineering. 2013;**57**:162-169

[55] Olson JS. Energy storage and the balance of producers and decomposers in ecological systems. Ecology. 1963;**44**(2):322-331

[56] Fornara DA, Tilman D, Hobbie SE. Linkages between plant functional composition, fine root processes and potential soil N mineralization rates. Journal of Ecology. 2009;**97**(1):48-56

[57] Romero LM, Smith TJ, Fourqurean JW. Changes in mass and nutrient content of wood during decomposition in a South Florida mangrove forest. Journal of Ecobiology. 2005;**93**(3):618-631

[58] Guo L, Dai J, Ranjitkar S, Xu J, Luedeling E. Response of chestnut phenology in China to climate variation and change. Agricultural and Forest Meteorology. 2013;**180**:164-172

[59] Guo L, Dai J, Wang M, Xu J, Luedeling E. Responses of spring phenology in temperate zone trees to climate warming: A case study of apricot flowering in China. Agricultural and Forest Meteorology. 2015;**201**:1-7

[60] Wold S. PLS for multivariate linear modeling. In: van der Waterbeemd H, editor. Chemometric Methods in Molecular Design: Methods and Principles in Medicinal Chemistry. Weinheim, Germany: Verlag-Chemie; 1995. pp. 195-218

[61] Han YR, Kun AL, Ma ZG, Masae SYM. Forty-eight-year climatology of air temperature and precipitation changes in Xilinhot, Xilingol steppe (Inner Mongolia), China. Grassland Science. 2011;**57**:168-172

[62] Yuan JY, Ouyang ZY, Zheng H, Xu WH. Effects of different grassland restoration approaches on soil properties in the southeastern Horqin sandy land, northern China. Applied Soil Ecology. 2012;**61**:34-39

[63] Zhang C, Xue S, Liu GB, Song ZL. A comparison of soil qualities of different revegetation types in the loess plateau, China. Plant and Soil. 2011;**347**:163-178

[64] Ungar I. Are biotic factors significant in influencing the distribution of halophytes in saline habitats? The Botanical Review. 1998;**64**(2):176-199

[65] Lambers H, Mougel C, Jaillard B, Hinsinger P. Plant–microbe–soil interactions in the rhizosphere: An

evolutionary perspective. Plant and Soil. 2009;**321**:83-115

[66] Oesterheld M, Loreti J, Semmartin M, Sala OE. Inter-annual variation in primary production of a semi-arid grassland related to previous-year production. Journal of Vegetation Science. 2001;**12**(1):137-142

[67] Sierra J. Temperature and soil moisture dependence of N mineralization in intact soil cores. Soil Biology and Biochemistry. 1997;**29**(9-10):1557-1563

[68] Nemani RR, Keeling CD, Hashimoto H, Jolly WM, Piper SC, Tucker CJ, et al. Climate-driven increases in global terrestrial net primary production from 1982 to 1999. Science. 2003;**300**(5625):1560-1563

[69] Bradford J, Lauenroth W, Burke I, Paruelo J. The influence of climate, soils, weather, and land use on primary production and biomass seasonality in the US Great Plains. Ecosystems. 2006;**9**(6):934-950

[70] De Boeck HJ, Bassin S, Verlinden M, Zeiter M, Hiltbrunner E. Simulated heat waves affected alpine grassland only in combination with drought. The New Phytologist. 2016;**209**(2):531-541

[71] Epstein HE, Lauenroth WK, Burke IC, Coffin DP. Ecological responses of dominant grasses along two climatic gradients in the Great Plains of the United States. Journal of Vegetation Science. 1996;**7**(6):777-788

[72] Knapp A, Burns C, Fynn RS, Kirkman K, Morris C, Smith M. Convergence and contingency in production–precipitation relationships in north American and south African C4 grasslands. Oecologia. 2006;**149**(3):456-464

[73] Koerner SE, Collins SL. Interactive effects of grazing, drought, and fire on grassland plant communities in North America and South Africa. Ecology. 2014;**95**(1):98-109

[74] de Valpine P, Harte J. Plant responses to experimental warming in a montane meadow. Ecology. 2001;**82**(3):637-648

[75] Dessureault-Rompré J, Zebarth BJ, Georgallas A, Burton DL, Grant CA, Drury CF. Temperature dependence of soil nitrogen mineralization rate: Comparison of mathematical models, reference temperatures and origin of thesoils. Geoderma. 2010;**157**(3-4):97-108

[76] Bokhorst SF, Bjerke JW, Tømmervik H, Callaghan TV, Phoenix GK. Winter warming events damage sub-Arctic vegetation: Consistent evidence from an experimental manipulation and a natural event. Journal of Ecology. 2009;**97**(6):1408-1415

[77] Yu H, Luedeling E, Xu J. Winter and spring warming result in delayed spring phenology on the Tibetan plateau. Proceedings of the National Academy of Sciences of the United States of America. 2010;**107**(51):22151-22156

[78] Cook BI, Wolkovich EM, Parmesan C. Divergent responses to spring and winter warming drive community level flowering trends. Proceedings of the National Academy of Sciences of the United States of America. 2012;**109**(23):9000-9005

[79] Fu YH, Zhao H, Piao S, Peaucelle M, Peng S, Zhou G, et al. Declining global warming effects on the phenology of spring leaf unfolding. Nature. 2015;**526**(7571):104-107

[80] Ladwig LM, Ratajczak ZR, Ocheltree TW, Hafich KA, Churchill AC, Frey SJK, et al. Beyond arctic and alpine: The influence of winter climate on temperate ecosystems. Ecology. 2016;**97**(2):372-382

[81] Qian S, Wang LY, Gong XF. Climate change and its effects on grassland productivity and carrying capacity of livestock in the main grasslands of China. Rangeland Journal. 2012;**34**(4):341-347

[82] Grimm NB, Chapin FS, Bierwagen B, Gonzalez P, Groffman PM, Luo Y, et al. The impacts of climate change on ecosystem structure and function. Frontiers in Ecology and the Environment. 2013;**11**(9):474-482

[83] Luedeling E, Guo L, Dai J, Leslie C, Blanke MM. Differential responses of trees to temperature variation during the chilling and forcing phases. Agricultural and Forest Meteorology. 2013;**181**:33-42

[84] Aerts R, Chapin FS. The mineral nutrition of wild plants revisited: A re-evaluation of processes and patterns. Advances in Ecological Research. 1999;**30**:1-67

# 4

# Plant Phenology and an Assessment of the Effects Regarding Heavy Metals, Nanoparticles and Nanotubes on Plant Development: Runner Bean, Artichoke and Chickpea Seedlings

*Feyza Candan*

**Abstract**

The relationship between environmental pollution and nutrition in particular, which forms the basis of health, is fundamentally important for protecting human health. Therefore, the data obtained from the examination of how plants and animals consumed as food are affected by environmental pollution can be seen as an indicator of their effects on humans. On the other hand, the role of technology and nanotechnology in life has been increasing in this century, and a considerable amount of heavy metals, nanoparticles (NPs), and nanotubes (NTs) are released to the environment. The results of morphological or anatomical examination of runner bean (*Phaseolus coccineus* L) and artichoke (*Cynara scolymus* L.) plants subjected to copper (Cu) and lead (Pb) heavy metals and chickpea (*Cicer arietinum* L) plants subjected to Au nanoparticles and $C_{70}$ single-walled carbon nanotubes (SWNTs) are presented with this study in the point of their phenological development process. The three taxa belonging to Fabaceae and Asteraceae families with high economic status and having flowers with characteristic features were chosen deliberately as representatives. This chapter presents a study that will shed light on future biomonitoring-based studies focusing on the impact of environmental pollution on plants phenology with economic value.

**Keywords:** heavy metal, nanoparticle, nanotube, runner bean (*Phaseolus coccineus* L), artichoke (*Cynara scolymus* L), chickpea (*Cicer arietinum* L), morphology, anatomy

## 1. Introduction

It is a known fact that environmental pollution constitutes an important problem in Turkey as well as in the rest of the world. Rapid industrialization and population growth have caused pollution in the atmosphere, pedosphere, and hydrosphere.

Therefore, it is seen that countries pay particular attention to pollution-related studies and health problems caused by pollution and allocate high amounts of resources to deal with the problem.

Heavy metals show toxic effects at certain concentrations for living organisms. However, low concentrations of some heavy metals are essential for normal and healthy plant growth. Furthermore, heavy metals and nanoparticles are causes of concern because they can penetrate into different parts and cells of plants at different rates, and by this way, they enter the food chain and reach the living beings.

There are about 22,000 bryophyte species and 20,000 algae species; however, vascular plants are the dominant plant group in the world with 255,000 species. Land plants, which perform their life cycles completely in the terrestrial environment, are mainly composed of bryophytes and vascular plants. Furthermore, at least a thin film of water is required for fertilization in all taxa except seed plants. Even in the two primitive genera seed plants, cycad and gingko, fertilization is a result of free-swimming spermatozoids released into the liquid medium in the archegonium chamber [1–3].

One of the most important features of vascular plants is the presence of buds at the ends of the trunk and side branches in the gymnosperms and generally in the angiosperms. The bud is an apical meristem coated with protective bud scales. Meristem is the region of cells to which new cells, tissues, and organs are added and has the potential for active cell division and contributes to plant growth. Therefore, despite the limited growth potential in animals, plant growth is limitless due to the presence of apical meristem. However, the development of plant parts, such as leaves, flowers, and fruits, is limited to their shapes and is genetically predetermined [1]. In short, when evaluated from a phenological point of view, plant parts do not show any further growth independent of the time they remain on the plant after completing their development.

Cell development and differentiation take place as the changes occurring in protoplast; for example with the fusion occurring in vacuoles to grow, via structures such as mitochondria, plastids and the golgi body, endoplasmic reticulum, microtubules, and microfilaments in cytoplasm. Cell walls differentiate and increase in thickness due to structural and environmental effects, and they may become permeable. Moreover, the walls may integrate with the lignin, which increases tensile forces. Tissues formed by the differentiation of apical meristem include parenchyma, collenchyma, sclerenchyma, and primary xylem and primary phloem, in which the pith and cortex are formed [1, 2].

Phenological stages are divided into eight possible principal stages: [1] bud development, [2] leaf development, [3] shoot/branch development, [4] inflorescence emergence, [5] flowering, [6] fruit development, [7] fruit maturity, and [8] senescence and the beginning of dormancy [3]. Secondary parts and secondary metabolites occur in the plant during the phenological cycle [1].
Genotype and environmental factors are involved in the emergence of seconder metabolites. In this case, based on the amount of soil, water, and air pollution in the environment in which the plant grows, various deteriorations may occur as a result of morphological and physiological changes whose effects on the plant can be seen with the naked eye or observed only through microscopic examinations.
In this chapter, general information about heavy metal and nanoparticles is given, and the effects of heavy metals and nanoparticles on the seedlings of runner bean (*Phaseolus coccinea*), chickpea (*Cicer arietinum*), and artichoke (*Cynara scolymus*) species, which are of economic importance, were examined morphologically and anatomically.

## 2. Effects of heavy metals, nanoparticles, and nanotubes on plant phenological development

The term heavy metal has been used by scientists with various definitions for about 60 years. An element with a density of more than 7 g/cm$^3$, in 1987 with a density of more than 4 g/cm$^3$, in 1992 with a density greater than 5 g/cm$^3$, and in 1995 with a density of 6 g/cm$^3$ with a metallic property was classified as heavy metal in 1964. Some scientists have classified heavy metals according to their atomic weights, atomic numbers, other chemical properties, and toxic properties. In biological terms, the term heavy metal is generally used for possible contamination of metals and metalloids on the environment and in terms of their toxicity or ecotoxicity [4].

Heavy metals are released into the atmosphere, pedosphere, and hydrosphere every day due to human activities besides natural causes, such as volcanic activities. Flying ashes from the chimneys of cement plants and thermal power plants; the use of heavy metal paint; the smoke emitted by motor vehicles as well as their plastic-based parts such as brake pads, garbage, and waste sludge incineration plants; and the release of industrial wastes, such as pesticides, fertilizers, paper, batteries, products, etc. are among the main causes of heavy metal pollution [5–7].

The discharge of heavy metal-containing particles released from the factory and plant chimneys onto agricultural lands, their dissolution in the soil by rain or irrigation, or the irrigation of agricultural land mixed with industrial wastewater leads to various diseases in crops grown on such lands and damages the agricultural economy [8–10].

Heavy metals have toxic effects for living organisms at certain concentrations. However, certain critical concentrations of some heavy metals are necessary for normal and healthy plant growth. Therefore, heavy metals are classified as essential elements and nonessential elements according to their participation in life processes. Cobalt (Co), copper (Cu), manganese (Mn), molybdenum (Mo), iron (Fe), nickel (Ni), and zinc (Zn) are heavy metals necessary for the growth and vitality of plants and are considered essential elements. Heavy metals such as barium (Ba), cadmium (Cd), mercury (Hg), antimony (Sb), lead (Pb), and chromium (Cr) are not essential for plants and other living organisms and are called nonessential elements [11].

Essential elements are found as a cofactor in many enzyme systems and as a structural component in biological processes in living organisms. For example, copper is an essential element for normal plant growth at certain concentrations. Copper is an essential cofactor for many metalloproteins in plants and plays a role in photosynthetic electron transport, mitochondrial respiration, cell wall metabolisms, and hormone signal transduction pathways [12, 13].

High concentrations of copper (depending on plant species) show toxicity in plants although it is an essential element. Lead is not an essential element and shows toxic properties for plants. The presence of excess copper and lead in the environment negatively affects phenological development in plants [14].

These heavy metals result in lipid peroxidation [15], degradation of cell and thylakoid membrane structure, and a decrease in chlorophyll amount due to the change in the chloroplast structure and thus chlorosis as a result of the oxidative damage they caused [16]. Heavy metals bind to sulfhydryl (-SH) groups of proteins and inhibit enzyme activity [17] and cause oxidative DNA damage [18, 19], chromosomal abnormalities [20], and lack of other essential elements [21–24].

More than 30 base lesions were characterized by DNA exposure to reactive oxygen species [25]. On the DNA, reactive oxygen species can cause single-nucleobase lesions, single-strand breaks, double-strand breaks, and various oxidative damages such as base connections in the strand [26–28].

Contamination of soils with heavy metals and the accumulation of heavy metals in high concentrations in plants grown here have a genotoxic effect in plants and lead to mutation-like changes in the DNA profile. Therefore, a connection can be established between these changes in the organism and the intensity of pollution in the soil [29].

Sresty and Rao (1999) examined the ultrastructural changes in the nucleolus, nucleus, endoplasmic reticulum, and vacuoles in pea plant stem cells in response to zinc and nickel stress [30].

Zengin and Munzuroğlu (2004) observed the root, stem, and leaf growth in bean (*Phaseolus vulgaris* L.) seedlings exposed to lead and copper stress and exam-ined which tissue was affected more in heavy metal stress [14].

Soudek et al. (2010) exposed flax (*Linum usitatissimum* L.) seeds to different concentrations of lead, nickel, copper, zinc, cadmium, cobalt, arsenic (As), and chromium heavy metals and examined the effects of heavy metal stress on plant germination and root development [31].

Öztürk Çalı and Candan have studied the effects of fungicide on the morphology and viability of pollens of tomato (*Lycopersicon esculentum* Mill.) [32]; the effect of activator application on the anatomy, morphology, and viability of tomato pollen [33]; and influence of activator on meiosis of tomato [34].

Candan and Öztürk Çalı (2015) have observed pollen micromorphology of four taxa of *Anemone coronaria* L. from western Turkey [35]. On the other hand, the authors have compared the pollen morphology and viability of four naturally distributed and commercial varieties of *Anemone coronaria* [36].

Some studies have used various methods based on single-cell gel electrophoresis (comet assay), micronucleus analysis, or cytogenetic analysis in order to investigate the genotoxic effects of pollution on plants.

Steinkellner et al. (1999) treated samples of *Tradescantia* sp. with water from seven regions of Austria where the water was exposed to industrial pollution, and they examined the chromosomal changes in the cells of the root region of the plant by micronucleus analysis. The authors reported negative changes in plant stem cells at the end of the study [37].

Menke et al. exposed the root area of *Arabidopsis thaliana* (L.) to different genotoxic effects. They examined the damage caused by the genotoxic effect in the plant by single-cell gel electrophoresis method and successfully demonstrated the mutagenic effect occurring in stem cell nuclei [38].

The *Comparison of Physiological, Biochemical and Molecular Parameters in Seedlings of Artichoke (Cynara scolymus L.) and Runner Bean (Phaseolus coccineus L.) Seeds Exposed to Lead (Pb) Heavy Metal Stress in the point of Ecological Pollution* was studied with the 2012-057 numbered project supported by Manisa Celal Bayar University [39]. Candan and Batır have presented this scientific important compar-ison at a conference after the project was completed [40]. On the other hand, Batır has studied on the thesis about determination of the DNA changes in the artichoke seedlings (*Cynara scolymus* L.) subjected to lead and copper stresses [41], and Batır et al. have written an article about that topic [42]. The original PCR photographs of some primers used related runner bean and artichoke samples are given below [39, 41] (**Figures 1–6**).

Today, especially due to increasing demand and changing climatic conditions, many studies have been carried out in plant biotechnology regarding more resistant agricultural plants against factors such as drought, salinity, freezing, and heavy metal contamination. Various biological, chemical, and physical methods are

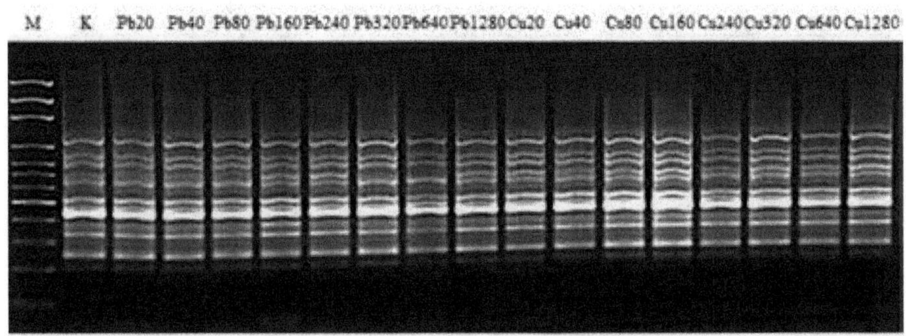


**Figure 1.**
*PCR gel photograph of OP A03 primer used related runner bean samples [39].*

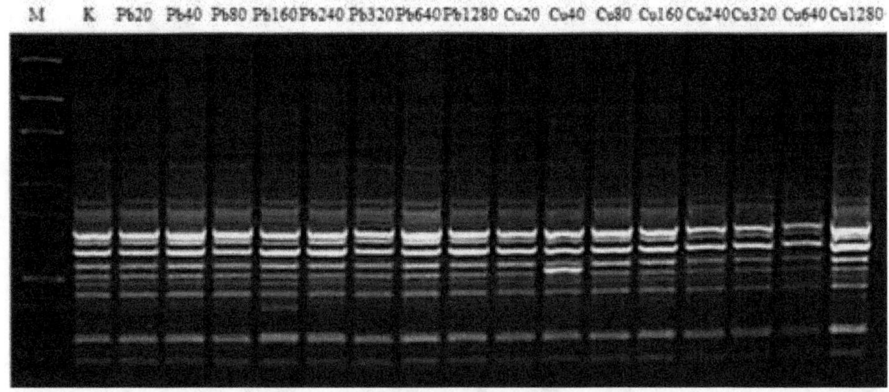


**Figure 2.**
*PCR gel photograph of OP C05 primer used related runner bean samples [39].*

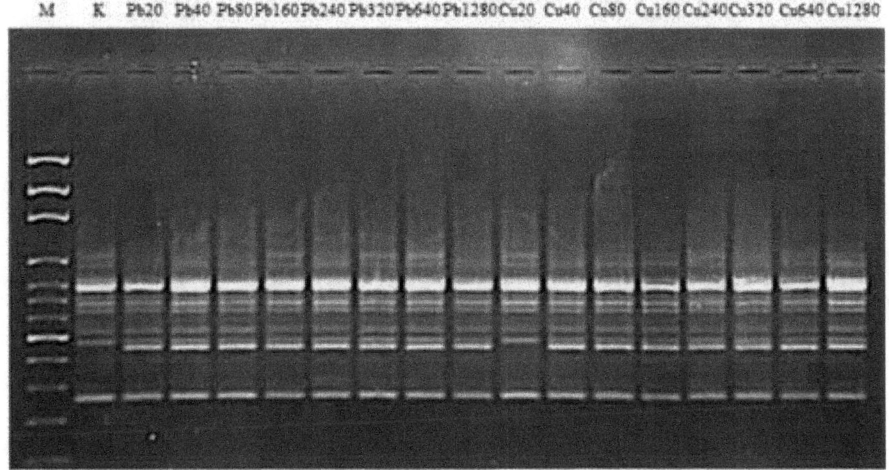


**Figure 3.**
*PCR gel photograph of OP C20 primer used related runner bean samples [39].*

available as regards obtaining biomolecules; for example, nanomaterials with their considerable reactions have much attention in biomass. The basis of this study is to determine the genotoxic effect levels of different stress factors on different plant

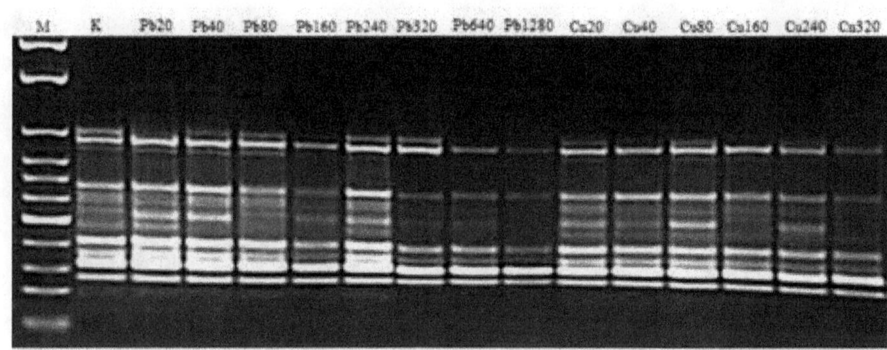


**Figure 4.**
*PCR gel photograph of OP Co3 primer used related artichoke samples [39, 41].*

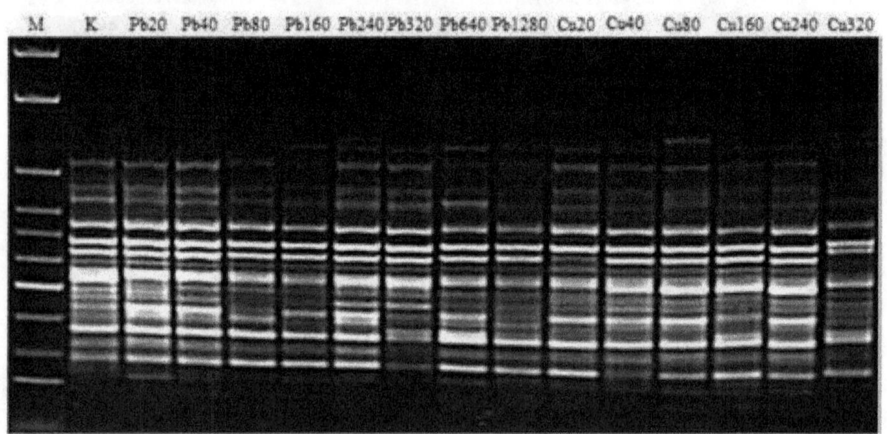


**Figure 5.**
*PCR gel photograph of OP Co5 primer used related artichoke samples [39, 41].*

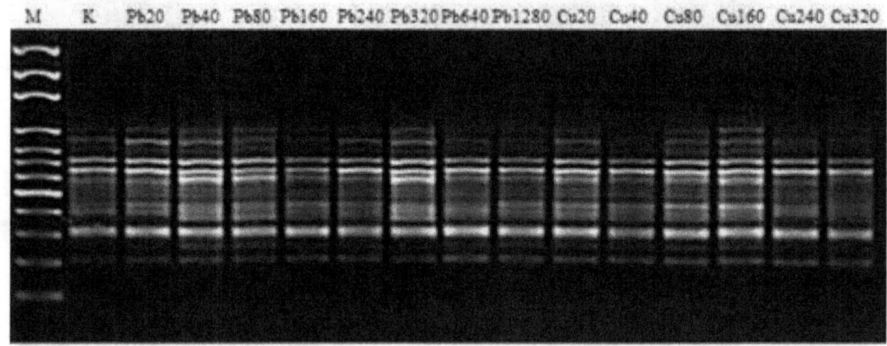


**Figure 6.**
*PCR gel photograph of OP C18 primer used related artichoke samples [39, 41].*

species. Until recently, the investigation of the effects of stress factors as heavy metals, nanoparticules, and nanoparticles on plant phenology remained at cellular, morphological, and anatomical levels.

## 3. Materials and methods

Fabaceae family to which runner bean and chickpea plants belong and Asteraceae family to which artichoke belongs were selected as the study material. Both of them are families with large numbers of economically important plants in Turkey as well as in the rest of the world. The flowers of the Fabaceae family are zygomorphic in shape and have legume and lomentum fruit. There are many flowers lined up on the flower tray (receptaculum) and head (capitulum) formed by the bracts surrounding these flowers, and there are achene type fruits in the Asteraceae family [43–46].

The aim of the present study was to investigate the effects of copper and lead heavy metals on runner bean (*Phaseolus coccineus*) and artichoke (*Cynara scolymus*) seedlings. In addition, the effects of Au nanoparticles and $C_{70}$ single-walled carbon nanotubes on chickpea (*Cicer arietinum*) seedlings were investigated morphologically. The tolerability of heavy metal and nanoparticle effects by these plants, the cultivation in heavy metal or nanoparticle-contaminated areas, the morphological and anatomical reflections of the changes in genomes, and to which extent the plant's general structure are preserved compared to controls were evaluated in this way.

### 3.1 Germination and cultivation of runner bean and artichoke seeds

Runner bean and artichoke seeds were sterilized and then planted for growing. At least 20 seeds were included and observed in the control group and for each heavy metal application. Seeds were planted and germinated in viols with fine-grained perlite [47].

$CuCl_2$ $2H_2O$ and Pb $(CH_3COO)_2$ $3H_2O$ solutions were applied in concentrations of 20, 40, 80, 160, 240, 320, 640, and 1280 ppm to the runner bean and artichoke seeds planted in groups of 3(Cu 20, Cu 40, Cu 80, Cu 160, Cu 320, Cu 640, Cu 1280 and Pb 20, Pb 40, Pb 80, Pb 160, Pb 320, Pb 640, Pb 1280). This procedure was repeated for 21 days. The seeds of the control group were planted and irrigated with distilled water. As a result, seedlings of the control group and those subjected to Cu-Pb heavy metal stress were obtained after 21 days [47] (**Figures 7, 8**).

a

b

**Figure 7.**
*Runner bean seedlings treated with Pb grown in viol. (a) general view at development phase. (b) general view after development.*

a

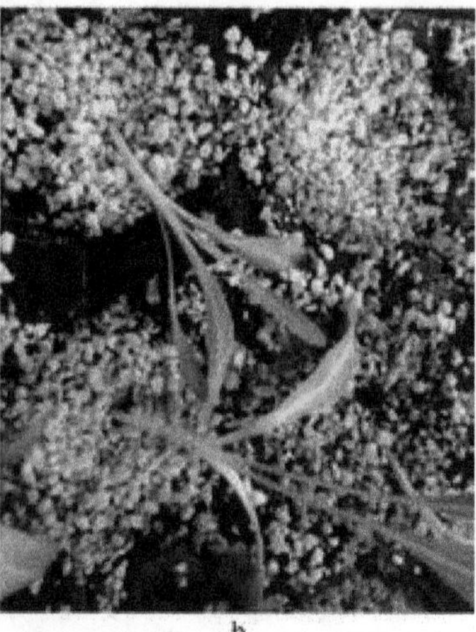

b

**Figure 8.**
*Artichoke seedlings treated with Pb grown in viol. (A) General view. (b) Close-up view of samples treated with 20 ppm Pb.*

**Figure 9.**
*General view of the control group grown in the climate cabinet, chickpea seedlings treated with Au NPs and $C_{70}$ SWNTs and pots with late germination.*

### 3.2 Germination and cultivation of chickpea seeds

Twenty chickpea seeds were exposed to 4 ml Au NPs and $C_{70}$ SWNTs and 15 ml deionized water mixture for 2 days, and they were grown in pots with perlite for 21 days in two groups. The control group of 20 seeds was also grown in other pots with perlite. All the plants in this group were watered every day in the morning only with water. Au NPs and $C_{70}$ single-walled carbon nanotubes exposed to chickpea plants and control group were taken from the pots after 3 weeks, and herbarium materials were made. On the other hand, some of them were stored at 70% alcohol for microscopical investigations [48] (**Figure 9**).

### 3.3 Methods used to obtain anatomical data

Cross sections were taken from taxa to determine and compare the characteristics of root and stem anatomy of runner bean and artichoke seedlings exposed to Cu and Pb heavy metal concentrations. While determining the samples from seedlings to take sections, 3–10 mm from the end of the roots and the middle part of the body above the ground were used. These fragments were used for sectioning with microtome by the paraffin method. All plant samples were subjected to various treatments to make the sections suitable for microtome removal [47].

These operations were carried out on the samples taken from the abovementioned parts of the plants retained in 70% alcohol. These parts were passed through 80, 90, and 100% alcohol and 2 alcohol/1 xylol, 1 xylol/1 alcohol, 1 alcohol/2 xylol, and 100% xylol solutions, in this order. The paraffin was allowed to penetrate the interior of the samples which were kept in the laboratory drying oven at 60° for 48 hours. Sections of 20, 25, 30, 35, and 40 µm used in the investigations were obtained via samples placed in paraffin blocks. The sections were placed on slides properly by using a hot water bath set at 40 °, and they were fixed onto the slides with adhesive. Sections were cleared off paraffin using 100% xylol, 1 xylol/1 alcohol, absolute alcohol, 95% alcohol, 80% alcohol, 70% alcohol, and purified water, in this order, for 5 minutes each, and then they were stained with safranin and fast green. Samples were kept in pure water, 70% alcohol, 80% alcohol, 95% alcohol, absolute alcohol, 1 xylol: 1 alcohol, and 100% xylol, for 1 minute each, so that water removal from the tissues was completed [49–51]. After removal of all the water, the preparations which were made permanent using Entellan and allowed to dry for 4–5 days at room temperature were examined in general. In this way, the reaction of the samples (with different concentrations of heavy metal in the cells) to dyes and the possible staining status were determined.

During the examination of the sections, the treatment of plant tissues with dye has caused a problem since the samples contain heavy metals, such as copper and lead, and they affect the physiology of the plant. However, it is known that in permanent preparations, an artificial appearance is obtained by losing some of the chemical content of the plant material and pigments due to the fact that the plant materials pass through a considerable amount of chemical stages. Therefore, it has been stated in some studies that permanent preparation methods are not suitable for some plants [52, 53]. It was also tried to make hand cross sections on artichoke and runner bean samples. The anatomical features of the taxa were evaluated and interpreted according to Carlquist, Fahn, and Yentür [2, 54, 55].

Dyes were prepared using different ratios of safranin and fast green, and all of them were tested for staining of heavy metal-treated anatomical specimens in accordance with the literature [50, 51], and examinations were performed on the specimens deemed appropriate.

Furthermore, objectives of 4, 10, 20, and 40 were used for examination and photographing the anatomical structures of the root and stem of the taxa. Runner bean and artichoke root and stem photographs were taken with 4, 10, and 20 objectives, and the unit of measurement was determined as 100 (µm).

## 4. Results

### 4.1 Morphological observations

Runner bean seeds were germinated at all concentrations of Cu and Pb. However, there wasn't any germination observed in 640 and 1280 ppm concentrations of Cu in the artichoke seeds, but germination occurred in all concentrations of Pb [39, 47].

It was determined in the morphological observations of runner bean and artichoke plants subjected to heavy metals that significant differences occurred in different doses of phytotoxic effects of Cu and Pb [47].

The control group samples of the runner bean and artichoke plants were photographed in order to compare the heavy metal phytotoxic effects on the morphological characteristics of the plants. Moreover, the general appearance and close-up photographs of the samples related to runner bean plant irrigated with Cu 20 ppm and Pb 160 and Pb 1280 ppm concentrations and artichoke plant irrigated with Cu 20 and 40 ppm and Pb 20 and 40 ppm were taken (**Figures 10–15**).

Cu and Pb heavy metals caused various phytotoxic effects in cases where the recommended dosage was exceeded or excessive pollution occurred in any other way was determined as the result of the study when the changes in the morphological structures of the plants were examined. Phytotoxicity seen in the morphological structure of the plant emerged as bending, shrinkage, and dark spots on the end of the leaves. On the other hand, while plant root, stem, and leaf lengths increase in low doses, high concentrations (640–280 ppm) cause size reduction and incomplete development [47] (**Figures 10–15**).

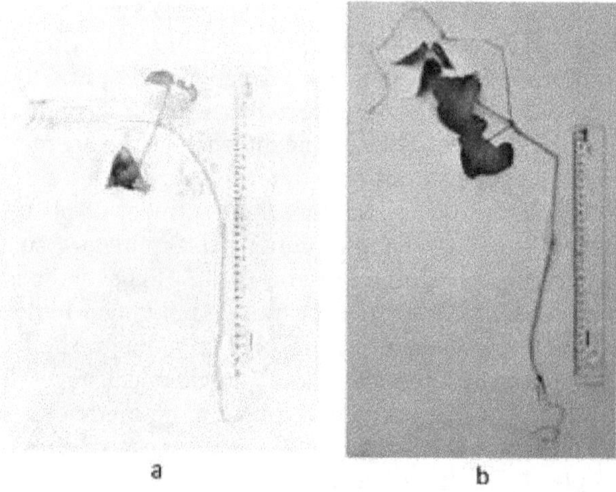

**Figure 10.**
*(a) Control group of runner bean seedling. (b) runner bean seedlings treated with Cu 20 ppm [47].*

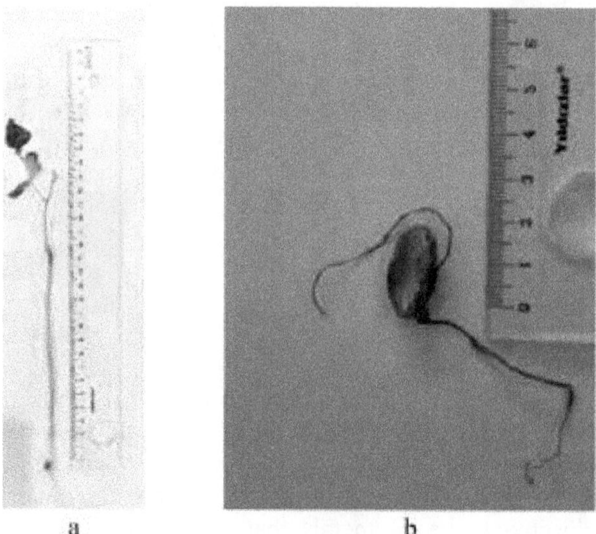

**Figure 11.**
*(a) Runner bean seedling treated with Pb 1280 ppm. (b) Undeveloped runner bean seed treated with Pb 160 ppm [47].*

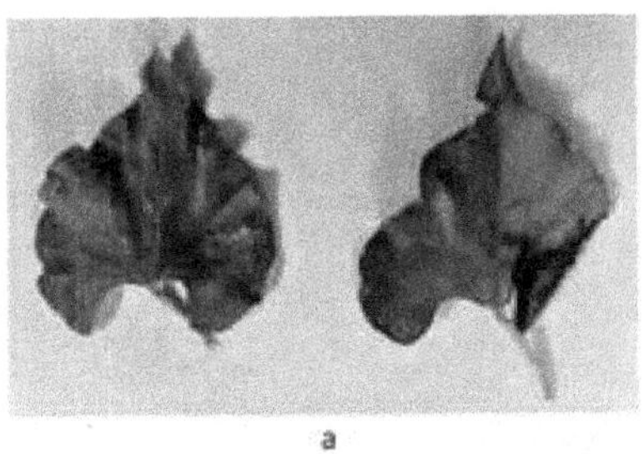

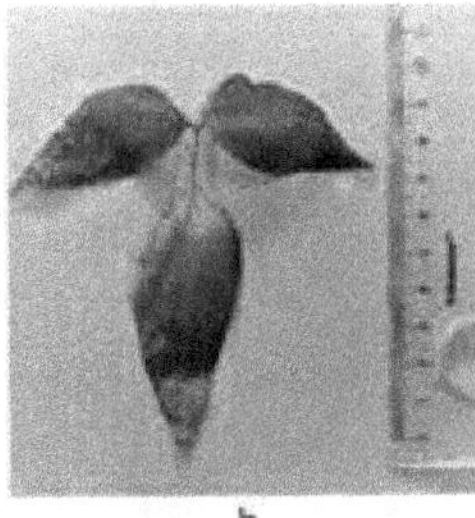

**Figure 12.**
*(a) Runner bean seedlings treated with Pb 1280 ppm. (b) General view of anomalies in terms of shape, chlorosis, and hole [47].*

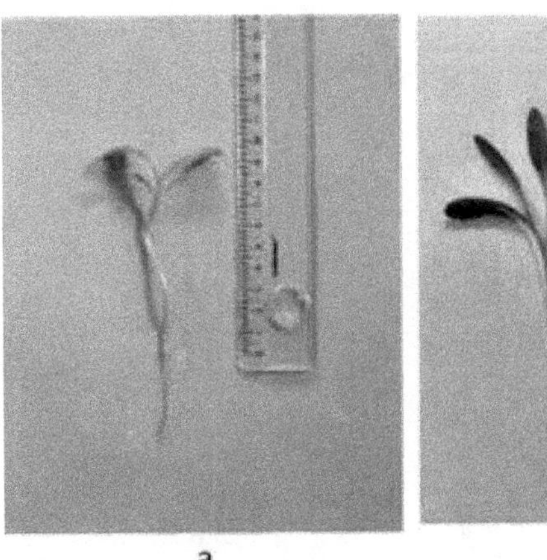

**Figure 13.**
*(a) Artichoke seedling of control group. (b) Artichoke seedling treated with Cu 20 ppm [47].*

**Figure 14.**
*Artichoke seedling treated with Cu 40 ppm. (a) General view. (b) Close-up view of the dried leaves [47].*

It was observed that there was an increase in plant root, stem, and leaf sizes in both treatment groups when chickpea plant control group and the plants subjected to Au Nps and $C_{70}$ SWNTs were compared. Furthermore, it was determined that there were increases in the number of fibrous roots, nodes, and subbranches in both groups (**Figure 16**).

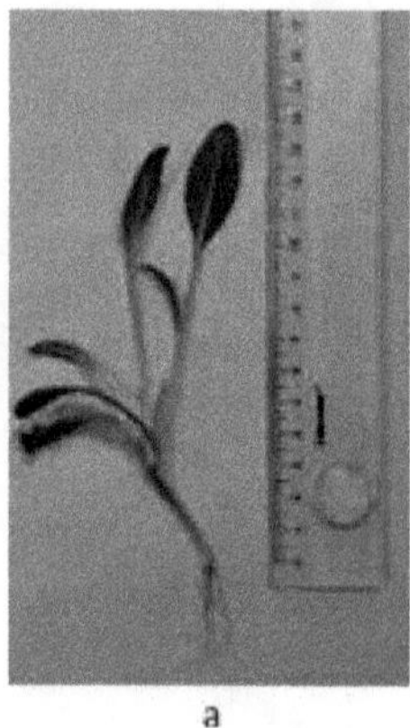

a b

**Figure 15.**
*Artichoke seedling treated with Pb 40 ppm. (a) general view. (b) close-up view of the dried leaves [47].*

a b

**Figure 16.**
*(a) Chickpea seedling of the control group and the sample treated with Au NPs. (b) Chickpea seedlings treated with $C_{70}$ SWNTs.*

## 4.2 Anatomical observations

Micrometer was selected as the unit for measurements taken from root and stem cross sections of the runner bean and artichoke seedlings. Cross sections taken from the roots and stem parts of the plants were considered suitable for evaluation. The roots and stems, epidermis, vascular bundle elements, secretary canals, sclerenchyma, starch sheath, cortex and pith cells, and cambium cells were measured, and the presence and variety of crystals were examined and compared [47].

Photographs of root and stem cross sections of the runner bean and artichoke plants were taken in order to compare the anatomical effects of heavy metal phytotoxic effects on the morphological characteristics of the plants. Runner bean root cross-sectional photographs were taken from root samples subjected to Cu 80 ppm and 640 ppm, and Pb 640 ppm concentrations and stem cross-section photographs were taken from the samples subjected to Cu 20, 80, and 640 ppm concentrations. Artichoke seedlings of root cross sections treated with Cu 160 ppm and Pb 320 and 640 ppm concentrations and stem cross sections treated with Cu 20 and 160 ppm and Pb 1280 ppm concentrations were examined [47] (**Figures 17–25**).

However, diseases caused by heavy metal stress in the plant, such as chlorosis and necrosis, and epidermal thickening, density of crystallization, increase in hairiness, and thinning in vascular bundles had negative effects on staining in anatomical studies and caused the tissues not to absorb the dye. Furthermore, the presence of heavy metals in the plant content and crystallization prevented the retention of the dye and made staining process difficult. Thus, a large number of experiments with different dyes and dye concentrations have been carried out for the tissue to absorb the dye into the cell [47].

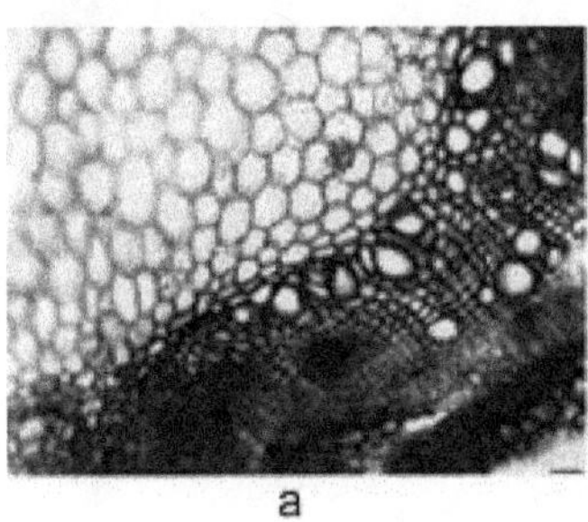


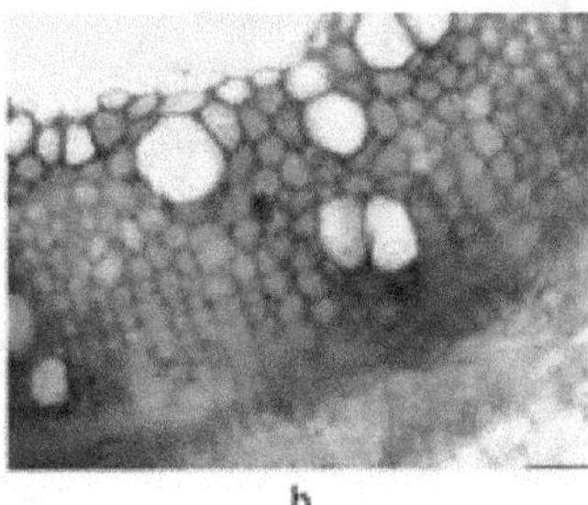


**Figure 17.**
*(a) Control group of runner bean seedling root cross section; vascular bundles, cambium, and glandular primordium. (b) Cross section of runner bean seedling treated with Cu 80 ppm; vascular bundles, cambium, sclerenchyma, endodermis, pericycle, and casparian strip [47].*

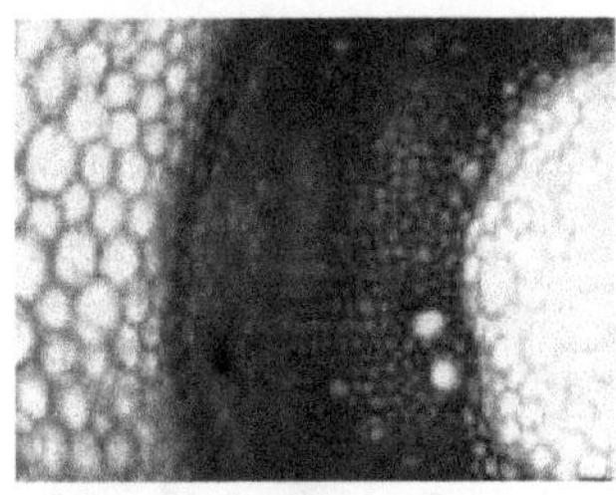

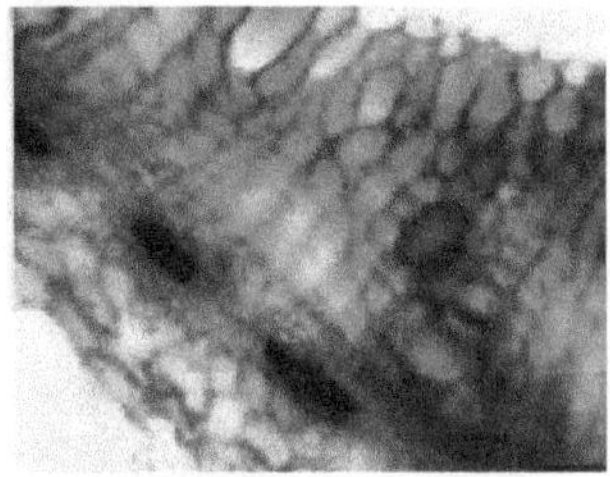

**Figure 18.**
*(a) Runner bean seedling treated with Cu 640 ppm: (a) general view of root cross section, vascular bundles, endodermis, pericycle, and cambium. (b) Close-up view of root cross section, secretion canals, cambium, endodermis, casparian strip [47].*

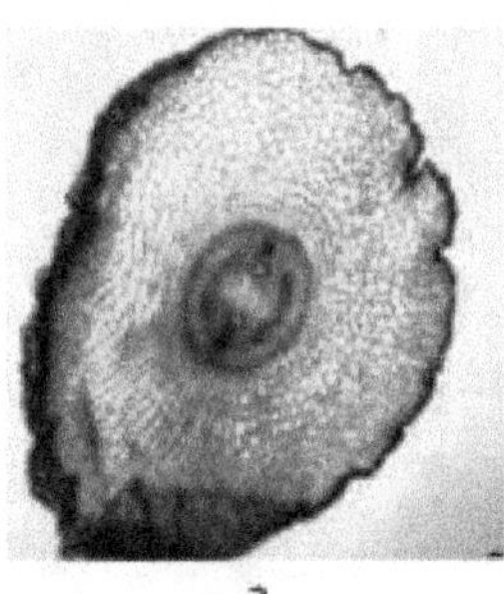


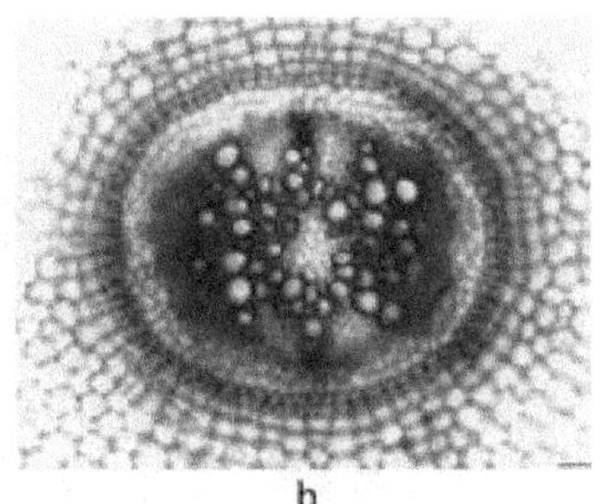


**Figure 19.**
*(a) Control group of artichoke seedling root cross section. (b) Cross section of artichoke seedling treated with Cu 160 ppm root cross-section central cylinder and conduction bundles [47].*

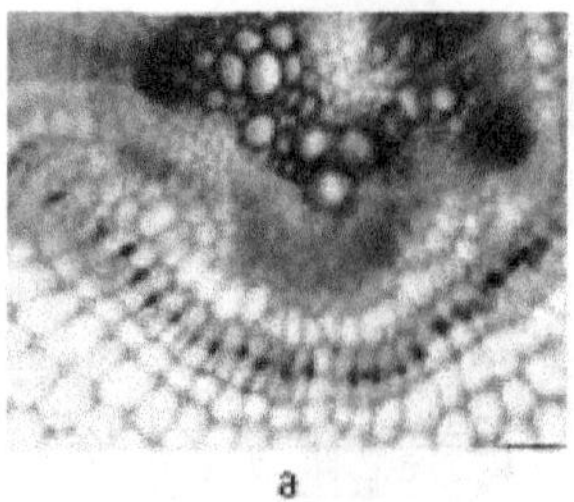
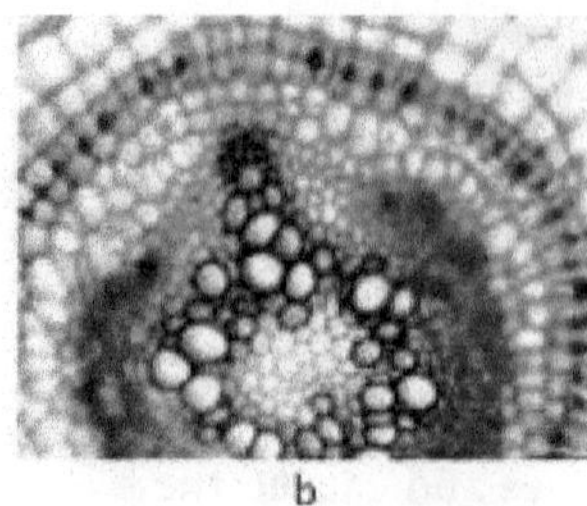

**Figure 20.**
*(a) Close-up view of artichoke seedling root treated with Pb 320 ppm: Central cylinder, endodermis, pericycle, and crystals. (b) Close-up view of root artichoke seedling treated with Pb 640 ppm: Vascular bundles and crystals [47].*

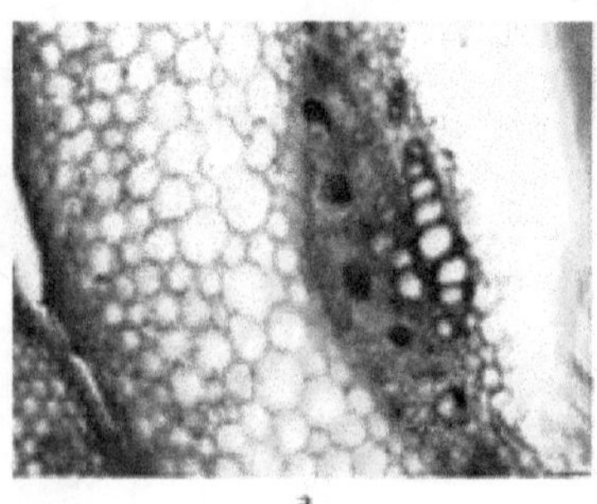
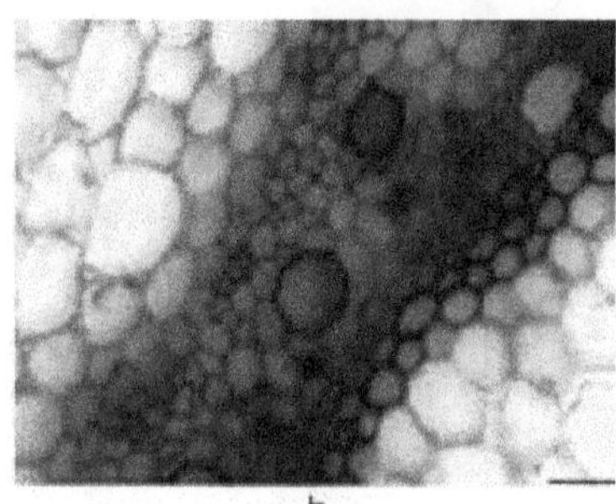

**Figure 21.**
*(a) General view of control group runner bean seedling stem cross section; close-up view of epidermis, cortex bundles, and secretion canals. (b) Close-up view of runner bean seedling stem treated with Cu 20 ppm: Xylem, phloem, secretion canals, and starch scabbard [47].*

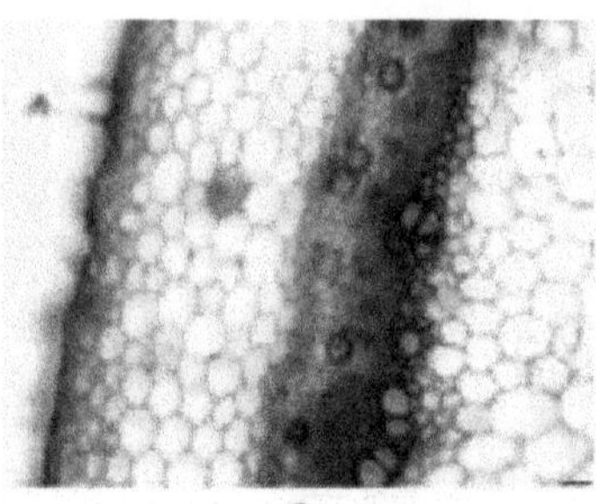
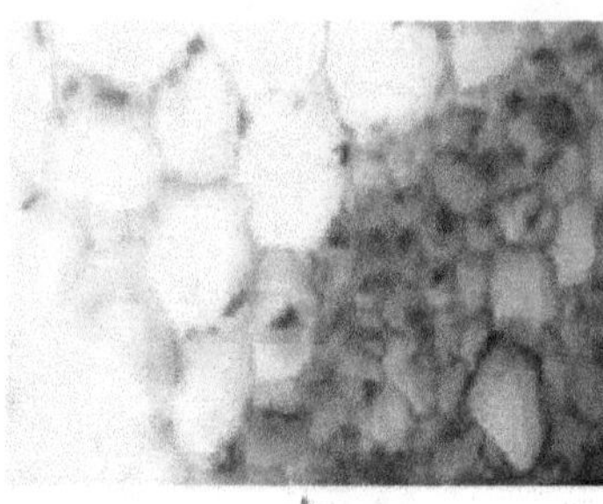

**Figure 22.**
*(a) General view of runner bean seedling stem treated with Cu 80 ppm: Vascular bundles, secretion canals, and crystals. (b) Close-up view of runner bean seedling stem treated with Cu 640 ppm: Secretion canals and crystals [47].*

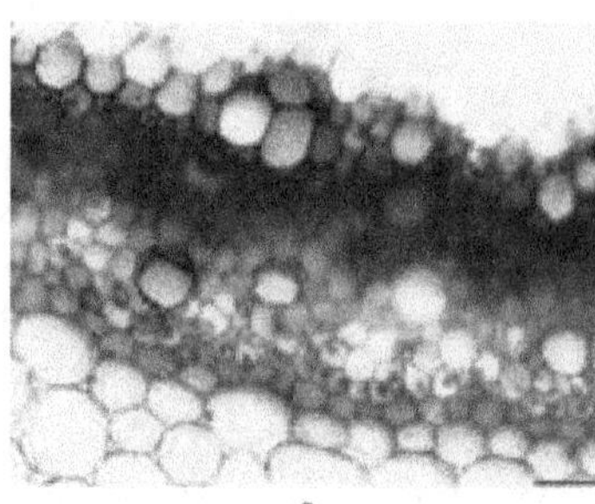
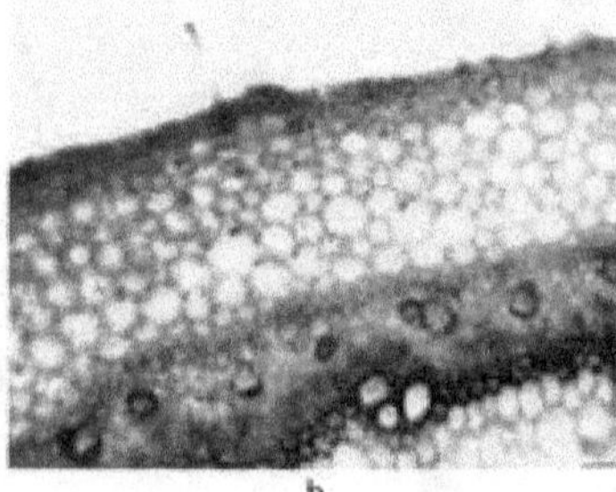

**Figure 23.**
*(a) Close-up view of runner bean seedling stem treated with Pb 40 ppm: Xylem, phloem, secretion canals, and crystals. (b) General view of runner bean seedling stem treated with Pb 640 ppm: Vascular bundle, secretion canals [47].*

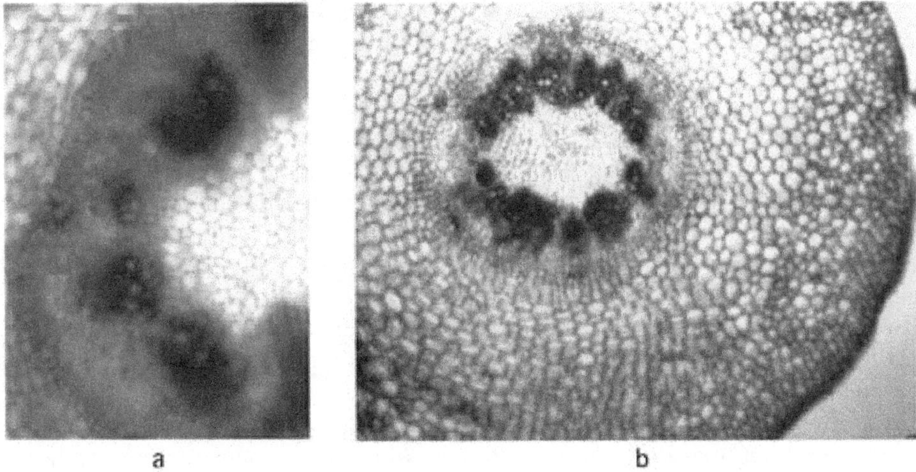

**Figure 24.**
*(a)general view of artichoke seedling stem treated with Cu 20 ppm stem: Vascular bundles and (b) general view of artichoke seedling stem treated with Cu 160 ppm.*

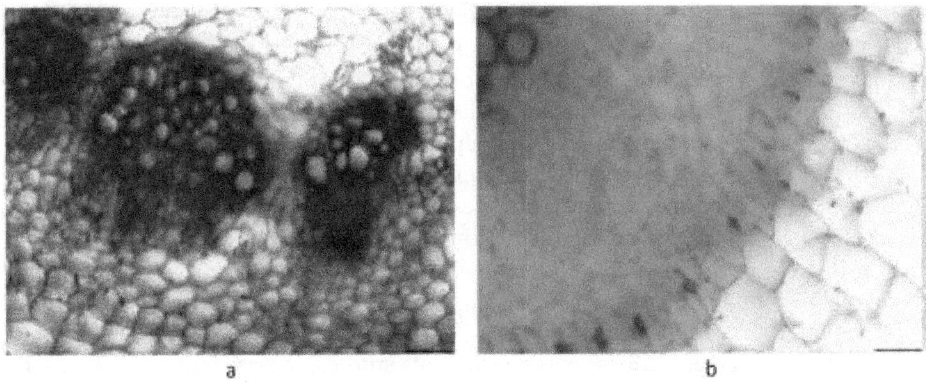

**Figure 25.**
*(a) Close up view of artichoke seedling stem treated with Cu 160 ppm:Xylem, phloem, and crystals and (b) close up view of artichoke seedling stem treated with Pb 1280 ppm cortex and crystals.*

## 5. Discussion

Heavy metal pollution in soil and water is one of the most important environmental problems in industrialized countries. Various heavy metals such as cadmium, lead, copper, mercury, and chromium from various industrial establishments such as leather, paint, fertilizer, textile, cement, and chemical industry are released onto soil and aquatic environments and cause environmental pollution [56–58]. Since most of the heavy metals do not undergo biodegradation in the environment, they can easily accumulate and increase their toxic effects on living things by forming very complex structures [59].

High-structured plants are equipped with advanced features that allow them to adapt to changes in nature, one of which is the retention of metals in the roots [60]. Retention and deposition of metals in roots have more negative effect on the root area and seed germination than stem and leaf growth. Zengin and Munzuroğlu (2004) reported that the most negative effect was in the root area of the bean (*Phaseolus vulgaris*) seedlings exposed to increasing concentration of lead and copper solutions; stem and leaf growth was negatively affected; however, they stated that the most negative effect was in the root area of the seedlings [14].

Soudek et al. (2010) treated linen (*Linum usitatissimum* L.) seeds with different concentrations of lead, nickel, copper, zinc, cadmium, cobalt, arsenic, and chromium heavy metals and reported that heavy metal stress had negative effects on the number of germinating seeds and seedling root development. The negative effect of heavy metal stress on root length in plants may result from the division of cells in the root region or the prolongation of the cell cycle [31]. The root, stem, and leaf structures of the runner bean and artichoke seedlings grown in high Cu and Pb concentrations (160, 320, 640, 1280 ppm) examined in this study were degraded as a result of oxidative damage. Therefore, the epidermal cells forming the surface of these parts were damaged, which negatively affected root, stem, and leaf growth. However, it also caused adverse conditions such as dryness, shrinkage, and necrosis in the leaves although Cu and Pb heavy metals applied at low concentrations generally stimulated growth and increase the number of leaves in the plant [39, 47]. Furthermore, browning caused by heavy metal stress was observed in the roots of the runner bean and artichoke seedlings in which high concentrations of heavy metals were applied. This color change occurs with the increase in the amount of suberin in stem cells. Therefore, suberin stem cells will limit the uptake of water, and plant growth inhibition occurs [61, 62].

The defense mechanisms developed by plants against heavy metal stress may vary in the level of family, genus, species, subspecies, and variety [54–57]. The defense mechanisms that allow plants to be tolerant to heavy metals have not yet been fully understood. However, the mechanisms of tolerance include vacuolar phenomena [58], enzymatic and nonenzymatic antioxidant systems [59, 60], metal-binding ligands such as metallothionein [61], and alternative oxidase pathways [62].

Although copper is an essential element, it is more toxic when it is present in high doses compared to cadmium, a nonessential element. This is explained by the direct influence of copper on the formation of reactive oxygen species [superoxide radical ($O_2 \cdot -$), hydrogen peroxide ($H_2O_2$), hydroxyl radical ($OH-$)] because it is a trace element. Since copper and iron transition metals are involved in oxidoreduction reactions, they act as catalysts that accelerate the formation of reactive oxygen species [63–66]. As in previous studies, it was found that artichoke seedlings are negatively affected mostly by copper in molecular terms [39, 41, 67, 68]. The highest negative effect was observed in groups subjected to copper solutions in terms of root length, root dry weight, total soluble protein amount, and genomic mold stability [39–42] (**Figures 17–22**).

Reactions to heavy metal stress in this study emerged as different anatomical results in both plants. However, it can be determined that the response of copper-treated samples was not always greater than that of the lead-treated samples even though they responded differently to different concentrations in terms of anatomical results. Previously the effect rate observed at the molecular level in Cu-treated samples has been found to be higher than the effect rate in Pb-treated samples in both plants [39–42]. However, it was not the case for the present anatomical study because the roots and stems of the plants were examined in terms of many parameters, so heavy metals did not show the same effect [47] (**Figures 17–25**).

It was observed that seedlings belonging to runner bean species showed high tolerance against lead and copper stress. The genome of the plants was preserved at 94–95%, and no significant reduction in total soluble protein was observed especially at 1280 ppm concentration of lead and copper solutions. This situation has led to the conclusion that runner bean species has a strong defense mechanism against heavy metal contamination [40–42, 49]. In this anatomically based study, it would be wrong to say that one heavy metal is always superior to another in terms of its effects because plant parts were examined anatomically in terms of many parameters and similar results were not observed in all. The reaction of the plants as crystallization is particularly important and has shown similar responses in both heavy metal treatments. The differences varied in terms of crystal density, location, and shapes (**Figures 17–25**) [47].

As a result, various anatomical differences determined in this study regarding the characteristic features, such as root and stem vascular bundles thicknesses, cell size and fragmentation in the pith region, formation or thickening of cambium and sclerenchyma, shape and size of secretary canals, differences in cortex cell sizes, the sizes of the vascular bundles, the formation of crystals and deposits in the phloem layer according to the heavy metal concentration applied, the number of epidermis cells per unit area, and the epidermal wall shapes, can be used to reveal the phytotoxic effects of Cu and Pb heavy metals.

The morphology of roots and shoots is extremely important for the growth and development of all plants, and each factor that changes their morphology has positive/negative effects [69].

On the other hand, Candan and Lu (2017) have shown that there are more differences on the pea green (*Pisum sativum*) anatomy under the effects of $C_{70}$ nanomaterial [70]. Candan and Markushin have studied about spectroscopic study of the gold nanoparticles (Au NPs) distribution in leaf, stem, and root of the pea green plant [71]. In this chapter, the effects of Au nanoparticles and $C_{70}$ nanotubes on the morphology of roots, leaves, and stems are investigated, and positive results on the development of chickpea have been observed (**Figure 16**).

## 6. Conclusion

It is the fact that the application of some materials which are not used consciously or at the recommended dosage and which contain heavy metals in order for the crop to be attractive for the consumers actually yields negative results. Therefore, this study is important about examining the extent of heavy metal phytotoxic effects related determining them on plants phenological development in the point of morphologically and anatomically changes. The information given in this study is valuable as it presents the negative molecular effect of heavy metal pollution on the plant in terms of morphological and anatomical aspects. Furthermore, this study will guide the researchers on the effects of environmental pollution in relation with the phenological development of economic plants and hence on human health.

Heavy metal and nanoparticule-nanotube-induced plants must be evaluated in the point of biochemistry and examined via scanning electron microscope (SEM) and transmission electron microscope (TEM) regarding their root, stem, and leaf struc-ture and apical tip, leaf-bud primordiums, and provascular tissue in detailed ways for interdisciplinary studies according to plants' phenological development progress.

## Acknowledgements

I would like to thank Manisa Celal Bayar University (Turkey) Scientific Research Projects Coordination Unit (BAP) for supporting most of this study as part of the projects numbered 2012-057 and 2014-120.

I am also thankful to Dr. Qi Lu, a member of the Department of Physics and Engineering, and Dr. Gulnihal Ozbay, a member of the Department of Agriculture and Natural Resources at Delaware State University (USA), for the support of some laboratory studies related to nanoparticles and nanotubes.

## Author details

Feyza Candan
Biology Department, Arts and Science Faculty, Manisa Celal Bayar University, Manisa, Turkey

*Address all correspondence to: feyzacandan2002@yahoo.com

## References


[1] Beck CB. An Introduction to Plant Structure and Development: Plant Anatomy for the Twenty-First Century. England: Cambridge University Press; 2010

[2] Carlquist S. Comparative Plant Anatomy. New York: Holt-Rinehart-Winston Inc.; 1961

[3] Mendes DS, Pereira MCT, Nietsche S, Silva JF, Rocha JS, Mendes AH, et al. Phenological characterization and temperature requirements of *Annona squamosa* L. in the Brazilian semiarid region. Annals of the Brazilian Academy of Sciences. 2017;**89**(3 Suppl):2293-2304

[4] Duffus JH. Heavy metals a meaningless term. Pure and Applied Chemistry. 2002;**74**(5):793-807

[5] Robert HM. Contaminants in the Upper Mississippi River: Boston. Stoneham, Massachusetts: Butterworth Publishers; 1995. pp. 195-230

[6] Dean JG, Bosqui FL, Lanouette VH. Removing heavy metals from waste water. Environmental Science and Technology. 1972;**6**(6):518-522

[7] Kahvecioğlu Ö, Kartal G, Güven A, Timur S. Metallerin çevresel etkileri-I. Metalurji Dergisi. 2004;**136**:47-53

[8] Vousta D, Grimanins A, Sammara C. Trace elements in vegetable grown in an industrial areas in relation to soil and air particulate matter. Environmental Pollution. 1996;**94**(3):325-335

[9] Alloway BJ, editor. Heavy Metal in soils: Trace metals and Metalloids in Soils and their Bioavailability. Dordrecht: Springer; 2010

[10] Nriagu JO, editor. Changing Metal Cycles and Human Health. Berlin, Dahlem Konferenzen: Springer-Verlag; 1984

[11] Bera AK, Bera A, Roy SB. Impact of heavy metal pollution in plants development in physiology. Biochemistry and Molecular Biology of Plants. 2005;**1**:105-124

[12] Marschner H. Mineral Nutrition of Higher Plants. London: Academic Press; 1995. p. 899

[13] Raven JA, Evans MCW, Korb RE. The role of trace metals in photosynthetic electron transport in $O_2$-evolving organisms. Photosynthesis Research. 1999;**60**:111-149

[14] Zengin FK, Munzuroğlu O. Effect of lead and copper (Cu) on the growth of root, shoot and leaf of bean (*Phaseolus vulgaris* L.) seedlings. Journal of Science. 2004;**17**:1-10

[15] Keller C, Hammer D. Metal availability and soil toxicity after repeated croppings of *Thlapsi caerulescens* in metal contaminated soils. Environmental Pollution. 2004;**131**: 243-254

[16] Braz J. Copper in plants. Brazilian Journal of Plant Physiology. 2005;**17**: 145-146

[17] Van Assche F, Clijsters H. Effects of metals on enzyme activity in plants. Plant, Cell & Environment. 1990;**13**: 195-206

[18] Hernandez-Jimenez MJ, Lucas MM, de Felipe MR. Antioxidant defence and damage in senescing lupin nodules. Plant Physiology and Biochemistry. 2002;**40**:645-657

[19] Banfalvi G, editor. Cellular Effects of Heavy Metals. Dordrecht: Springer; 2011. p. 348

[20] Candan F. Some observations on plant karyology and investigation methods. In: Silva-Opps M, editor.

Current Progress in Biological Research. Croatia: InTech-Open Press; 2013

[21] Meharg AA. Integrated tolerance mechanisms constitutive and adaptive plant responses to elevated metal concentrations in the environment. Plant, Cell & Environment. 1994;**17**: 989-993

[22] Ouzounidou G, Ilias I, Tranopoulou H, Karataglis S. Amelioration of copper toxicity by iron on spinach physiology. Journal of Plant Nutrition. 1998;**21**:2089-2101

[23] Gill SS, Tuteja N. Reactive oxygen species and antioxidant machinery in abiotic stress tolerance in crop plants. Plant Physiology and Biochemistry. 2010;**48**(12):909-930

[24] Mourato M, Reis R, Martins LL. Characterization of Plant antioxidative system in response to abiotic stresses: A focus on heavy metal toxicity. In: Montanaro G, Dichio B, editors. Advances in Selected Plant Physiology Aspects. Croatia: InTech-Open Press; 2012. pp. 23-44

[25] Kuraoka I, Robins P, Masutani C, Hanaoka F, Gasparutto D, Cadet J, et al. Oxygen free radical damage to DNA. The Journal of Biological Chemistry. 2001;**276**(52):49283-49288

[26] Cadet J, Loft S, Olinski R, Evans MD, Bialkowski K, Wagner JR, et al. Biologically relevant oxidants and terminology, classifi cation and nomenclature of oxidatively generated damage to nucleobases and 2-deoxyribose in nucleic acids. Free Radical Research. 2012;**46**(4):367-381

[27] Robert JB. Genetics: Analysis & Principles. 4th ed. New York: McGraw-Hill Publishers; 2011

[28] Dedon PC. The chemical toxicology of 2-deoxyribose oxidation in DNA. Chemical Research in Toxicology. 2008; **21**:206-219

[29] Aras S, Kanlıtepe Ç, Cansaran D, Halıcı MG, Beyaztaş T, et al. Journal of Environmental Monitoring. 2010;**12**: 536-543

[30] Sresty TVS, Rao KVM. Ultrastructural alterations in response to zinc and nickel stress in the root cells of pigeonpea. Environmental and Experimental Botany. 1999;**41**:3-13

[31] Soudek P, Katrusakova A, Sedlacek L, Petrova S, Koci V, Marsik P, et al. Effect of heavy metals on inhibition of root elongation in 23 cultivars of flax (*Linum usitatissimum* L.). Arch. Environmental Contamination and Toxicology. 2010; **59**:194-203

[32] Öztürk Çalı İ , Candan F. Effects of a Funguside on the morphology and viability of pollens of tomato (*Lycopersicon esculentum* mill.). Bangladesh Journal of Botany. 2009; **38**(2):115-118

[33] Öztürk Çalıİ, Candan F. The effect of activator application on the anatomy, morphology and viability of *Lycopersicon esculentum* mill. Pollen. Turkish Journal of Biology. 2010;**34**:281-286

[34] Öztürk Çalı İ , Candan F. Influence of activator on meiosis of tomato (*Lycopersicon esculentum* mill.). Bang. Journal of Botany. 2013;**42**(2):361-365

[35] Candan F, Öztürk Çali İ. Pollen micromorphology of four taxa of *Anemone coronaria* L. from western Turkey. Bang. Journal of Botany. 2015; **44**(1):1-36

[36] Candan F, Öztürk Çİ. Studies on the comparison of pollen morphology and viability of four naturally distributed and commercial varieties of *Anemone coronaria* L. Pakistan Journal of Botany. 2015;**47**(2):517-522

[37] Steinkellner H, Kassie F, Knasmuller S. *Tradescantia-*

micronucleus assay for the assessment of the clastogenicity of Austrian water. Mutation Research. 1999;**426**: 113-116

[38] Menke M, Chena P, Angelis KJ, Schubert I. DNA damage and repair in *Arabidopsis thaliana* (L.) as measured by the comet assay after treatment with different classes of genotoxins. Mutation Research. 2001;**493**:87-93

[39] Candan F, Batır MB, Büyük İ. Determination of DNA Changes in Seedlings of Artichoke (*Cynara scolymus* L.) and Runner Bean (*Phaseolus coccineus* L.) Seeds Exposed To Copper (Cu) And Lead (Pb) Heavy Metal Stress. Manisa: Manisa Celal Bayar University Scientific Research Projects Coordination Unit, Manisa Celal Bayar University BAP Project, Project No: 2012-057; 2013

[40] Candan F, Batır BM. The comparison of physiological, biochemical and molecular parameters in seedlings of artichoke (*Cynara scolymus* L.) and runner bean (*Phaseolus coccineus* L.) seeds exposed to Lead (Pb) heavy metal stress in the point of ecological pollution. In: Proce–edings of International Conference on Agricultural, Ecological and Medical Science, AEMS, 7–8 April 2015, Phuket/ Thailand; 2015

[41] Batır MB. Determination of DNA changes in seedlings of artichoke (*Cynara scolymus* L.) seeds exposed to copper (Cu) and lead (Pb) heavy metal stress [master thesis] Eskişehir Osmangazi University, Science Institute; 2014

[42] Batır BM, Candan F, Büyük İ. Determination of the DNA changes in the artichoke seedlings (*Cynara scolymus* L.) subjected to lead and copper stresses. Plant, Soil and Environment. 2016; **62**(3):143-149

[43] Davis PH. *Fabaceae*.In:TheFlora of Turkey and The East Aegean Islands Vol. 3. Edinburgh: Edinburgh University Press; 1970

[44] Davis PH. *Asteraceae*. In: The Flora of Turkey and The East Aegean Islands. Vol. 5. Edinburgh: Edinburgh University Press; 1975

[45] Seçmen Ö, Gemici Y, Leblebici E, Görk G, Bekat L. Tohumlu Bitkiler Sistematiği. Turkey, İzmir: Ege Üniversitesi Fen Fakültesi Kitaplar Serisi; 1998. p. 396 s

[46] Zeybek N, Zeybek U. Farmasotik Botanik (Kapalı Tohumlu Bitkiler Sistematiği) ve Önemli Maddeleri. 2. Baskı. Turkey, İzmir: Ege Üniversitesi Eczacılık Fakültesi Yayınları No: 2; 1995. p. 436 s

[47] Candan F. Determination of Root, Stem and Leaf Anatomical Changes Regarding Seedlings of Artichoke (*Cynara scolymus* L.) and Runner Bean (*Phaseolus coccineus* L.) Seeds Exposed to Copper (Cu) and Lead (Pb) Heavy Metal Stress. Manisa: Manisa Celal Bayar University Scientific Research Projects Coordination Unit, Manisa Celal Bayar University Scientific Research Project (BAP Project), Project Number: 2014-120; 2018

[48] Candan F, Lu Q. Comparative effects of gold (Au) and carbon (C70) Nanomaterials translocation on Chick pea (*Cicer arietinum* L.) plant morphology. In: Proceedings of VIII. International symposium on ecology and environmental problems, ISEEP, 4-7 October 2017, Çanakkale/Turkey; 2017

[49] Algan G. Bitkisel Dokular İçin Mikroteknik. Turkey: Fırat University, Science Faculty Press; Bot-No: 1. 1981. pp. 3-34

[50] Vardar Y. Botanikte Preperasyon Tekniği. Izmir/Turkey: Ege University, Science Faculty Books; No: 1. 1987

[51] İnce H. Bitki Preperasyon Teknikleri. Izmir/Turkey: Ege University, Science Faculty Press; No: 127. 1989

[52] Johansen DA. Plant microtechnique. New York and London: McGraw-Hill Book Co. Inc.; 1940

[53] Omosun G, Edeoga HA, Markson AA. Anatomical changes due to crude oil pollution and its heavy metals component in three *Mucuna* species. Recent Research in Science and Technology. 2009;**1**(6):264-269

[54] Fahn A. Plant Anatomy. Oxford: Pergamon Press; 1982

[55] Yentür S. Bitki Anatomisi. Istanbul/Turkey: Istanbul University Press, Science Faculty; No: 209. 1989

[56] Bradford MM. A rapid and sensitive method for the quantitation of microgram quantities of protein utilizing the principle of protein-dye binding. Analytical Biochemistry. 1976; **72**:254-284

[57] Nriagu JO, Pacyna JM. Quantitative assessment of worldwide contamination of air, water and soils by trace metals. Nature. 1988;**333**:134-139

[58] Boyer HA. Trace elements in the water, sediments, and fish of the Upper Mississippi River, Twin Cities metropolitan area. In: Wiener JG, Anderson RV, McConville DR, editors. Contaminants in the Upper Mississippi River. USA, Boston: Butterworth Publishers; 1984. pp. 195-230

[59] Kabatas PA, Pendias H. Trace elements in soils and plants. In: Nnagu JO, editor. Changing Metal Cycles and Human Health. Springer; 1984. pp. 87-91

[60] Fernandes JC, Henriques FS. Biochemical, physiological and structural effect of excess copper in plants. The Botanical Review. 1991;**57**: 246-273

[61] Mohanpuria P, Rana NK, Yadav SK. Cadmium induced oxidative stres influence on glutathione metabolic genes of *Camellia sinensis* (L.). Environmental Toxicology. 2007;**22**: 368-374

[62] Steudle E. Water uptake by roots: Effects of water deficit. Journal of Experimetal Botany. 2000;**51**(350): 1531-1542

[63] Mediouni C, Benzarti O, Tray B, Mohamed HG, Jemal F. Cadmium and copper toxicity for tomato seedlings. Argonomy for Sustainable Development. 2006;**26**:227-232

[64] Doncheva S, Nikolov B, Ogneva V. Effect of excess copper on themorphology of the nucleus in maize root meristem cells. Physiologia Plantarum. 1996;**96**(1):118-122

[65] Ralph PJ, Burchett MD. Photosynthetic response of *Halophila ovalis* to heavy metal stres. Environmental Pollution. 1998;**103**: 91-101

[66] Batır MB, Candan F, Buyuk İ, Aras S. The determination of physiological and DNA changes in seedlings of maize (*Zea mays* L.) seeds exposed to the waters of the Gediz River and copper heavy metal stress. Environmental Monitoring and Assessment. 2015;**187**:169

[67] Candan F. Effects of copper (Cu) heavy metal stress on artichoke (*Cynara scolymus* L.) plants morphology. In: Proceedings of XIII. Congress of Ecology and Environment with International Participation, UKECEK, 12–15 September 2017a, Edirne/Turkey; 2017a

[68] Candan F. Determination of morphological changes regarding seedlings of artichoke (*Cynara scolymus* L.) seeds exposed to lead (Pb) heavy metal stress. In: Proceedings of XIII.

Congress of Ecology and Environment with International Participation, UKECEK, 12–15 September 2017b, Edirne/Turkey; 2017b

[69] Talebi SM. Nanoparticle-induced morphological responses of roots and shoots of plants. In: Tripathi DK, Ahmad P, Sharma S, Chauhan DK, Dubey NK, editors. Nanomaterials in Plants, Algae, and Microorganisms. Cambridge: Academic Press; 2018. pp. 119-141

[70] Candan F, Lu Q. Anatomical study on translocation of carbon Nanomaterials distribution in leaf and stem of the pea green (*Pisum sativum)* plant. In: Proceedings of the 3rd International Symposium on EuroAsian Biodiversity, SEAB, 7–8 July 2017a, Minsk/Belarus; 2017a

[71] Candan F, Markushin Y. Spectroscopic study of the gold nanoparticles (AuNPs) distribution in leaf, stem, and root of the pea green (*Pisum sativum*) plant. In: Procedings of the 3rd International Symposium on EuroAsian Biodiversity, SEAB, 7–8 July 2017b, Minsk/Belarus; 2017b

5

# Major Natural Vegetation in Coastal and Marine Wetlands: Edible Seaweeds

*Ilknur Babahan, Birsen Kirim and Hamideh Mehr*

**Abstract**

For thousands of years, seaweeds grown in coastal and marine have been used as food, materials and medicines by the people. Edible seaweeds directly consumed, especially in Asian, are used for preparing food due to the their components contain-ing minerals, essential trace elements, and various natural compounds. At the last decades, they have been getting more and more attention in food and pharmaceutical industries because of their biological activities such as anti-cancer, anti-obesity, anti-diabetes, anti-microbial, and anti-oxidant activity. Therefore, in the present study, we have worked on to understand the structure of edible seaweeds. It is worthy to mention that they can be considered as source of some proteins, polyunsaturated fatty acids, minerals, vitamins, dietary fibers, antioxidants, and phytochemicals.

**Keywords:** edible seaweeds, polysaccharides, anti-cancer, anti-obesity, anti-diabetes, anti-oxidant, anti-microbial

## 1. Introduction

Macroalgae or called seaweeds are multicellular, marine species and are con-sidered as non-vascular plants. Although term seaweed is widely used but these species are characteristically far from "weeds" but the fact is, seaweeds are the main productive species in the oceans and food chain basis. Seaweeds are used directly or indirectly in food and household products without being tasted or smelled. They are vastly used in food industry due to their valuable elements, vitamin, and proteins [1]. Sea or brackish water is the main habitat for seaweeds (macroalgae), and are referred as benthic marine algae or sea vegetables due their choice of habitat in the sea [2–4]. These simple unique organisms are one of the major productions of Asian industries, and main goal for these industries is to use the production to the maximum extent [2, 5]. Since seaweeds are main part of the diet in east Asia and to some extent being used as snacks and delicacies in other countries. Although seaweeds are as part of food in Far East countries, western countries use them as sources of phycocolloids, thickening and gelling agents for various industrial applications including food. Different applications are due to various chemical composition of seaweeds with habitats, maturity, salinity, environmental habitat, and temperature [2, 6].

Seaweeds are produced more than million tons per year, while microalgae are being produced almost 20,000 tons annually. This has to be mentioned the macroalgae have higher biomass sells in comparison with seaweeds. Seaweeds are mostly cultivated in near shores in China, Philippines and Japan [7, 8]. The total aquatic production in 2004 passed 15.36 million tons while 93% of the contribution belonged to seaweeds. Among this, 6000 species of seaweeds are harvested which fall into three categories as, green (Chlorophytes), brown (Phaeophytes) and red (Rhodophytes). Regarding the abundancy, the human consumption is mainly on brown algae (66.5%), red algae (33%), and green algae (5%) in Asia [2, 9]. The main producers in Asia are focusing on specific algae such as, in China, Japan and Korea are Nori (Porphyra, red algae), Konbu (Laminaria, brown algae) and Wakame (Undaria, brown algae). Among 6000 species discovered about 150 species are known as food source and 100 for phycocolloid production. The total revenue for edible algae passed 1 billion US dollars only in Japan and this estimation is the value of 1.4 kg seaweed per person consumption.

Marine algae, in addition of being used in food, dairy, pharmaceutical, cosmetic and medicine industries, they can also be used in biodiesel, bioethanol, and hydrogen gases preparation. They can also be applied as antioxidant, antibiotics, and virostatic agents [10, 11] application of algae in food industry either for human or animal consumption has brought some negative perspectives due to some toxic elements such as cadmium or fucotoxins. The amount of toxin in algae is related to the contents of fiber and bioactive compounds present. This has direct impact on digestibility and application in food industry. Digestibility has connections mainly with the nitrogen consumption before and after digestion by using specific enzymes called pepsin [10, 12].

### 1.1 Marine seaweeds

Three main categories of marine seaweeds are Chlorophyta (green algae), Rhodophyta (red algae) and Phaeophyta (brown algae) [13]. Each class is explained in different section in this review.

#### *1.1.1 Brown seaweeds*

Brown algae also known as phaeophyta are the seaweeds mainly grow in cold waters at Northern Hemisphere. Marine is their main habitat and it plays a great role in their properties. Macrocystic is a kelp which grows underwater forests and may extend 60 m (200 ft) in length and has high level of biodiversity [14, 15]. Sargassum, another brown algae is an example of singular floating mats of seaweed in tropical waters. Many other brown algae grow along rocky shores and they have been used as food by humans since 2000 years [14].

*Padina boerengsnii*

*Dictyota ciliolata*

**Figure 1.**
*Some example of brown seaweeds [13].*

Brown algae is a great source of iodine and is the most commonly used alginates. They have thickening property, which is greatly used in food products such as salad dressings, in oil industry for oil-drilling muds and in coatings. The color of brown algae is from green and yellow pigments coming from xanthophyll and chlorophyll, respectively. The ranges of colors in brown algae species are due to the variable blending of these two pigments [13, 16, 17]. Some example of brown seaweeds are shown in **Figure 1**.

#### *1.1.2 Red seaweeds*

Red algae are also classified as phylum Rhodophyta. They are the most abundant and commercial value of sea algae. The term Rhodophyta refers to the algae group with red pigments due to phycoerythrin and phycocyanin. These pigments mask chlorophyll a, which do not contain chlorophyll b, β-carotene and a number of unique xanthophylls [13, 18].

Red algae are found on rocky shores. Some species are much deeper than brown or green algae. It is known that there are 550 species of red seaweed in the world and so they are the largest seaweed group [13, 17]. Two of them are shown in **Figure 2**. For example, a total of 128 varieties of red algae had been recorded in Red Sea so far [13, 16]. The main reserves in red seaweeds are typically floridean starch, and floridoside. The walls of these algae are made of cellulose and agar and carrageen-ans, both of which are long chained polysaccharides. Cellulose, agar and carrageen-ans are widely used commercially and have a large number of uses. Some red algae cells are different because, they are not good enough in amoeboidly, but none of the algae contain flagella, so none can swim quickly [13, 17].

#### *1.1.3 Green seaweeds*

The green algae are classified in the phylum Chlorophyta and are usually grown in the intertidal zone which has the high and low water marks, and in shallow water where there is plenty of sunlight. Due to the similarity of pigments, it is thought that they are the most closely related algae. Many types of green algae found on the surface of the ocean or near rocky surfaces. Some species of green seaweeds, *Halimeda macroloba*, *Ulva lactuca*, *Enteromorpha clathrata* and *Caulerpa trifara* are shown in **Figure 3** [13]. Approximately 140 species were recorded on the shores of the world. On the coast of Eritrea, there are about 50 species in the Red Sea.

Sea lettuce is one of the most widely known species. Green algae with bright green leaves up to 30 cm are called sea lettuce (*Ulva lactuca*). Green algae are deeply bound to the lower layers and are not usually cast by waves on the beach. Sometimes exceptionally, some green algae can tear through their substrate during storms and with heavy wave motion.

### 1.2 Edible seaweeds

Seaweed as a fundamental diet matter are known in Asia since prehistoric times. Some 21 species are used in daily cuisine in Japan and even six of them have been used since the eighth century. Kaiso, one of the edible seaweeds, accounted for 10% of the Japanese cuisine until recently. Using seaweeds in the kitchens extend an average of 3.5 kg per household in 1973, an increase of 20% over 10 years [19–22]. The 12 largest countries, using seaweeds in the kitchens in the world, are China, France, UK, Japan, Chile, Philippines, Korea, Indonesia, Norway, USA, Canada and Ireland. The seaweed production and the health of the sea algae is growing

**Figure 2.**
*Some species of red seaweeds [13].*

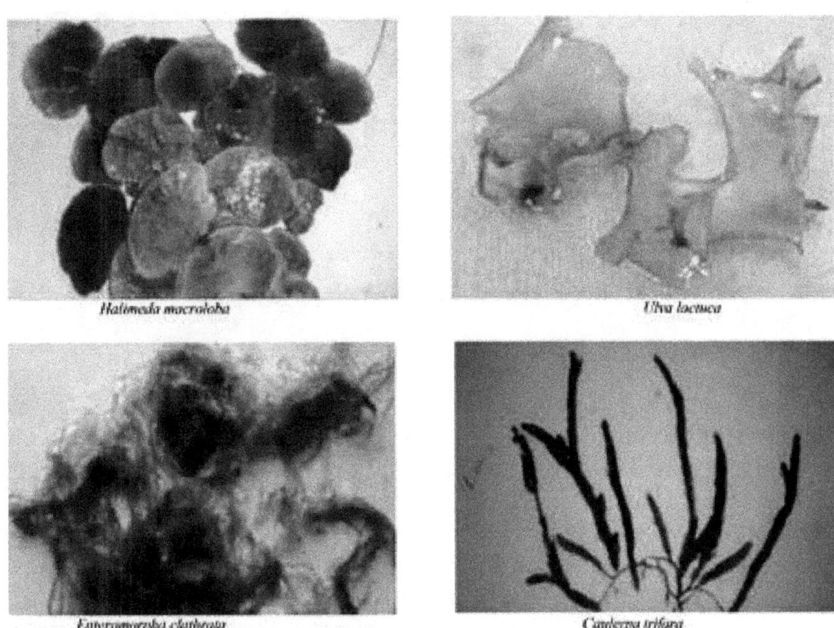


**Figure 3.**
*Some species of green seaweeds [13].*

increasingly important, the world seaweed production in 2000, including wild and softened, has reached about 10 million tons [21, 22].

As it was mentioned above, seaweeds are categorized into three classes of Rhodophyta (red), Phaeophyta (brown) and Chlorophyta (green) marine macroalgae. Some reported common edible Brown algae (Phaeophycae) are divided as Kelp, Fucales and Ectocarpales, given in **Table 1** [23]. Common edible Red algae (Rhodophyta) are Carola (*Callophyllis* spp.), Carrageen moss (*Mastocarpus stellatus*), Dulse (*Palmaria palmata*), *Eucheuma* (*Eucheuma spinosum* and *Eucheuma cottonii*), Gelidiella (*Gelidiella acerosa*), Ogonori (*Gracilaria*), Grapestone *Mastocarpus papillatus*, *Hypnea*, Irish moss (*Chondrus crispus*), Laverbread (*Porphyra laciniata/ Porphyra umbilicalis*), Gim (*Pyropia*, *Porphyra*) and Nori (*Porphyra*) [23]. Common edible Green algaes are Chlorella (*Chlorella* sp.), Gutweed (*Ulva intestina-lis*), Sea grapes or *green caviar* (*Caulerpa lentillifera*), Sea lettuce (*Ulva* spp.) [23].

### *1.2.1 Biological activity of edible seaweeds*

There two categories of algae as macro-algae (macroscopic) and micro-algae (microscopic) [10, 24]. Although algeas are great source and producers of vitamins,

| Kelps | Fucales | Ectocarpales |
|---|---|---|
| Arame (*Eisenia bicyclis*) | Bladderwrack (*Fucus vesiculosus*) | Mozuku (*Cladosiphon okamuranus*) |
| Badderlocks (*Alaria esculenta*) | Channelled wrack (*Pelvetia canaliculata*) | |
| Cochayuyo (*Durvillaea antarctica*) | Hijiki or Hiziki (*Sargassum fusiforme*) | |
| *Ecklonia cava* | Limu Kala (*Sargassum echinocarpum*) | |
| Kombu (*Saccharina japonica*) | *Sargassum* | |
| Oarweed (*Laminaria digitata*) | Spiral wrack (*Fucus spiralis*) | |
| Sea palm (*Postelsia palmaeformis*) | Thongweed (*Himanthalia elongata*) | |
| Sea whip (Nereocystis luetkeana) | | |
| Sugar kelp (*Saccharina latissima*) | | |
| Wakame (*Undaria pinnatifida*) | | |
| Hiromi (*Undaria undarioides*) | | |

**Table 1.**
*The list of the common edible Brown algae types (Kelps, Fucales and Ectocarpales).*

minerals and proteins and fatty acids but not great efforts have been allocated on the research of these plant like organisms [10, 25, 26]. Seaweeds are considered as source of soluble dietary fibers, proteins, minerals, vitamins, antioxidants, phytochemicals, and polyunsaturated fatty acids, with low caloric value. These nutrient factors are directly influenced by external environmental factors such as geographic location, temperature and season [27, 28]. Although their main application is gelling agent, thickened and stabilizers in food industries but currently studies are focused on medicinal usage and their anticancer, diabetes, inflammation, obesity and other ailments treatments [27].

Edible seaweeds are fundamental part of the cuisine for people living by the seas in areas such as Asia, Hawaii, South America and Africa, as well as marine products obtained from the sea, which are the source of protein from the sea, and in recent years a focus of interest in Europe and America due to the increasing interest in healthy nutrition. Edible seaweeds are a very good source of vitamins such as A, $B_1$, $B_2$, $B_6$, $B_{12}$, niacin and C and also rich in iodine, potassium, iron, magnesium and calcium. They are nutritious as a component in food [4, 13, 21, 22, 29–32].

In addition to being source of food, seaweeds have antibacterial, antiviral and antifungal properties [2, 33]. In ancient researches, there are tracks of seaweed applications specially in 2500 years old Chinese literature [34, 35]. For instance, Japanese were using seaweed as one of the main ingredients in recipe for Nori in addition to raw fish and sticky rice. It is also well known in Europe and North America that seaweeds have therapeutic powers in treatment of tuberculosis, arthritis, colds and influenza. Very early discovery in 1990s, marine bacteria, invertebrates and algae were used in bioactive compounds [34, 36]. Major milestone in pharmaceutical industries during 1980–1995 was research on seaweed [34, 37]. The discoveries showed that many types of seaweed have anti-inflammatory and anti-microbial agents. These agents are able to be used in treatment of wounds, burns and rashes and some evidences have suggested the algae have been used in treatment of breast cancer in ancient Egypt [38, 39].

Seaweeds are known to contain strong natural anti-oxidants, since algae contain a lot of secondary metabolites such as tocopherol, carotenoids, polyphenols, flavonoids, tannins, lignans, and mycosporine-like amino acids (MAA), vitamin C, and glutathione [40].

The studies have shown that seaweeds possess anticancer agents and there are hopes they can be effective in treatment of tumors and leukemia [34]. As the efforts have continued, scientists successfully isolated chemical compounds from brown seaweed with anticancer and antitumor activities [38, 39]. It has been reported that fucoidan from *U. pinnatifida* shows very good anti-cancer activity against human lung cancer cell line which is known A549 cell line [41].

Some studies in last decades show that fucoxanthin and fucoxanthinol from *U. pinnatifida* shows also anti-obesity activity. As obesity is known a serious health issue and has cost significant economic problem, that edible seaweeds possess anti-obesity activity, is very notable because; it is well known that obesity cost to some chronic diseases, such as liver steatosis, cardiovascular disease, osteoarthritis, type 2 diabetes, and some types of cancer. Alongside having anti-obesity activity, some reports show that fucoxanthin and fucoxanthinol from *U. pinnatifida* possesses anti-diabetic activity [41]. Therefore, it can be considered that anti-obesitity activity and anti-diabetic activity are related each other. It is expected that if seaweeds have anti-obesity activity, they can able to show anti-diabetic activity.

## 2. Chemical components of edible seaweeds

### 2.1 Polysaccharides

Polysaccharides are the main components of green, brown and red seaweed. Algae cell walls contain numerous polysaccharides such as, alginates, alginic acid, carrageenans, agar, laminarans, fucoidans, ulvans and derivatives with storage and structural functions (Perez et al., 2016; [42, 43]). As it is shown in **Figure 4** [44], agar polysaccharides have complex molecular structure with alternating composition of 3-linked-D-galacropyranose (G unit) and 4-linked-3,6-anhydro- L-galactopyranose (LA unit) [45]. Substitution of hydroxyl group by ester sulfate, methyl groups and pyruvic acid at various positions have direct impact on physical and rheological properties of polysaccharides [44, 46–50].

Polysaccharides have noticeable effects in immunomodulatory and anti-cancer as one of the most important macromolecules. These effects are driving force for wide research in biochemical and medical areas. As it was mentioned above, polysaccharides are abundant in cell walls and their composition is under the influence of season, age, species and geographical location. Their main goal in plants are food reservoir, however they can provide strength, and flexibility encountering wave actions and also balancing the ionic equilibrium inside the cell. Other structural benefits of polysaccharides, such as regularity of the hydroxyl group, can increase the ion interactions our of cell walls and interchain hydrogen bonding and causing gelation.

Depending on seaweeds, different polysaccharides can be produced by alginates, fucoidans, and laminarans. Laminarans, fucoidans as water soluble and high molecular mass alginic acids as alkali soluble polysaccharides are main products of brown seaweeds [51]. Main components of brown algae wall are cellulose microfibrils merged in amorphous polysaccharide while they relate to each other via proteins. There are two kinds of acid polysaccharides in extracellular structure of brown algae, sulfated fucans and alginic acid (Perez et al., 2016).

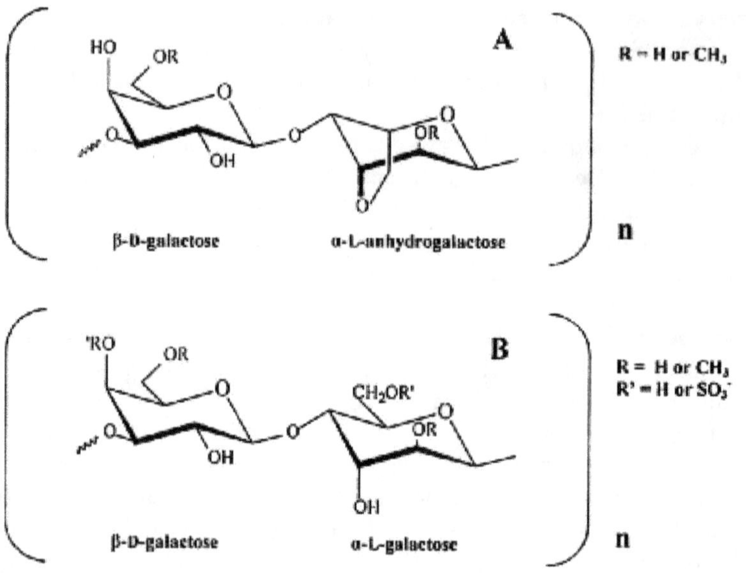

**Figure 4.**
*Chemical structure of agar polysaccharides with the different types of monomers [44].*

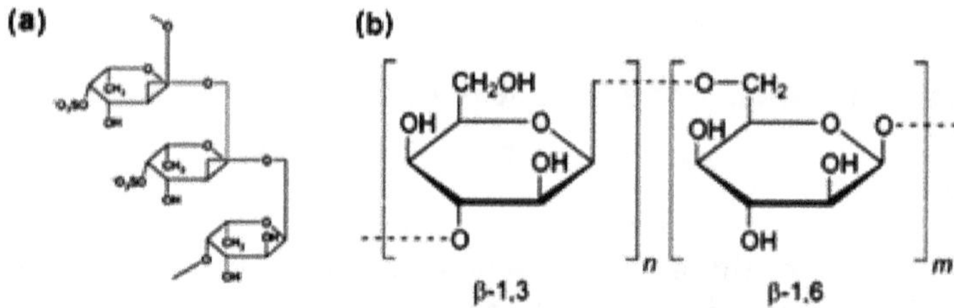


**Figure 5.**
*(a) Structure of fucoidan [51] and (b) structure of laminaran [51].*

### *2.1.1 Fucans*

Fucans as one of the acid polysaccharides present in extracellular structure of brown algae (**Figure 5**) are categorized into three major groups: fucoidans, xylofucoglycuronans and glycorunogalactofucans [51].

### *2.1.2 Fucoidans*

Fucoidan is a branched sulfate ester polysaccharide with branching. The major branches in this polysaccharide are L-fucose-4 sulfate or sulfate ester at $C_3$. Fucoidan has molecular weight ranging from 100 kDa [52] to 1600 kDa [53]. Main components of Fucoidan are fucose, uronic acids, galactose, xylose and sulfated fucose. Fucoidan structure contains sulfated fucans backbone, which is made of different sugars, fucose, or uronic acid. The backbone also has different degrees of branching. This structure is highly dependent on the algae's species. Due to the complex structure, especially due to branching, it is very difficult to study the whole molecule [54].

As a known fact, fucoidan has solubility in water and acid solution [53] and acid hydrolysis can result various amounts of D-xylose, D-galactose, and uronic acid. Algal fucoidans as very common sulfated polysaccharide present in all brown algae are mainly found in Fucales and Laminariales, also present in Chordariales,

Dictyotales, Dictyosiphonales, Ectocarpales, and Scytosiphonales. Although algal is present in brown algae but it seems to be absent in green algae, red algae, as well as in freshwater algae and terrestrial plants [55].

Study the structural composition of polysaccharides showed that xylofucoglycuronans or ascophyllans have polyuronide backbone, fundamentally poly-b-(1,4)-D-mannuronic acid branched with 3-O-D-xylosyl-L-fucose-4-sulfate or sometimes uronic acid. While, glycuronogalactofucans are composed of linear chains of (1,4)-D-galactose branched at $C_5$ with L-fuco-syl-3 sulfateoroccasionallyuronicacid [56]. This backbone consists of (1 → 3)-linked α-L-fucopyranose residues (type 1, **Figure 6A**) or alternating (1 → 3)-linked α-L-fucopyranose, (1 → 4)-linked α-L-fucopyranose residues (type 2, **Figure 6B**), and fucose and sulfate branching (**Figure 6C**) [54].

### 2.1.3 Carrageenans

Carrageenan as linear sulfated polysaccharides are extracted from edible red seaweeds. Carrageenan name is from *Chondrus crispus* species of seaweed known as Carrageen Moss or Irish Moss in England, and Carraigin in Ireland [57]. This large and highly flexible polysaccharide contain 15–40% of ester-sulfate as the main component of sulfated polygalactan with average molecular weight of 100 kDa. The structural composition shows alternate units of anhydrogalactose (3,6-AG) and D-galactose. These units are joint by α-1,3 and β-1,4-glycosidic linkage. There are different classes of carrageenan such as λ, κ, ι, ε, μ. All these classes have sulfate groups in range of 22–35%. The number and position of the ester sulfate is the key for the primary differences in different types of carragenans. It must be mentioned that these nomenclatures have no reflect on the chemical structures. Kappa and Iota type have ester sulfate content around 25–30% and 3,6-AG content of about 25–30%. While, Lambda type has higher ester sulfate content of about 32–39% and no content of 3,6-AG (**Figure** 7) [57, 58].

### 2.1.4 Alginic acids

Alginic acid or algin is a linear polysaccharide with 1,4-linked, b-D-mannuronic and a-L-guluronic acid (**Figure 8**) as building blocks which are arranged in non-regular and different sequences fashion [59]. Alginic acid is derived from brown seaweed in form of sodium and calcium alginate with main application in food and pharmaceutical industries. Structural functionalities make them able to bind with

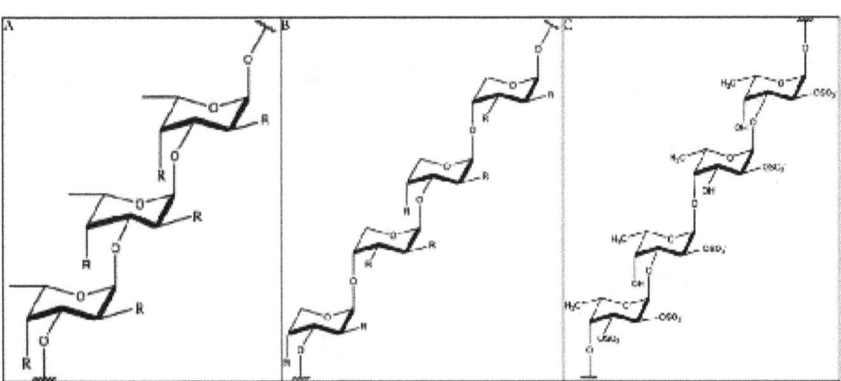


**Figure 6.**
*Structure of fucoidans [54].*

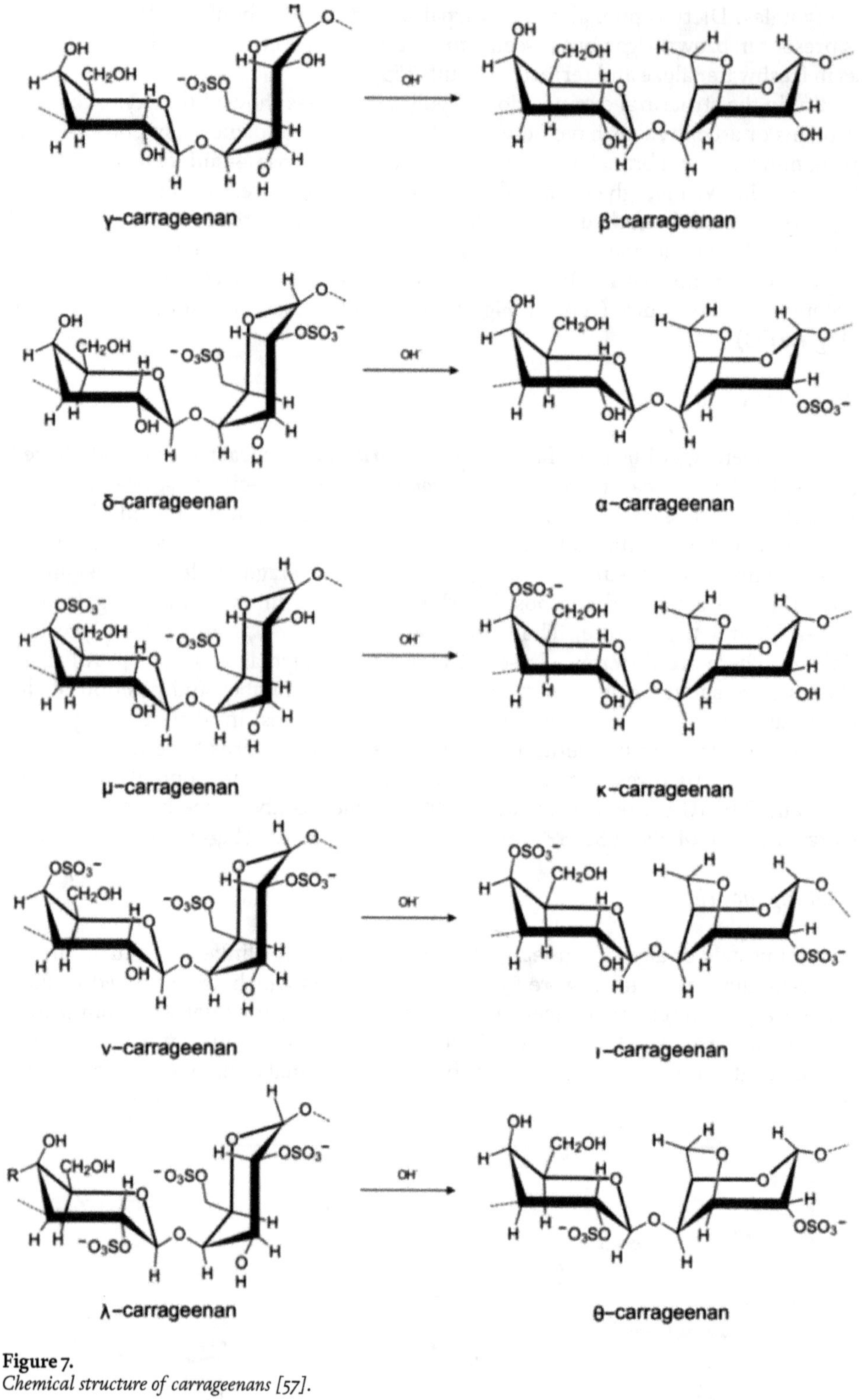

**Figure 7.**
*Chemical structure of carrageenans [57].*

metal ions and obtained very viscous solutions when hydrated. This water absorption property makes alginate suitable for different applications specially in biologi-cal studies with potential application as anti-coagulant, anti-tumor, antiviral and anti-oxidant [55, 60, 61].

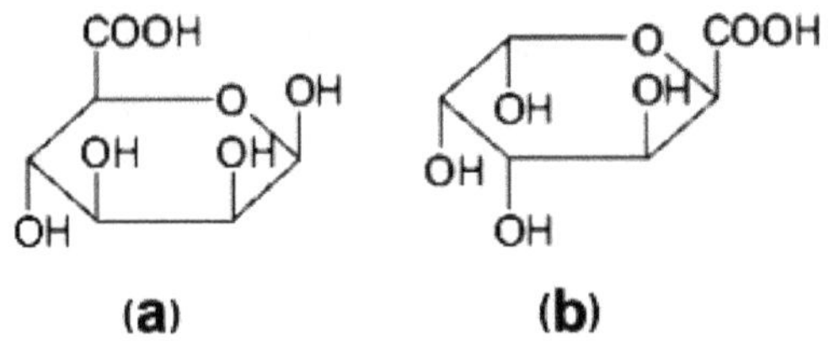

**Figure 8.**
*(a) b-D-mannuronic acid in alginic acid and (b) a-L-guluronic acid in alginic acid (a and b adapted from [51, 62]).*

### *2.1.5 Laminarans*

Laminarans, the nutritional reserve of all brown algae, was first detected in Laminaria species. The molecular weight of the laminaran is about 5000 Da depending on the degree of polymerization. The main sugar, structure and composition of the laminaria species is the laminar, which varies according to the algae species. Laminaran is a polysaccharide, which is soluable in water and consisting 20–25 glucose units including of (1,3)-b-D-glucan including of (1,3)-b-D-glucan, b (1,6) branched (**Figure 8b**). There are two kinds of laminar chain, called M or G, which are different at the reduction ends. While the M chains ends with a mannitol residue, the G chains end with a glucose residue. Most of the laminates, which are impervious to hydrolysis in the upper gastrointestinal tract (GIT) and which are considered to be dietary fiber, are stabilized by cross-chain hydrogen bonds [63]. The activity of structure of laminarans, which are affected by environmental factors such as water temperature, nutrient salt, salinity, waves, sea flow and plunge depth, vary. Besides the role of laminar as a prebiotic and dietary fiber, it is also interesting to have anti-microbial and anti-cancer activities [63].

## 2.2 Alkaloids

Alkaloids are organic compounds, which contains nitrogen atom in their structures. Various structures of amines, cyclic nitrogen and halogenated containing organic compounds exist in the plants and natural materials. Cyclic nitrogen containing alkaloids are only be found in marine organisms and marine algae and are classifies in three main categories [64].

### *2.2.1 Phenylethylamine alkaloids (PEA)*

β/2-phenylethylamine, phenethylamine also known as PEA is made of benzene ring with different ethylamine side chains (**Figure 9a**). These important alkaloids are precursors for making natural and synthetic compounds. Many pharma-ceutical precursors can be achieved from substituted PEAs present in plant and animals, such as, simple phenylamine (tyramine, hordenine) and catecholamine (dopamine) [64].

Some type of brown algae, *Gracilaria bursa-pastoris*, *Halymenia floresii*, *Phyllophora crispa*, *Polysiphonia morrowii* and *Polysiphonia tripinnata*, have PEA in their structures [65].

### *2.2.2 Indole and halogenated indole alkaloids*

Morales-Ríos et al. recorded that the alkaloids produced by *Flustra foliacea*, possessing an unusual pyrroloindoline skeleton, are divided into simple indoles (1–6)

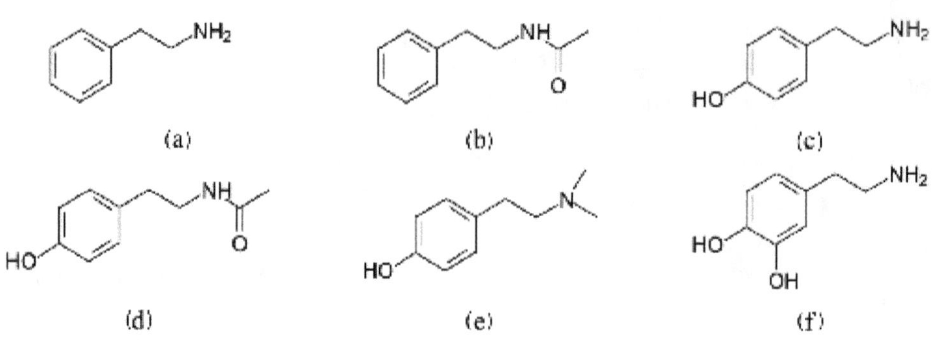

**Figure 9.**
*Structures of phenylethylamine derivatives: (a) PEA; (b) N-ACPEA; (c) TYR; (d) N-ACTYR; (e) HORD; (f) DOP (adapted from [65]).*

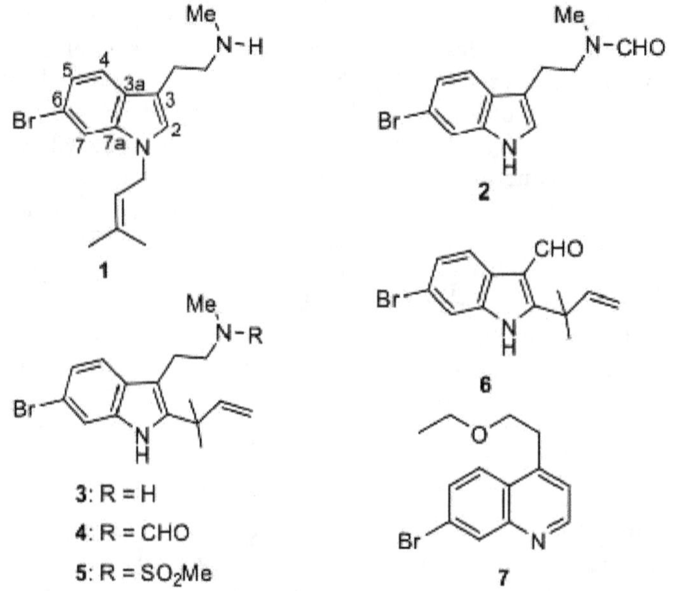

**Figure 10.**
*Structure of indoles (1–6) and quinoline (7) extracted from F. foliacea (adapted from [66]).*

and a quinoline 7 (**Figure 10**), and those with a pyrrolo[2,3-b]indole framework (8–23), including hexahydro-1,2-oxazino[5,6-b]indole (24) (**Figure 11**). Main metabolites in marine seaweeds such bryozoan *Flustra foliacea* are brominated indoles. The structure of these seaweeds have number of brominated indoles with prenyl or isoprenyl substituents at different positions [66].

The marine cheilostome bryozoan *Flustra foliacea* contain an order of brominated pyrroloindolines and indoles, terpenes, and a kind of quinoline, having a variety of biological activities, including anti-microbial, anti-tumor and some biological activities, as secondary metabolites [66].

### *2.2.3 Other alkaloids*

Main alkaloids achieved form algae are from family of 2-phenylethylamine and indole with different substitutions and functionalities. 2,7-naphthyridine derivatives are also alkaloids. Substitutions such as bromide and chloride are specifically seen in Chlorophyta (Perez et al., 2016; [67]). Regarding to the medical properties

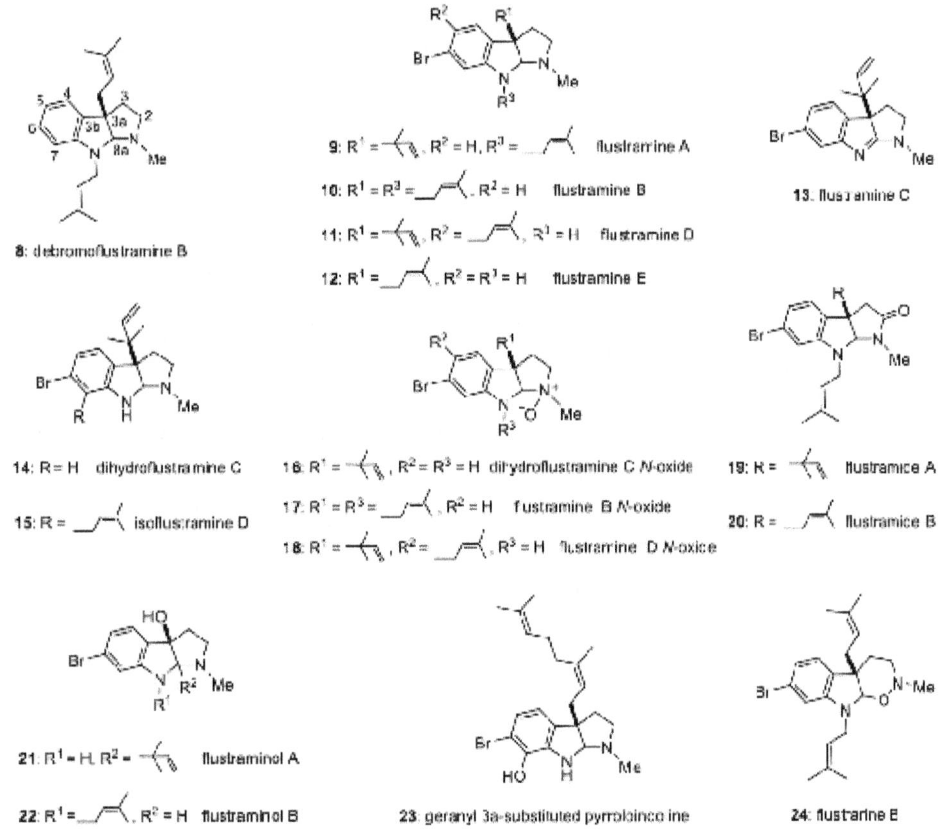

**Figure 11.**
*Indolines 8–24, including indolenine (13), isolated from F. foliacea (adapted from [66]).*

of marine alkaloids, further research and study successfully separated sufficient amount of pure organic derivatives for biological testing [66].

## 2.3 Terpenes

Terpenes known as main algae metabolites, have chemical structure including five-carbon precursor. They are classified into, hemiterpenes, including five carbons (C5); monoterpenes, including ten carbons (C10); sesquiterpenes, including fifteen carbons (C15); diterpenes, including twenty carbons (C20); sesterterpenes, including twenty-five carbons (C25); triterpenes, including thirty carbons (C30) and polyterpenes, including above thirty carbons (>C30). It is known that some sea-weeds contains terpenes. Chlorophyceae is one of them. It contains cyclic and linear sesqui-, di-, and triterpenes. The other one is Rhodophyceae and contains high structural diversity of halogenated secondary metabolites whose polyhalogenated monoterpenes show a variety of antibacterial properties (Perez et al., 2016; [68]).

# 3. Conclusion

In recent decades, seaweed was thought as an abundant and renewable natural resource. Especially, edible seaweeds are rich in polysaccharides, unsaturated fatty acids, protein composition, vitamins, and minerals, as well as natural bioactive compounds such as alkaloids. Their main component, polysaccharides may vary depending on seaweeds and growth conditions. Because of their components, edible seaweeds possess various bioactivities such as anti-oxidant, anti-cancer, anti-obesity, and anti-diabetes activity.

## Author details

Ilknur Babahan[1*], Birsen Kirim[2] and Hamideh Mehr[3]

1 Department of Chemistry, Faculty of Arts and Sciences, Adnan Menderes University, Aydin, Turkey

2 Department of Aquaculture Engineering, Faculty of Agriculture, Adnan Menderes University, Aydin, Turkey

3 Department of Polymer Engineering, University of Akron, Akron, Ohio, USA

*Address all correspondence to: ilknurbabahan@yahoo.com

## References


[1] Maneveldt GW. You Use Seaweeds for that? Veld & Flora; 2006. pp. 24-25

[2] Chandini SK, Ganesan P, Suresh PV, Bhaskar N. Seaweeds as a source of nutritionally beneficial compounds—A review. Journal of Food Science and Technology. 2008;**45**(1):1-13

[3] Dhargalkar VK, Neelam P. Seaweed: Promising plant of the millennium. Science and Culture. 2005;**5**:60-66

[4] Wong KH, Cheung PCK. Nutritional evaluation of some subtropical red and green seaweeds. Part-1. Proximate composition, amino acid profiles and some physico chemical properties. Food Chemistry. 2000;**71**:475-482

[5] David T. Seaweeds. London: Natural History Museum; 2002. Available from: www.seaweed.ie/books

[6] Floreto EAT, Teshima S. The fatty acid composition of seaweeds exposed to different levels of light intensity and salinity. Botanica Marina. 1998;**4**:467-481

[7] Bruton T, Lyons H, Lerat Y, Stanley M, Rasmussen MB. A Review of the Potential of Marine Algae as a Source of Biofuel in Ireland. Dublin: Sustainable Energy Ireland; 2009

[8] Rothe J, Hays D, Benemann J. Macroalgae (Seaweeds). New Fuels: Macroalgae Future Transportation Fuels Study, National Petroleum Council. 2012. p. 12. Available from: https://www.npc.org/FTF_Topic_papers/12-Macroalgae.pdf

[9] Dawes CJ. Marine Botany. New York: John Wiley & Sons, Inc.; 1998. p. 480

[10] Pooja S. Algae used as medicine and food. Journal of Pharmaceutical Sciences and Research. 2014;**6**(1):33-35

[11] Raja A, Vipin C, Aiyappan A. Biological importance of marine algae —An overview. International Journal of Current Microbiology and Applied Sciences. 2013;**2**(5):222-227

[12] Mabeau S, Fleurence J. Seaweed in food products: Biochemical and nutritional aspects. Trends inFood Science & Technology. 1993;**4**:103-107

[13] Kasimala MB, Mebrahtu L, Magoha PP, Asgedom DG. A review on biochemical composition and nutritional aspects of seaweeds. Caribbean Journal of Science and Technology. 2015;**3**:789-797

[14] Cock JM, Peters AF, Coelho SM. Brown algae. Current Biology. 2011;**21**(15):R573-R575

[15] Hoek C, den Hoeck HV, Mann D, Jahns HM. Algae: An Introduction to Phycology. Cambridge University Press; 1995. p. 166. ISBN 9780521316873. OCLC 443576944

[16] Ateweberhan M, Prud"homme Van Reine WF. A taxonomic survey of seaweeds from Eritrea. Blumea. 2005;**50**:65-111

[17] Waaland JR. Common Seaweeds of the Pacific Coast. USA: Pacific Search Press; 1977. ISBN 0-914718-19-3

[18] Lee B. Seaweed Potential as Marine Vegetable and Other Opportunities. Australia: Can Print; 2008. ISBN 1 74151 598

[19] Indergaard M. The aquatic resource. I. The wild marine plants: A global bioresource. In: Cote W, editor. Biomass Utilization. New York: Plenum Publishing Corporation; 1983. pp. 137-168

[20] Nisizawa K. Seaweeds Kaiso – Bountiful harvest from the seas. In: Critchley A, Ohno M, Largo D, editors. World Seaweed Resources—An

Authoritative Reference System: ETI Information Services Ltd. Hybrid Windows and Mac DVD-ROM; 2006. ISBN: 90-75000-80-4

[21] Pereira L. A review of the nutrient composition of selected edible seaweeds. In: Pomin VH, editor. Seaweed: Ecology, Nutrient Composition and Medicinal Uses, Hauppauge, NY. USA: Nova Science Publishers, Inc.; 2011. pp. 15-47

[22] Pereira L. A Review of the Nutrient Composition of Selected Edible Seaweeds. Nova Science Publishers, Inc.; 2011. ISBN 978-1-61470-878-0

[23] Harrison M. Edible Seaweeds around the British Isles. Wild Food School; 2008. Retrieved 11 November 2011

[24] Singh S, Kate BN, Banerjee UC. Bioactive compounds from cyanobacteria and microalgae: An overview. Critical Reviews in Biotechnology. 2005;**25**(3):73-95

[25] Blagojević DK, Simeunović JS, Babić OB, Milovanović IM. Algae in food and feed. Food and Feed Research. 2013;**40**(1):21-31

[26] Pulz O, Gross W. Valuable products from biotechnology of microalgae. Applied Microbiology and Biotechnology. 2004;**65**(6):635-648

[27] Mohamed S, Hashim SN, Rahman HA. Seaweeds: A sustainable functional food for complementary and alternative therapy. Trends in Food Science & Technology. 2012;**23**:83-96

[28] Renaud SM, Luong-Van JT. Seasonal variation in the chemical composition of tropical Australian marine macroalgae. Journal of Applied Phycology. 2006;**18**:381-387

[29] Hotchkiss S, Trius A. Seaweed: The most nutritious form of vegetation on the planet? Food Ingredients—Health and Nutrition. 2007;**1**:22-33

[30] Mondragon J, Mondragon J. Seaweeds of the Pacific Coast. Monterey, California: Sea Challengers Publications; 2003. ISBN 0-930118-29-4

[31] Norziah MH, Ching CY. Nutritional composition of edible seaweed *Gracilaria changi*. Food Chemistry. 2000;**68**:69-76

[32] Pereira L. Seaweed: An unsuspected gastronomic treasury. Chaîne de Rôtisseurs Magazine. 2010;**2**:50

[33] Trono GC Jr. Diversity of the seaweed flora of the Philippines and its utilization. Hydrobiologia. 1999;**398/399**:1-6

[34] Patı MP, Sarma SD, Nayak L, Panda CR. Uses of seaweed and its application to human welfare: A review. International Journal of Pharmacy and Pharmaceutical Sciences. 2016;**8**(10):12-20. ISSN: 0975-1491

[35] Tseng CK. The past, present and future of phycology in China. Hydrobiologia. 2004;**512**:11-20

[36] Mayer AMS, Lehmann VKB. Marine compounds with antibacterial, anticoagulant, antifungal, anti-inflammatory, anthelmintic, antiplatelet, antiprotozoal, and antiviral activities; with actions on the cardiovascular, endocrine, immune, and nervous systems; and other miscellaneous mechanisms of action. Pharmacologist. 2000;**42**:62-69

[37] Ireland CM, Copp BR, Foster MP, McDonald LA, Radisky DC, Swersey JC. Biomedical potential of marine natural products. In: Marine Biotechnology, Pharmaceutical, and Bioactive Natural Products. Vol. 1. NY: Plenum Press; 1993. pp. 1-43

[38] Pal A, Kamthania MC, Kumar A. Bioactive compounds and properties of seaweeds. 2014;**1**:e752. DOI: 10.13140/2.1.1534.7845

[39] Uppangala N. Seaweeds Show Anti-Cancer Activity: Alternative Cancer Therapy. Industry News; 2010

[40] Michalak I, Chojnacka K. Algae as production systems of bioactive compounds. Engineering in Life Sciences. 2015;**15**:160-176

[41] Wanga L, Parkb Y-J, Jeona Y-J, Ryua B. Bioactivities of the edible brown seaweed, *Undaria pinnatifida*. Aquaculture. 2018;**495**:873-880

[42] Usov AI. Chemical structures of algal polysaccharides. In: Domínguez H, editor. Functional Ingredients from Algae for Foods and Nutraceuticals. Cambridge, UK: Woodhead Publishing; 2013. pp. 23-86

[43] Vera J, Castro J, González A, Moenne A. Seaweed polysaccharides and derived oligosaccharides stimulate defense responses and protection against pathogens in plants. Marine Drugs. 2011;**9**:2514-2525

[44] Barros FCN, Silva DC, Sombrab VG, Macielb JS, Feitosab JPA, Freitasa ALP, et al. Structural characterization of polysaccharide obtained from red seaweed, *Gracilaria caudata* (J Agardh). Carbohydrate Polymers. 2013;**92**:598-603

[45] Araki C. Some recent studies on the polysaccharides of agarophytes. Proceedings of the International Seaweed Symposium. 1966;**5**:3-17

[46] Andriamanantoanina H, Chambat G, Rinaudo M. Fractionation of extracted Madagascan Gracilaria corticata polysaccharides: Structure and properties. Carbohydrate Polymers. 2007;**68**(1):77-88

[47] Freile-Pelegrin Y, Murano E. Agars from three species of Gracilaria (Rhodophyta) from Yucatan Peninsula. Bioresource Technology. 2005;**96**(3):295-302

[48] Lahaye M. Developments on gelling algal galactans, their structure and physico-chemistry. Journal of Applied Phycology. 2001;**13**(2):173-184

[49] Melo MRS, Feitosa JPA, Freitas ALP, de Paula RCM. Isolation and characterization of soluble sulfated polysaccharide from the red seaweed Gracilaria cornea. Carbohydrate Polymers. 2002;**49**(4):491-498

[50] Usov AI. Structural analysis of red seaweed galactans of agar and carrageenan groups. Food Hydrocolloids. 1998;**12**(3):301-308

[51] Gupta S, Abu-Ghannam N. Bioactive potential and possible health effects of edible brown seaweeds. Trends in Food Science & Technology. 2011;**22**:315-326

[52] Patankar MS, Oehninger S, Barnett T, Williams RL, Clark GF. A revised structure for fucoidan may explain some of its biological activities. The Journal of Biological Chemistry. 1993;**268**:21770-21776

[53] Ruperez P, Ahrazem O, Leal JA. Potential antioxidant capacity of sulphated polysaccharides from edible brown seaweed Fucus vesiculosus. Journal of Agricultural and Food Chemistry. 2002;**50**:840-845

[54] Weelden GV, Bobinski M, Okła K, van Weelden WJV WJ, Romano A, JMA P. Fucoidan structure and activity in relation to anti-cancer mechanisms. Marine Drugs. 2019;**17**:32

[55] Shanmugam M, Mody KH. Heparinoid-active sulphated polysaccharides from marine algae as potential blood anticoagulant agents. Current Science. 2000;**79**:1672-1683

[56] Jime-enez-Escrig A, Sanchez-Muniz FJ. Dietary fibre from edible Seaweeds: Chemical structure, physicochemical properties and effects on cholesterol

metabolism. Nutrition Research. 2000;**20**:585-598

[57] Bartosikova NJL. Carrageenan: A review. Veterinární Medicína. 2013;**58**:187-205

[58] Barbeyron T, Michel G, Potin P, Henrissat B, Kloareg B. ι-Carrageenases constitute a novel family of glycoside hydrolases, unrelated to that of κ-carrageenases. Journal of Biological Chemistry. 2000;**275**:35499-35505

[59] Andrade LR, Salgado LT, Farina M, Pereira MS, Mourao PAS, Amado-Filho GM. Ultrastructure of acidic polysaccharides from the cell walls of brown algae. Journal of Structural Biology. 2004;**145**:216-225

[60] Koyanagi S, Tanigawa N, Nakagawa H, Soeda S, Shimeno H. Oversulfation of fucoidan enhances its anti-angiogenic and antitumor activities. Biochemical Pharamacology. 2003;**65**:173-179

[61] Ponce NMA, Pujol CA, Damonte EB, Flores ML, Stoerz CA. Fucoidans from the brown seaweed *Adenocystis utricularis*: Extraction methods, antiviral activity and structural studies. Carbohydrate Research. 2003;**338**:153-165

[62] Davis TA, Volesky B, Mucci M. A review of the bio-chemistry of heavy metal biosorption by brown algae. Water Research. 2003;**37**:4311-4330

[63] Neyrinck AM, Mouson A, Delzenne NM. Dietary supplementation with laminarin, a fermentable marine b (1e3) glucan, protects against hepatotoxicity induced by LPS in rat by modulating immune response in the hepatic tissue. International Immunopharmacology. 2007;**7**:1497-1506

[64] Güven KS, Percot A, Sezik E. Alkaloids in Marine. Algae Marine Drugs. 2010;**8**:269-284

[65] Percot A, Yalçın A, Aysel V, Erdugan H, Dural B, Güven KC. β-Phenylethylamine content in marine algae around Turkish coasts. Botanica Marina. 2009;**52**:87-90

[66] Morales-Ríos M, Suárez-Castillo OR. Synthesis of marine indole alkaloids from Flustra foliacea. Natural Product Communications. 2008;**3**(4):629-642

[67] Barbosa M, Valentão P, Andrade PB. Bioactive compounds from macroalgae in the new millennium: Implications for neurodegenerative diseases. Marine Drugs. 2014;**12**:4934-4972

[68] Bedoux G, Hardouin K, Burlot AS, Bourgougnon N. Bioactive compounds from seaweeds: Cosmetic applications and future development. Advances in Botanical Research. 2014;**71**:345-379

# 6

# Vegetation Dynamics: Natural versus Cultural and the Regeneration Potential — The Example of Sahara-Sahel

*Erhard Schulz, Aboubacar Adamou, Sani Ibrahim, Issa Ousseini and Ludger Herrmann*

**Abstract**

There is a principal and controversial debate on the so-called 'Greening-Regreening' of the Sahel. There still is the old philosophy of an expanding/shrinking ecosystem Sahara versus Sahel. In some concepts, it is presented as annual. Another concept is based on a general degradation of the Sahelian savannas – in some cases with a decline to a lower state of ecological equilibrium after a short period of resilience. Anyhow, there are also signs of still ongoing regeneration processes of vegetation and soil. The main problem, however, lies in the principal lack of terrestrial observation and in the confusion of terms. This mostly concern on vegetation units and their dynamics. The goal of this article is to explain the general nature of the Sahara and the Sahel based on maps and graphs. We try to analyse the dynamics of boundaries during the last 200 years. The main results are the tripartite nature of the Sahara, divided into semidesert, desert and Saharan savanna with relatively stable boundaries. A reconstruction of the vegetation for the last 200 years confirmed the position of these borderlines even under different states of the plant cover. It also revealed the nature of Sahelian savannas as cultural landscapes – in higher diversity and density. It is also possible that the North Sahelian savannas had been for long times under the dynamics of elephant landscapes. A high-resolution sediment and pollen record from the Middle Sahel of Niger evidenced the high diversity and resilience up to the severe drought of the 1970s. It was a definite stroke from which these savannas never reached again their former diversity despite a slide recovery named 'Regreening'. The various projects for regeneration or conserva-tion in Sahara or Sahel differ in two types of projects. The one is the installation of Nature Reserves/National Parks with special reserves for emblematic animals as keystone organisms and an auto-regeneration of vegetation and soil. The other type consists of pasture rotation projects such as in the Malian Gourma or in the Central Air Mts. The first initiative resulted in the decade-long protection against the severe degradations, which were typical for the surrounding regions. The rotation system was based on timewise open wells and of observed pasture status. It was conceived together with the local populations and has been respected until the invasion of northern cattle keepers during the peak of drought in 1984. After severe quarrels, the system collapsed and the savannas degraded heavily. A comparable project worked in the central Air Mts. for 5 years. Remarkable results have been, but the rebellion of the 1990s, put a sudden end on it. The general insecurity of the last decades caused by civil war and/or various terrrorist groups led to a re-evaluation of a great number of regeneration initiatives including the pharaonic 'Great Green Wall', a continent wide forest belt. However, smaller projects on the village level may better develop as they are under the responsability of local population, which can reactivate their long experience. The 'regreening' might be restricted to the region of the southern Sahara and the northern Sahel as well as to the traditional park systems. Anyhow, even if a long-time amelioration of production systems will happen, the former must be regarded on the background of a rapidly increasing demography.

**Keywords:** Sahara, Sahel, vegetation, landscape types, present situation, historical development, stability of limits, cultural, landscapes, degradation, regeneration potential

## 1. Introduction

In the last years, a 'Greening' or 'Regreening' of the Sahel was a most disputed topic. It mutated to a general discussion of regeneration potential of the ecosystems and the possibilities to find production modes for the necessary food production. Moreover, conservation and nature protection were discussed and great projects were initiated [1–13].

On the other side, the general political insecurity of the last 15 years supressed fieldwork and made an end to several initiatives. Many of the conservation projects are now classified as 'in suspense.' This stands especially for the big National Natural Parks in the Sahara of Niger and Chad [14–16] and more or less for the 'Great Green Wall' too [17, 18].

Thus, the reasoning on degradation or regeneration is often based on pure remote sensing without the necessary ground check or field work. In addition, for the case of Sahara-Sahel-complex, there is still a deep confusion on the nature and dynamics of ecosystems and landscapes as well as on their definitions. Limits and boundaries seem to be free floating – sometimes on an annual scale.

On this background, we will characterise the main ecosystems – landscapes of Sahara-Sahel by a general vegetation map in order to avoid further confusions. This should also work as a base to interpret palaeorecords. Furtheron, we will try to reconstruct the landscape evolution during the last 200 years. Finally, we will discuss the chances of measures of regeneration and conservation.

## 2. The 'bandoneon desert'. Concepts and nature of the Sahara

It is fascinating to see that the old concept of an extension of one large ecosystem on the cost of another – here the advancing/encroaching desert into the savannas is still taken as valid. The alarm of Stebbing [19] of an advancing desert in the Niger-Nigeria border region was rapidly disproved by a common French-English – Forester expedition [20]. More than half a century later, Tucker et al. [21] presented the model of an expanding and retracting Sahara, which he considered as desert for the whole in the scale of years. Their conclusions were based on interpreted vegeta-tion changes with help of satellite images; however, without any differentiation between permanent and short-time plant cover. Another less meaningful approach was presented by Thomas and Nigaru [22], who claimed a 10% expansion of the Sahara/ desert since 1920 both to the North and to the South. The authors based their conclusion on changes in precipitation as they defined ecosystems/landscapes exclusively by mean annual precipitation.

Thus, we have to deal with a variety of methods and concepts in the analysis of landscapes/ecosystems in northern and western Africa. We take the term 'landscape' we take in a broad sense as a characteristic part of the earth's surface, which is defined by various features such as vegetation, relief or the intensive human impact, which developed in time, and which is visibly different from neighbouring regions.

**A.** Field observation and subsequent definition of landscape – or vegetation types. This was the procedure at the beginning of the twentieth century in the aftermath of the colonisation with Chevalier [23] as an example.

**B.** The characterisation of recognised landscapes and definition of leading features of their vegetation or geomorphology.

**C.** A difference in the concept of consistent or transition zones. 'Sahara' as the transition from the Mediterranean to the 'Sahel' [24] or the 'Sahel' as transition from the 'Sahara' to the (real) savannas [25].

**D.** Reduction of landscapes/ecosystem (and climate too) to a single feature such as rainfall. It is the main cause of confusion on the dynamics of the large ecosystems in northern and western Africa [26].

**E.** A main problem is the emotional component in the term 'desert', which impedes often a neutral recognition. Mostly, the terms 'Sahara' and 'desert' are used as synonyms – see [24].

Thus, it is necessary to explain clearly the terms in order to avoid confusion and to define them from direct observation in the field – or at least from clear descriptions.

## 3. What are we talking about?

A vegetation map of northern and western Africa was established in order to explain clearly the large vegetation types and their repartition. It is based on direct observation during several expeditions (see the small included map) and on published vegetation maps [27]. It deals with the physiognomic units such as forest, shrub-land or grass-land and gives the main floristic components. There is no differentiation between natural formation, near to nature formation or cultural units; however, their dynamics are shortly discussed. Here, we will concentrate on the South-Mediterranean steppe, on the Sahara and on the Sahel to give a background to the discussion about limits, their dynamics and their regeneration potential. Block diagrams and designs will support it. Thus, we will try to avoid the various confusions on terms such as 'steppe or prairie'. The map is to document visible units and their limits and to work as a modern model when reconstructing the past. The question of limits might be regarded as an academic one. However, it is an assessment of resource areas-mainly of pasture.

### 3.1 The southernmost formation of the Mediterranean realm is the 'steppe'

The term 'steppe' is freely used in literature – comparable to the term 'savanna'. For both the statement of Cole [28] is still valid: 'Most discussed and least understood' (see **Figure 1** nr. 10 and 2). Thus, 'steppe' is rarely referred to the origi-nal definition as a tussock-grassland of the genus *Stipa* under continental winter cold conditions [29, 30]. In this area, it is mainly characterised by *Stipa tenacissima* and *Lygeum spartium* on fine-grained substrates such as loess. It stretches over the plateaus of the Atlas Mts., and it is severely exploited for pasture, agriculture or paper production.

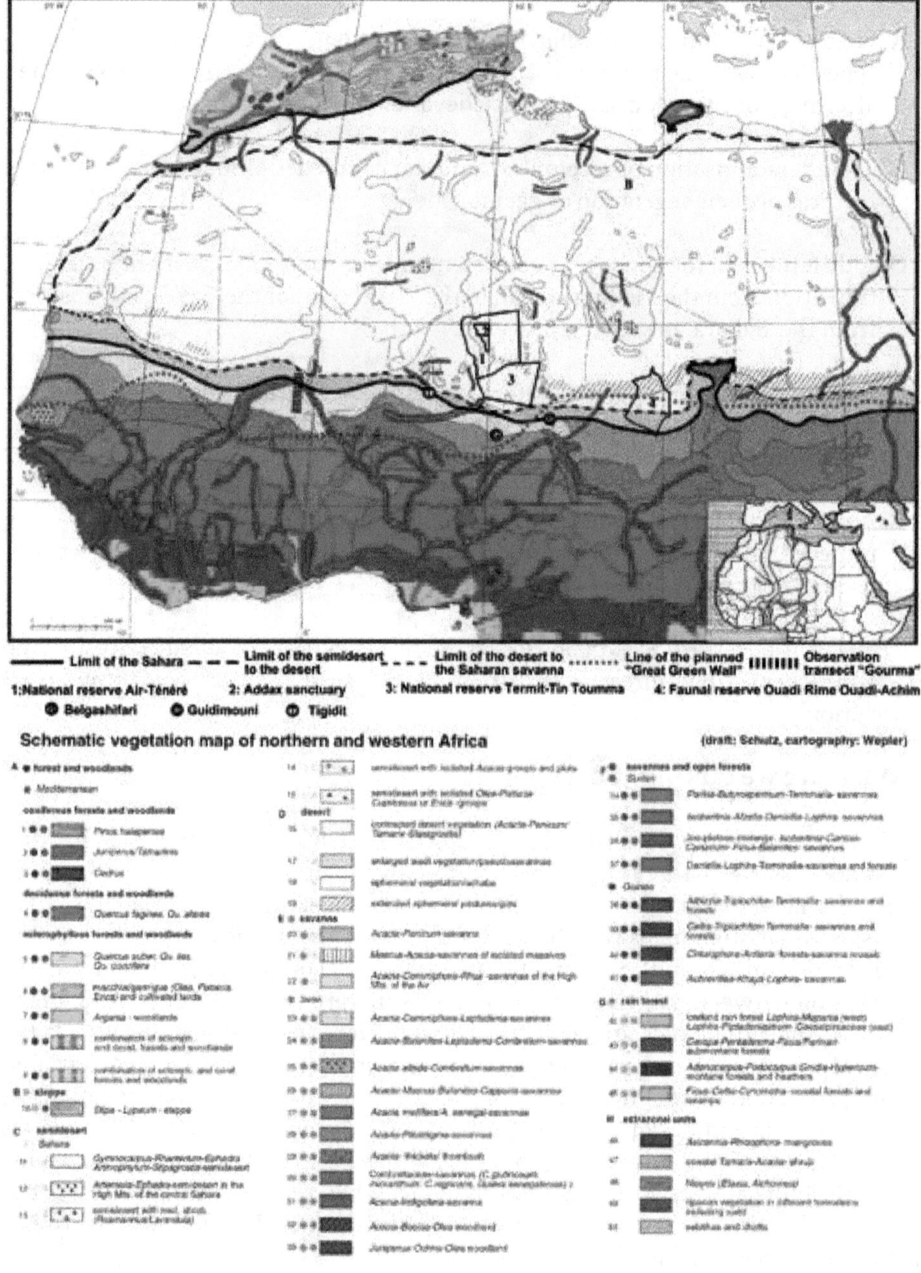


**Figure 1.**
*Schematic presents vegetation map of northern and western Africa. Also shown are the national parks-national reserves in the Sahara of Niger and Chad and the location of the planned 'Great Green Wall'. From [27], modified and enlarged.*

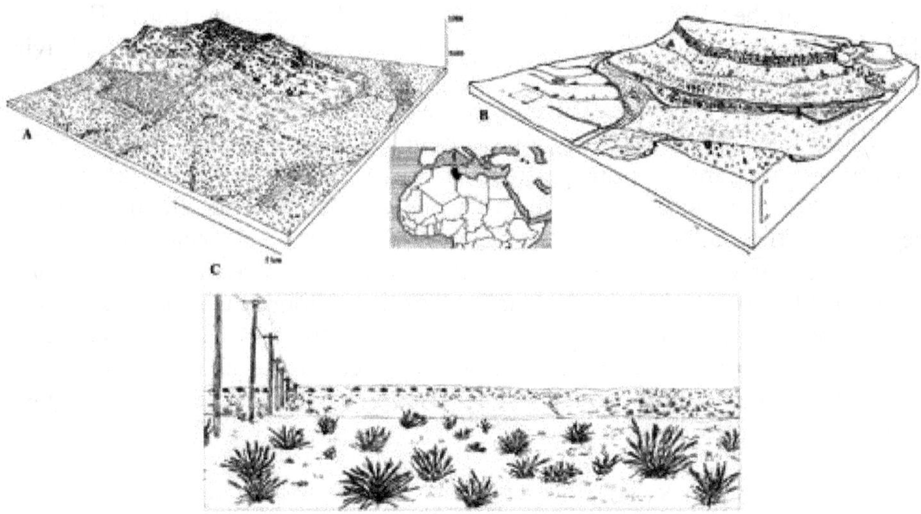

**Figure 2.**
*Aspects of the steppe (cf.* **Figure 1**, *10). (A) Block diagram of Djebel Chaambi in Central Tunisia. In the upper part, the southernmost stand of* Quercus-*forests, in the lower part the* Juniperus-Rosmarinus-*shrubs and on the plain the* Stipa-Lygeum Artemisia-*steppe. (B) Djebel Dahar, Southeast Tunisia. The southernmost outpost of steppe on the loess plateaus. (C) Aspects of the* Stipa-Lygeum-*steppe near Kasserine, Central Tunisia. Drawing Schulz.*

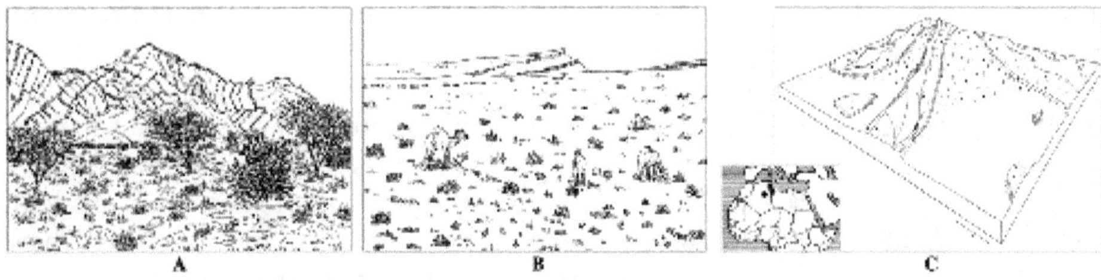

**Figure 3.**
*Aspects of the semidesert. (A)* Acacia-Rhantherium-*stands in the Bou Hedma, southern Tunisia. (B)* Rhantherium-*semidesert South of Remada, southern Tunisia. (C) The southern limit of semi desert with* Calligonum-Ephedra *at 30°N, South of El Golea, Algeria. Drawing Schulz.*

To the North, the steppe interfingers with Mediterranean *Rosmarinus-Juniperus* formations and a clear limit is hardly visible. However, at the southern part of the Sahara-Atlas, it ends with the loess cover. But there is an outpost of steppe on the loess plateaus of the Dahar Mts. in southeast Tunisia.

This follows the basic 'law of relative constancy' [30]. It means that plants or animals change the type of their habitat in the border region of their main area in order to guarantee the basic needs of the respective organism. Finally, it fits well to the original definition as a grassland under continental and wintercold conditions (**Figures 2, 3**).

## 4. The landscape system of the Sahara

Descriptions and characterisations of the Sahara are manyfold, see [31, 32]. Mostly it is taken as the greatest desert on earth with an extension of about 2000 × 5000 km. The area is structured by a system of wide basins and ridges often topped by mountains of more than 4000 m. Climatically, it is characterised by the interaction of the Westafrican monsoon and the tradewinds – see below. However, the most important feature is the general lack of water – a fact, which all living organisms have to cope with.

For a useful partition of the Saharan area, we need criteria, which are applicable to the whole area. Moreover, they must summarise the ecological effects of the respective region and in principle it must be visible and recognisable even in a reduced form, and it is not useful to choose volatile elements. In that way, the vegetation is the most appropriate way to characterise the whole region and to divide it in several parts. Thus, it has an indicator function. On a second level, the plant cover shall be described by its floristic content. In addition, the plant cover can be understood from detailed descrip-tions – even by non-specialists. And we should not forget that vegetation is the most important resource for various organisms. In that way, we will describe and divide the Sahara in units, which are easy to recognise– also from ancient descriptions.

*Sahara est. omnis divisa in partes tres.......*

## 4.1 The semidesert

South of Atlas Mts. there is a double change in landscape. It is from grassland (steppe) to shrub land and from the Mediterranean realm to the Saharan one (see **Figure 1**, 11–15). Vegetation is still diffuse, but rarely exceeds 30% of soil cover, and the greater part of biomass is below the surface. Saharan floristic elements like *Fagonia arabica, Rhantherium suaveolens, Gymnocarpus decander* or *Stipagrostis pungens* on dunes dominate in the small- or dwarf shrub lands. It is the northern part of the Saharan landscapes – the semidesert.

The authors [33–37] claimed that the double stress by frost and drought impedes a tree development. However, the double strategy of life in the Sahara is already vis-ible. Only a restricted number of organisms are equipped against drought and frost. On the other hand, there is the strategy of mass and accident. Aleatoric rainfall may activate the seed bank of herbs and grasses. These therophytes must fulfil their lifecycle in the short time of limited rainfall.

These accidental floras are an important resource for nomadic animal keeping.

Anyway, we must not forget that *Acacia raddiana*-stands still exist in southern Tunisia (Dj. Bou Hedma) or in southwestern Morocco. Perhaps, future records will convince us to rethink the dynamic of the northern Sahara [38–42].

The southern limit of semidesert is easy to recognise. Around 30°N (31° N in the East or along the Atlantic coast of Morocco), it changes from diffuse stands of *Calligonum* or *Ephedra* to another mode, (conracted or linear) of the *Acacia-Panicum*-type. This characterises the change from semidesert to desert.

## 4.2 The desert

The desert is extremely difficult to define because of the emotional component of the term (see **Figure 1**, 16-19, **Figures 4, 5**). Here, we follow the definition of Monod [44]. He stressed the difference of diffuse modes of semidesert or savanna to the con-tracted one – the desert. The desert is the region where permanent life is only possible in favourable places such as wadis (dry valleys) or depressions where groundwater and run off are available. Thus, permanent vegetation is contracted or linear.

It follows the oasis system, as few places, where the basic needs are guaranteed. There are several modes to cope with the scarce water resources such as the Acacia-strategy. Aleatoric rainfalls may induce germination of the seeds – perhaps already prepared by the intestines of animals. After germination, all resources are mobilised to develop a tap root to reach ground water. In that case, the plant gets independent

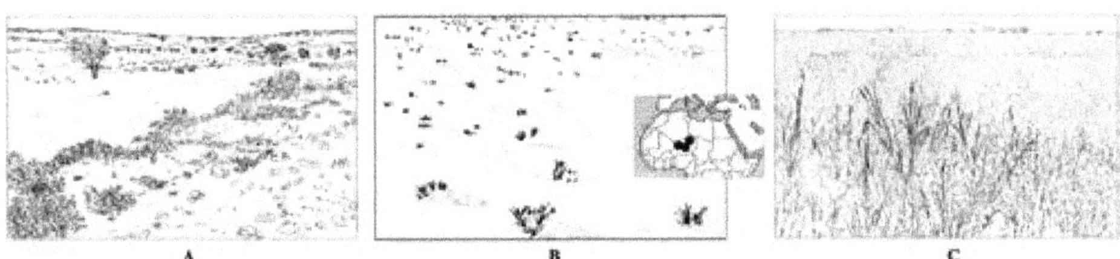

**Figure 4.**
*Aspects of the desert. (A) Contracted vegetation in the Wadi Achelouma, northeastern Niger. (B) Achab in the Ténéré, northern Niger. (C) Wild cereal fields in the southwestern foreland of the Air Mts., N-Niger. From [27] modified.*

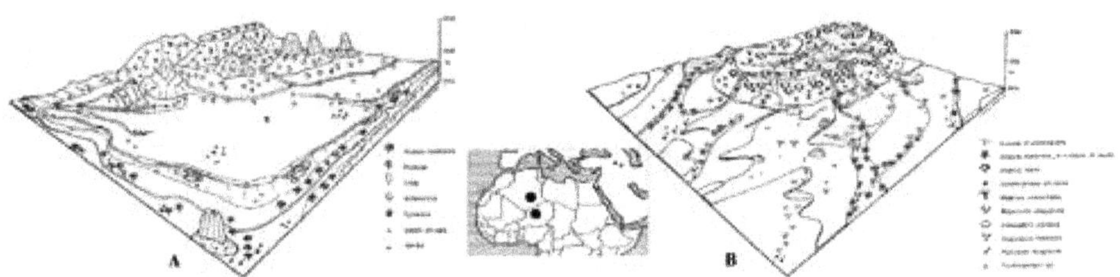

**Figure 5.**
*The modes of altitudinal change in the Sahara. (A) From desert to semidesert (Ahaggar/Algeria and Tibesti/Chad) (B) From desert to savannah (Air Mts./Niger). From [43], modified.*

from climate. However, there is the other strategy of life – that of achabs, already discussed in regard to the semidesert. The seed bank rapidly reacts on aleatoric rainfalls with a short time-flora. It may be the case once in 3 years or several times a year. There are also wild cereals as part of the achabs – an important resource for human food (see **Figure 4C**).

The contracted vegetation, mainly of the *Acacia-Panicum-* (tree-tussock grass) type, is typical for the wadis of mountain areas and their forelands (see **Figure 1**, 16, 17). Large wadis in the forelands – especially in the Southwest of Adrar des Iforas and Air Mts. might touch for a short distance but they separate afterwards. In that way the impression of a diffuse plant cover may exist. It is perfectly demonstrated by Voss et al. [45, 46] for the western forelands of the Adrar des Iforas in northern Mali.

### 4.3 Altitudinal change

As in other regions, the plant cover changes with altitude in the Sahara (see **Figure 1**, 12, 22, **Figures 5, 6**). There is an altitudinal change of vegetation in the High Mts. of the Sahara (**Figure 5**). In the Ahaggar Mts/South Algeria, the characteristic *Acacia-Panicum* vegetation of the desert wadis changes from about 2000 m into a diffuse *Artemisia*-shrub vegetation-a semidesert of a Mediterranean affiliation. In small gorges, some tree groups of *Olea lapperinii* or *Pistacia atlantica* exist. The Tibesti Mts. show similar features, however, on the highest peaks, some stands of *Erica arborea* survived (**Figure 7**). This is the Mediterranean type of altitudinal change [48].

The Air Mountains are different. Above 1800 m the contracted *Acacia-Panicum*-plant cover changes to a diffuse *Acacia-Commiphora-Rhus*-savanna (savanna seen as a tree grass – vegetation under a tropical climate). Thus, it is a Sahelian type of altitudinal change. **Figure 6** gives a general overview of the Air Mts. (A) with the locations of the change to High-Mts.-savannas and the upper catchment of the wadi Anou Mekkerene (**Figure 8** see below, see also [49]).

1 Cuestas, margins of the mountains.
2 Isolated granite massifs.
3 Lavas and basalts.
4 Main wadis.
5 Contracted desert vegetation. (*Acacia*, *Panicum*, achabs).
6 Inundation areas in the mountains and plains with extended grass floras.
7 Large areas of wild cereals (*Panicum laetum*, *Sorghum aethiopum*).
8 Pseudodiffuse vegetation of the southwestern forelands. Extended wadi vegetation.
9 Remains of former alluvial woods.
10 *Acacia-Commiphora-Rhus*- savannas.
11 *Acacia-Panicum*-savannas
12 *Commiphora-Acacia*-savannas
13 Aera of the annual salt pasture.

**Figure 6.**
*Vegetation of the Air Mts./Niger and the limits of the desert as an example for the southern Sahara. Also shown are Upper Wadi Anou Mekkerene (A, see also* **Figure 8***) and the Air-Ténéré-National Park (B) with its Addax sanctuary (C). From [47], modified.*

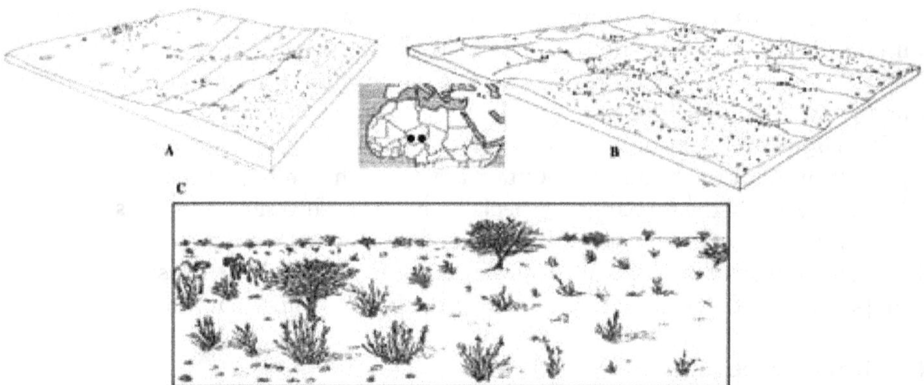


**Figure 7.**
*The southern limit of the desert and the Saharan savanna. (A) The passage from desert to (Saharan) savanna at the Tigidit escarpment, northern Niger. (B) The change from the linear desert vegetation to the savanna a the Belgaschifari well NE-Niger. (C) The general aspect of the Saharan Acacia-*Panicum savanna. *From [13], modified and complemented.*

## 4.4 The southern limit of the Sahara

In the southern forelands of the Air Mts. around 16°N/16°30′N, the aspect changes again in two steps. The first step is visible by a diffuse *Maerua crassifolia-Acacia ehrenbergiana*-savanna on the fissured sandstone-plateaus of Tigidit and also Agadem-Homodji in Southeast Niger (see **Figure 1**, 21). These savannas depend on the cistern effects of the fissures, which collect and hold water from runoff and dew [27]. On the plains, however, one observes a densification of the tree lines and the transition into a savanna of the same elements within a short distance (see **Figure 1**, 20).

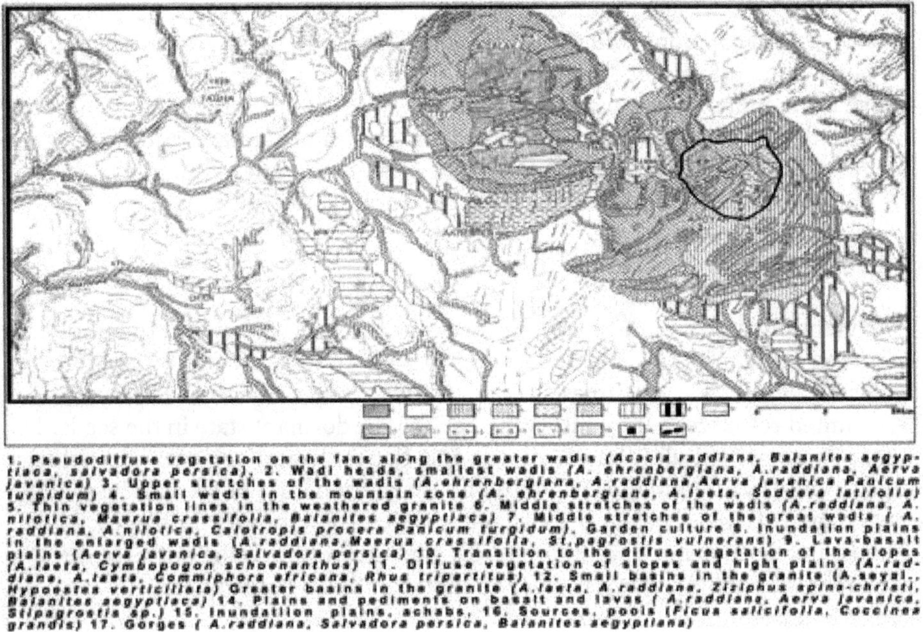


**Figure 8.**
*Wadi Anou Mekkerene and the Agalak-Aroyane Mts. of the central Air Mts. The area of the Guide pasture reserve is indicated. The difference of the diffuse mountainous savannas and the linear desert vegetation is clearly visible as well as the densification of the alluvial vegetation following down the wadis. From [43], modified.*

Thus, there is the definite transition from the desert to savanna within the Saharan realm. Similar features are confirmed for northern Mauretania and Mali [50, 51]. In northern Chad, this transition is modified by substrata [52]. Large inundation plains are quasi devoid of plants, which appear only on sand ridges. On the sandy plains at about 16°N, the change into a tree-tussock grass savannah occurs similarly as it is the case for Niger. Akthar-Schuster [53] reports a comparable transition belt for the northern Sudan too. This boundary is the most disputed limit between landscape zones, as it caused the misunderstanding of degradation-desertification, etc.

Finally, the Sahara is a tripartite landscape system, where the desert takes the greatest part but has its borders to the semidesert in the North and the savanna in the South. Thus, the main change in the landscape system, that of desert to savannah, takes place within the Saharan realm.

*.... quarum unam dominat semideserta, aliam deserta et tertiam savanna saharica.*

## 4.5 The climatic implications

At this point of description, we should also deal with climatic conditions. In the aftermath of Dubief [54], the main boundaries are often paralleled to - or defined by mean annual precipitation. However, there are also dew, runoff and especially the access to groundwater which determines plants and vegetation. So, various components are summed up. Note, that two main systems interact: the summer rains of the monsoon and the Mediterranean winter rains and trade winds. We also have to consider the aleatoric rainfalls during the whole year derived from monsoon or cold airdrops from the North. They are responsible for achabs and the short time floras demonstrate their existence. The northern boundary of the Sahara is usually assigned to an annual precipitation of about 100 mm – mainly in winter. More to the centre of the Sahara mean values are fictional. Rainfall becomes aleatoric and accident is the main component in the ecosystem. The southern limit of the semidesert may be attributed to about 50 mm/y and the southern border of the desert within the Sahara is more or less parallel to 150 mm/y of summer rain. As mentioned above, both limits largely depend on the combination of rainfall, runoff, dew and storage of humidity in soil. Anyhow, these clear boundaries are among the few pure climatic ones. They are visible across the whole continent.

### 4.6 Life strategies

As mentioned above, there are two basic strategies to cope with the uncertain resources. These are the 'achab-strategy,' to answer with a mass of unprotected organisms to aleatoric resources – here rainfall. They fulfil their life cycle with these limited resources before returning back to the dormant state in the seeds. The '*Acacia*-strategy' includes the use of tap-and flat roots and vegetative/generative propagation. Useful rainfall is exploited by the germinating of seeds. The saplings grow in the first years below the surface and develop taproots until they reach a groundwater lens or horizon. Afterwards, they grow above the surface, develop lateral roots and are more or less independent from the actual climate.

### 4.7 Differences in concepts and analyses/interpretations

Different concepts may produce different interpretations. The vegetation map (**Figure 1**) differs in several points from the concepts of other colleagues especially in type and position of the southern boundary of the Sahara. We do not follow the interpretation given by Medail-Quezel [24] or White [55] for the North-extensions of the Sahel in the southwester forelands of the Adrar des Iforas (N-Mali) and of the Air Mts. (Niger) as well as for the southern half of the Air Mts. [56].

The forelands are not seen as part of the Sahel but as regions of enlarged wadis see [45, 46]. The Air Mts. are considered as Saharan desert-mountains with a Sahelian altitudinal change – as for example, the Ahaggar Mts. or the Tibesti, which do not belong to the Mediterranean out of their high altitude vegetation. White [54] takes the northernmost savannas as part of the Sahel. Another point is the statistical approach as shown by Linder et al. [57]. They define various borderlines of Sahel versus Sahara out of all zoological and floristic elements. Most of those boundaries reach several hundreds of km more to the North – into the region of plain desert. This represents the principal difference of field analysis and pure statistical analysis without any ground check. Another point is the difference and extension of the Sudan- and Guinea-zones. The concepts of the Kew and Toulouse schools [58, 59] differ at the Nigeria-Cameroon border. In that case, we follow the 'Toulouse' school.

## 5. The Sahel and its savannas

A few km to the South, the aspect of landscape changes again (see **Figure 1**, 23-33, **Figures 9, 10**). The savanna remains but the floristic composition differs. Beside *Acacia, Commiphora* becomes characteristic and the annual grasses like *Aristida mutabilis*or *Cenchrus biflorus* are dominant. It is the definite change from the Sahara to the Sahel. Phytosociologically, it is defined by the transition from the*Acacio-Panicion* to the *Acacio-Aristidion* [47].

As **Figures 9** and **10** demonstrate, the Sahelian savannas are intensively exploited. The northern ones are pasture areas, and millet growing dominates in the *Acacia-Piliostigma* savannas. These are cultural landscapes and they demonstrate

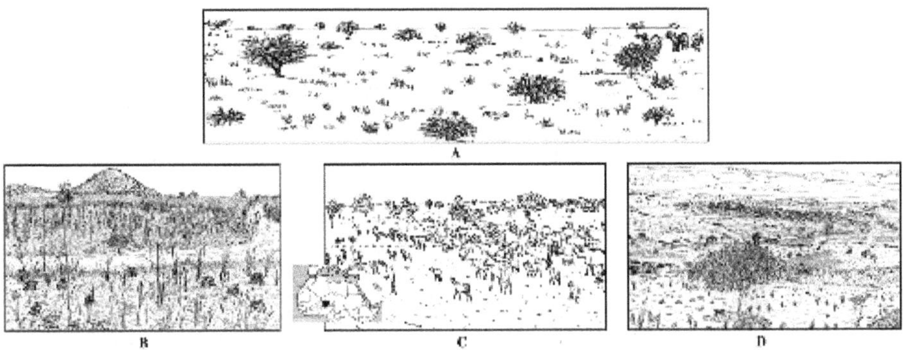

**Figure 9.**
*Aspects of the Sahel. (A) The Sahelian savanna on the Tigidit plateau, Central Niger* Acacia, Commiphora, Maerua *and annual grasses. (B) Millet fields near Birni-n-Konni, southern Niger. (C) Animal keeping near Abalak, northern Niger. (D) Desertification. The overexploited area of Ader, near Koutous, central Niger. Drawing Schulz.*

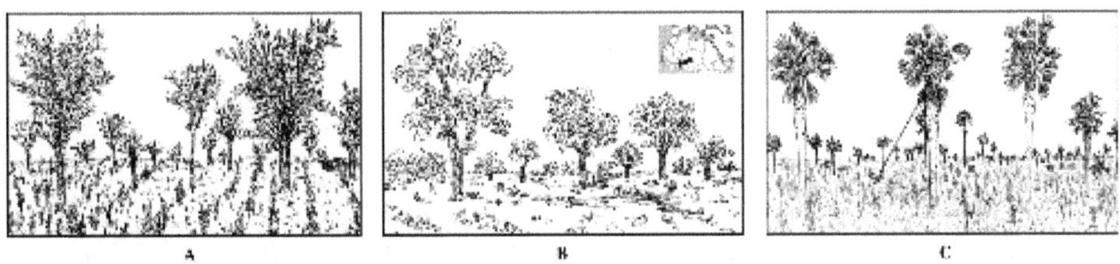

**Figure 10.**
*The aspects of agro-forestry. (A)* Faidherbia albida*-Park for animal keeping and agriculture in southern Niger. (B) Karité-Park (*Vetiveria paradoxa*) in northern Togo. Tree cultivation for fat and agriculture with the general employment of fire. (C) Ronier palm-Park (*Borassus aethiopum*), southern Niger, for various exploitations of the trees and agriculture. Former defence parks. Drawing Schulz.*

the variety of degradation. The southern Sahel is also a region of the old rooted agroforestry systems – the parks [60–62]. They are dual and integrated systems of animal (cattle) keeping and agriculture. The main feature is the two storey aspects of trees of a restricted species composition and only one or two generations. The Gao (*Faidherbia*)-parks, however, often show several generations of shrubs and trees. The intention of these parks is the production of vegetal or animal fat and agrarian products. They have been constituted by selection from a pre-existing vegetation (*Vitellaria* and *Parkia*-parks), by tolerance and assistance- as for the *Faidherbia*-parks – or by former defence plantation as it is the case of *Borassus*-parks [61]. Fire is still a part of the agricultural management.

Either it is a tool to clear land for new fields – few areas where fallow – either shifting cultivation is still practised or it is used for cleaning or sanitary purposes [63–66]. The Sahel is a savanna region and climatically it is influenced by tropical summer rains (monsoon) with a gradient from about 800 to 150/ 200 mm/y and with a rainy season of 3–5 months.

For long periods, the Sahel was only regarded as a transition zone to the real (Sudanian)-savannas [33, 50]. From the 1970s, this region was accepted as one of the consistent savannas [67] even widely transformed to cultural landscapes [68].

## 6. Lessons from the past. The last 200 years

Type and dynamic of landscape may often be read and understood from its history (see **Figures 11–16**). A series of more or less precise descriptions is on our

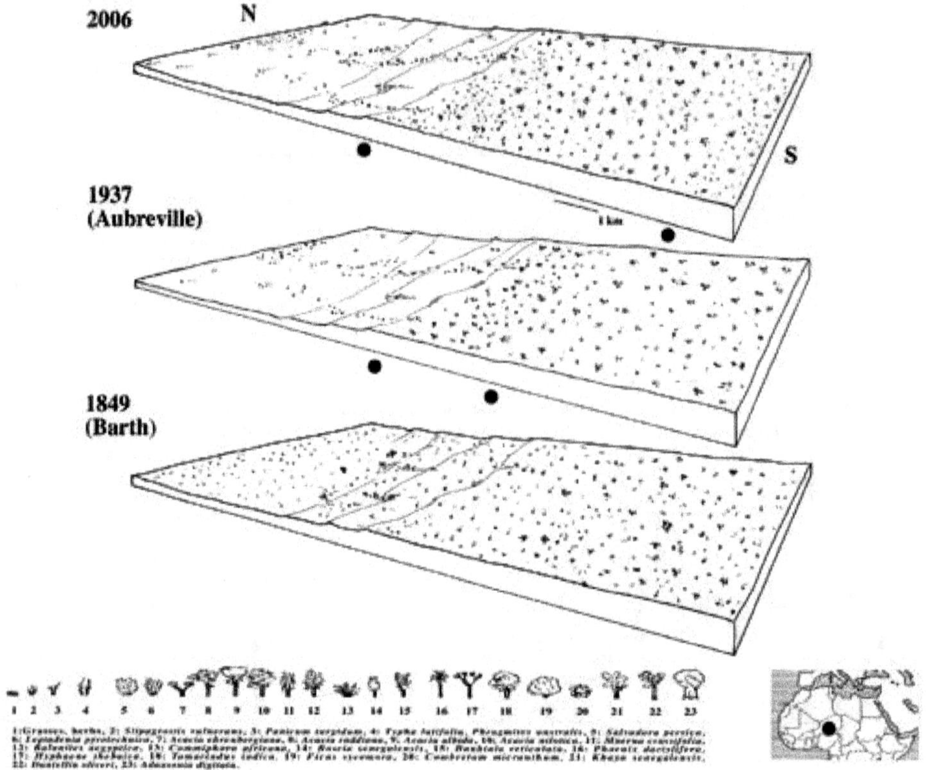


**Figure 11.**
*The history of the southern limit of the desert and the Sahara at the Tigidit plateau, northern Niger (from [69], modified).*

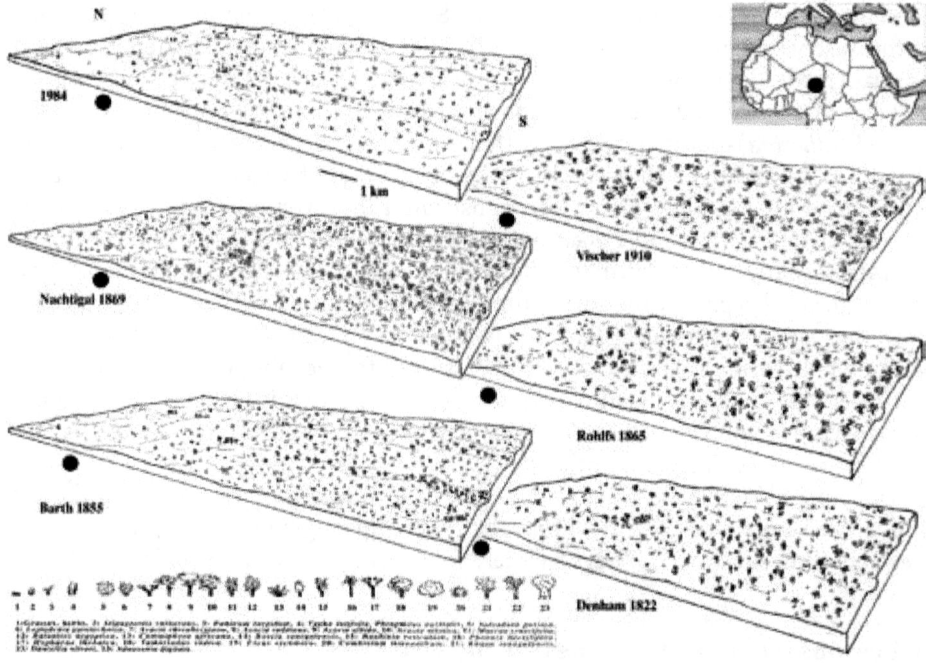


**Figure 12.**
*The history of the southern limit of the desert and the Sahara at the Belgashifari well, NE Niger (from [69], modified).*

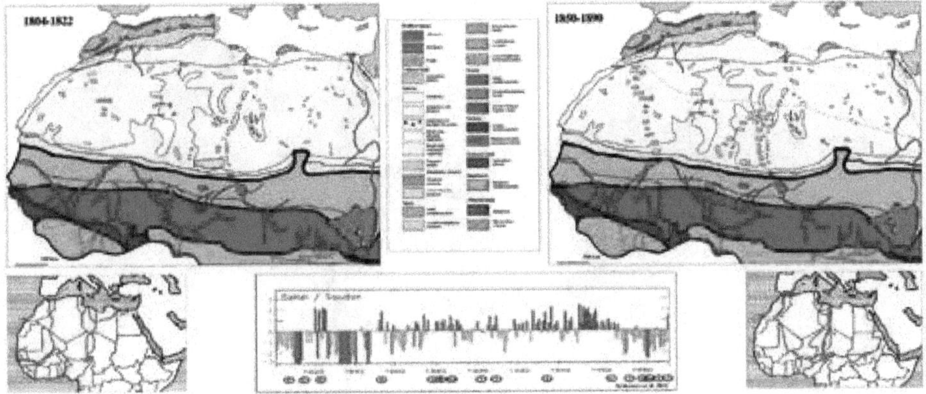

**Figure 13.**
*Northern and western Africain the nineteenth century. Vegetation maps and a reconstruction of precipitation based on the reports of the early explorers [70]. The importance of the achabs is visible in the second half of the nineteenth century. Cartography Schulz.*

disposition centred on the traditional transsaharan trade routes from the 1820s on. For the present case, the historic 'Borno-Road' – Tripolis-Kukawa and its deviation via Ghadames-Rhat-Agadez – served as a perfect source of information. It was the most frequented caravan-route in the nineteenth and early twentieth century, whereas the Tombouctou-Fez (Morocco) road was already less used. From 1822 on, we have for every 30 years a report of the voyagers [20, 71–76] on the nature of the landscapes. As wells were crucial points for the caravans, they also served as reference points in all the reports. Vegetation has always been an important topic in their reports which relied on the vernacular names of plant species – in Arabic or in other languages. Thus, we have a suitable base to reconstruct the plant cover for the nineteenth and for the first half of the twentieth centuries as we can use the indicator values of the modern vegetation.

### 6.1 At first, we will present the landscape changes at the desert-savanna-transition: at the reference points Tigidit cuesta and at the Belgashifari well (see above)

#### *6.1.1 The Tigidit cuesta (16°25′N, 7°55′E)*

As mentioned above, the contrast between the contracted mode of the *Acacia-Panicum*-vegetation (desert) in the foreland of the cuesta and its diffuse mode (savanna) on its top is clearly visible (cf. **Figure 17**). The dots depict the extension of the Saharan savanna and the change to those of the Sahel. At 1937, the situation was similar but the belt of the *Acacia-Panicum*-savanna was smaller and the extension of the Sahelian *Commiphora*-savanna was greater [20]. In the middle of the nineteenth century, the situation was different. A large grass cover masked the main transition and the Saharan savanna was much more extended [13].

#### *6.1.2 The transition at the Belgashifari well (16°2 N, 13°14′E)*

In 1984, the change from contracted to diffuse (permanent) vegetation was as clear as at Tigidit (see above). However, the Saharan savanna was much more extended (see **Figure 12**). In 2014, the situation was comparable, but trees were much more scarce. It was in 1822, when Denham [71] gave the first of the historical descriptions: he reported the change from desert to savanna near its present posi-tion. After a belt of a lush savanna, he described a clear change to a dense savanna.

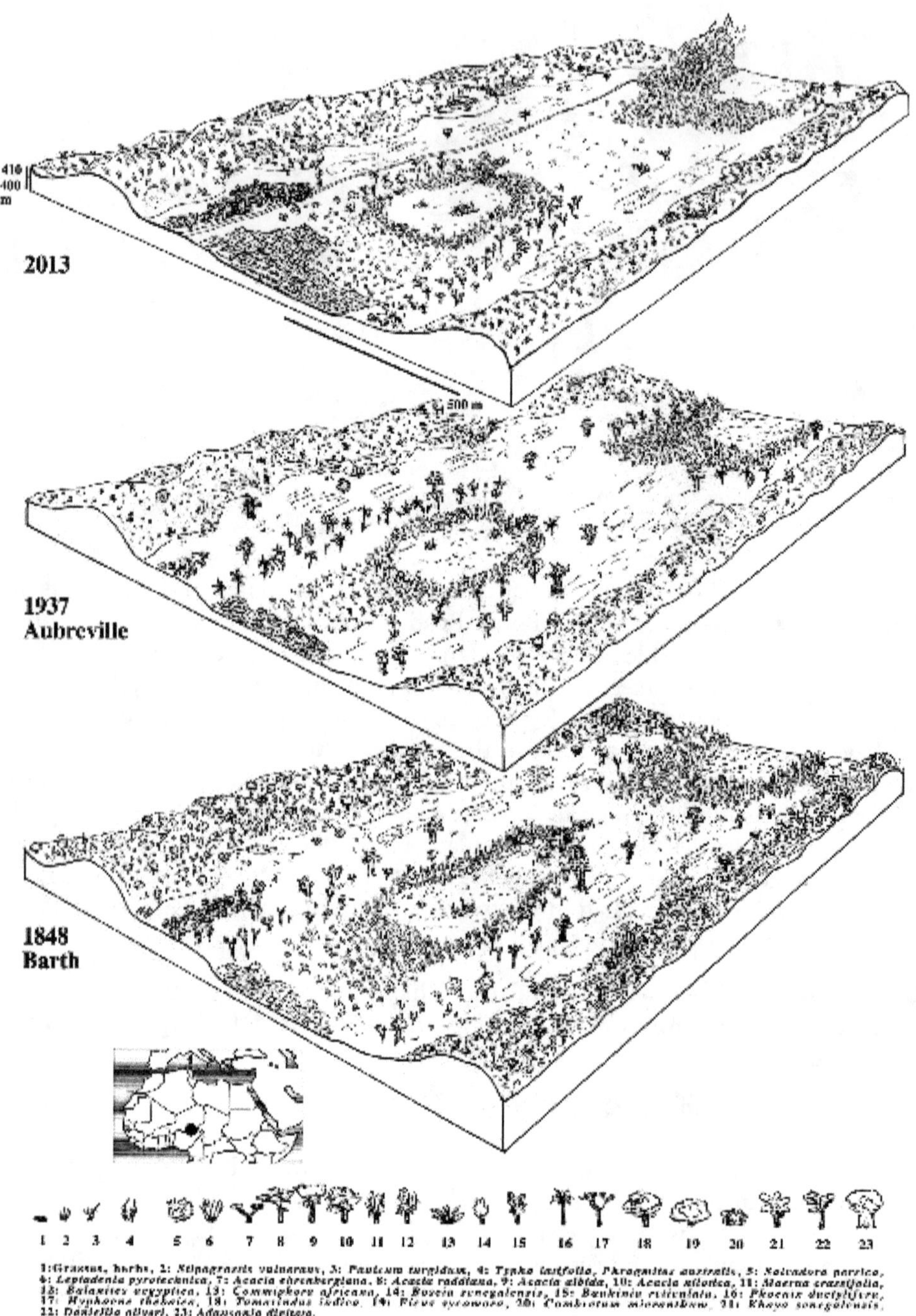


**Figure 14.**
*The present situation and the historical development of the Guidimouni depression/SE-Niger (from [69], modified).*

Thirty years later, Barth [72] saw again the desert-savanna-boundary in a similar position as at present; however, he noted a dense herb and grass cover and an important tree-vegetation in the dune depressions. Rohlfs [74] described a dense grass and herb cover that masked the main transition, and for the South of Belgashifari well, he noted a dense savanna with Sudanian trees in the dune valley. Nachtigal [73] confirmed this mosaic too. Thirty years later, Vischer [75] described a loose grass and herb cover with the desert boundary near the present position.

However, the tree cover south of it was less dense than described by his predecessors. In conclusion, we state that the main boundaries did not change their position very much, but during the 1860s, the plant cover was much more dense and diversi-fied with a remarkable Sudanian tree vegetation reaching far to the North in the dune valleys.

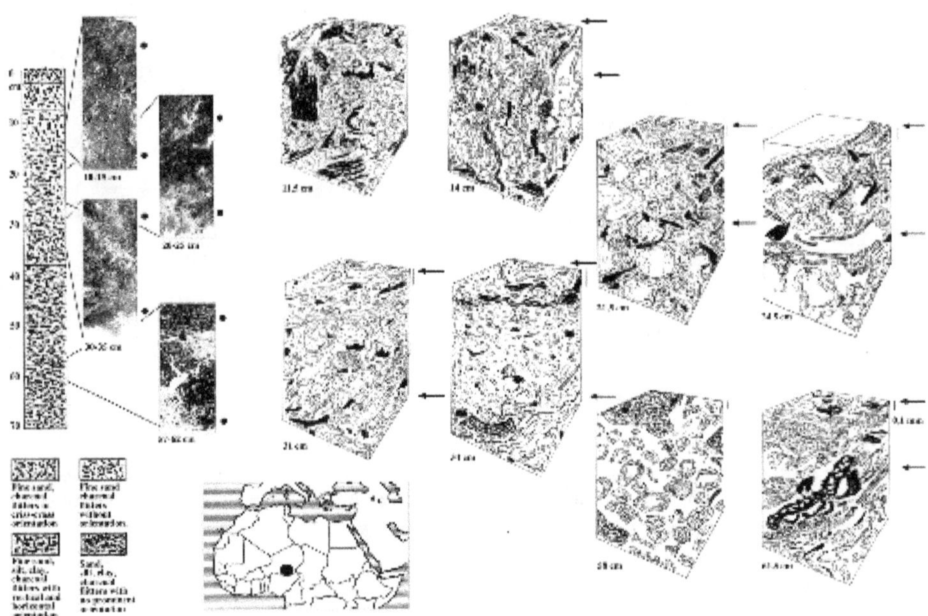

**Figure 15.**
*The sediment structure of the Guidimouni record/Southeast Niger. It demonstrates the stability of the sediments by the formation of algea-layer sand also the steady presence offire asproved by the charred material (from [69], modified).*

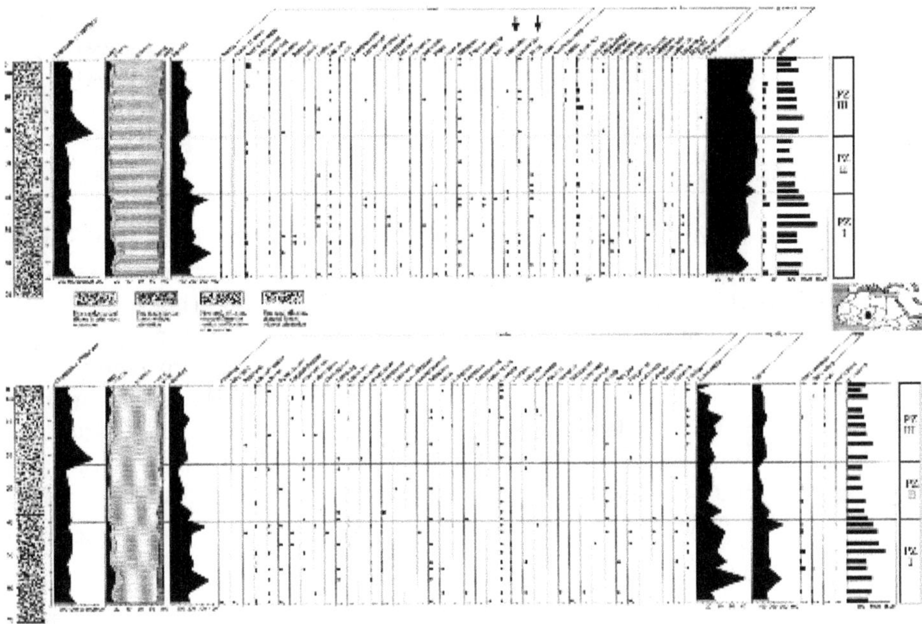

**Figure 16.**
*The Guidimouni pollen record/ SE-Niger (from [69], modified).*

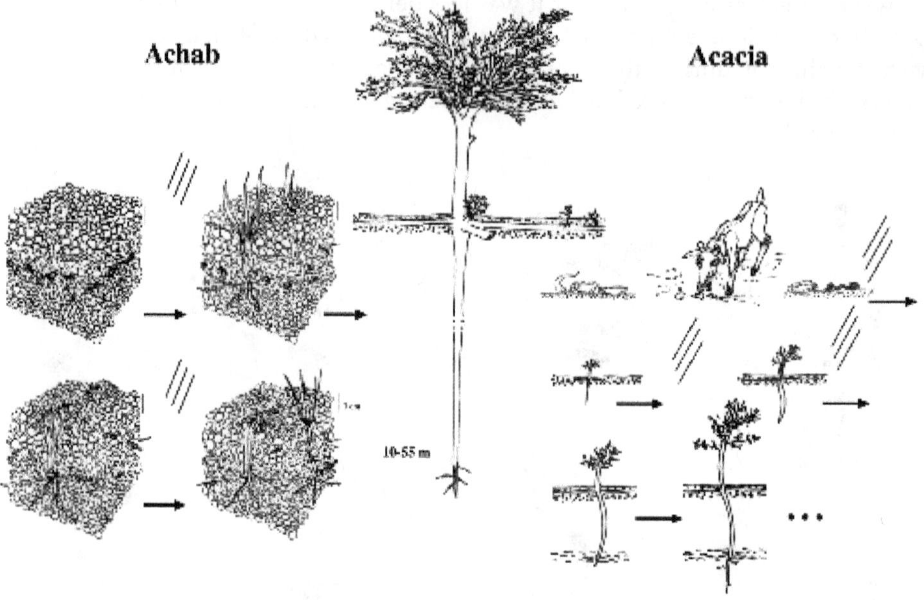


**Figure 17.**
*The strategies of life in the desert. (A) Achab. Development of therophytes after aleatoric rainfall and the formation of a root carpet. (B) Vegetative and generative strategy of* Acacia*. Deep and lateral roots with root suckers, seed germination and development of deep roots for groundwater. Drawing Schulz.*

## 6.2 Mapping the past

These reports made it possible to establish vegetation maps for the first and for the second half of the nineteenth century. The small maps give the expedition routes. It is clear that the information does not cover the whole area, and the interpretation is certainly limited. The maps rely on the written reports, and the descriptions for the central part of the region were the most precise ones. For the southern parts of the visited regions, the descriptions are mostly based on trees, which were mentioned by their vernacular names. Due to several robberies or damages during the transport, most collected plants were lost. Finally, the bombing and burning of the Berlin Herbarium in the Second World War destroyed the last preserved plant specimens [77].

### *6.2.1 Climate history*

Nicholson et al. [70] reconstructed the mean precipitation for the last two centuries based on the landscape descriptions of early voyagers, early measurements and interpretations out of lake level- or sediment records. The diagram (**Figure 13**) is also marked for the expeditions of the early voyagers. The record depicts a long drought period from the beginning of the nineteenth century to about 1850 with a short humid spell around 1820. During the 1850s, some humid years occured followed by a series of dry years up to the 1870s. Afterwards, a humid period lasted with some inter-ruptions until 1910. After another dry period up till 1920, the twentieth century, a long humid period occurred until the end of the 1960s. Suddenly, the climate changed to a long series of droughts, which only at the end of the 1980s seemed to diminish.

### *6.2.2 The vegetation*

Mapping is based on the present vegetation map (**Figure 1**) and documents the differences, which could be read from the historical reports. The two maps show similarities and differences. They correspond in their regular presence of trees in the northern semidesert and in the position of the southern boundary of the Sahara. The Sahel region was described for extended *Acacia*-thickets and forests in the northern part and Combretaceae-savannas in the South. The Sudanian region was similar to the present one; it was, however, denser. A further differentiation was not possible. But the presence of parks was regularly mentioned. The rain forests were much more extended and consequently the Guinean zone much more restricted – compared to the present situation. The most important information lies in the regular presence of achabs in the Sahara during the second half of the nineteenth century. We conceive quasi-permanent pastures by repeated rainfall in the Sahara up to 25°N. Certainly, this was the base of a strong nomad economy providing the base for their dominance over sedentary people.

All voyagers agreed on the rich and diverse game in Sahel and Sudan. They men-tioned in particularthe large elephant populations. Together with the large exten-sion of *Acacia*-forests-thickets, we have to think on elephant-landscapes [78–82]. Elephants are known as landscape engineers. They produce a twofold landscape. For the one they transform forests to medium high thickets by positive and negative selection of tree species and the elimination of high trees and for the other they also create new structures. They open the forests for their tracks giving chance to grasses, herbs or shrubs with a ruderal behaviour. In this way, they form thickets and provide dry and combustible material. So, elephants also create fire-prone landscapes too. Elephants are bound to water and they make or enlarge pools for drinking, bathing and also the uptake of minerals. As they are social animals with a long life period, they school their young generations to maintain and preserve these types of landscapes. The *Acacia*-thickets around lake Chad or in northern Cameroon give an – even poor – model of these former landscapes.

### 6.3 The landscape history of the Sahelian savannas during the last 100 years. Guidimouni – a key locality for the Sahelian savannas

In addition to the historical descriptions (see above), we also dispose on physical archives,which describe the landscape evolution during the last 100 years. They come from southeast Niger. The dune depression of Guidimouni in southeastern Niger has been described several times in the last 200 years [20, 72]. Moreover, it was possible to core the upper part of the lake sediments [76].

#### *6.3.1 The physical situation of the Guidimound depression*

A long interdune depression in SE-Niger (13°42′N/9°32′E) represents the situation of the Middle Sahalian savannas (see **Figure 1**). The region is part of that area, which was supposed to be endangered by an enchroaching desert [19]. The depression has two lakes which are fed by fresh water sources assuring a more or less permanent water body.

**Figure 14** depicts the present situation of the depression and its recent history. The upper diagram shows the whole depression in its present situation. A degraded Middle-Sahelian savanna surrounds the lake, mainly consisting of *Acacia*- and *Balanites*-trees, *Leptadenia*-bushes and grasses. At present, it is still a regular habit to burn the reed in spring in order to have space for gardens and fields. In parallel, the *Leptadenia*-bushes on the dunes are cut and the branches afterwards burned (slash and burn) to prepare new fields after a fallow period of several years. Soils in this region belong to the arenosol-, regosol- or chambic-arenosol groups [83].

In 1848, Barth [72] described the Guidimouni depression as densely vegetated by grasses and herbs the dunes bearing an *Acacia-Commiphora-Leptadenia*-savanna. The depression itself had an *Hyphaene-Phoenix*-belt around the *Typha*-reeds. Also some *Adansionia* trees were planted. A comparable picture was given in 1936 by Aubreville et al. [20]. However, the savannas on the dunes were not as dense as Barth described it but some Sudanian trees were still present, such as *Daniellia sp.*This situation was one of the strongest arguments against the idea of an enchroaching desert.

### *6.3.2 The Guidimouni sediment record*

The sediment core was taken in 2013 in order to reconstruct the recent landscape and vegetation history [69]. The lakes of the Guidimouni depression are shallow lakes or ponds, and they are not more than 2 m in depth. However, their surface varies much during the year. In drought periods, the lakes may dry out (see [83]). Thus, one has also to think on the risk of disturbance by wind and breaking waves and also of deposition gaps caused by desiccation. A 70 cm-long tube could be enforced into the sediments of the western lake. The sediment record consists of silty or sandy gyttias with a variable content of organic matter. Four thin sections were made in the Mineralogy Department of Szeged University, Hungary. They should help to understand the sedimentation processes and also detect possible zones of reworked sediments.

At a first look, the sediments seem to be uniform or amorphous. However, at 400× magnification, it was possible to discriminate into two mayor features, which are explained by **Figure 16**. Under a disturbed section of about 12 cm, the deposits are organised in fine – millimetric – layers, which are separated by algae/bacteria films, respectively, by their jellies. These are always densely coloured by Fe-oxides. The uppermost sediment is mixed and does not show a distinct structure, but it depicts the presence of diatoms. Thus, the sedimentation starts with an inwash/inflow of sandy-silty material and alterated organic matter. On this layer, a film of algae/bacteria-jelly is formed indicating a eutrophic and energy-rich shallow water body. It fixes the sediment beneath. Small arrows indicate the positions of these films. However, the water-rich and unstable layers of the upper cm are exposed to wave action, slumping phenomena or other disturbances. So, they may be contorted, displaced or mixed again. The upper two columns of **Figure 16** show these phenomena. The central part of the record (about 20–53 cm), however, is mainly made of sands or silt, but still separated by the algae layers. There is information that this part belongs to the drought period of the 1970/80s. During this time, the lakes became almost dry as reported by locals (Adamou, frdl. comm.). Anyway, a certain amount of water still must have persisted to allow the formation of the algae/bacteria films. The lowest thin section depicts an in-wash of weathered middle and coarse sand and a dense organic rich gyttia, which again is divided by algae/bacteria layers. The general formation of bacteria/algae films will counteract the disturbance effects of waves in the shallow water. Considering these facts, a sampling with a distance less than 5 cm seemed not to be useful – out of the disturbance risk.

An important feature is the regular presence of charred material. It is made of grass coal-flitters consisting of cuticulae, leaves or parenchyma remains. Charcoal from wood seems to be very rare. These flitters are kept in the thin layers and are oriented along the algea-films. Thus, during the time of the deposition of the record, fire always was an important part of landscape dynamics. At present, the inhabitants regularly use fire to clear the dune area and the reeds in order to prepare their fields. So, it is likely to adopt this model also for the past. It is indicated by the regular presence of grass-coal flitters. Coarse ones will not have been transported over long distances.

### *6.3.3 Vegetation history of the last 100 years*

The detection of the stabilising bacteria/algae films visible in the thin sections allowed exploiting the record for pollen analysis. The diagram (**Figure 16**) was constructed on the base of all pollen but aquatics were excluded. The most of the arboreal and non-arboreal elements show only values of less than 1%. Thus, they are only represented for their presence in the diagram. The pollen diagram is characterised by the elements of an open Sahelian savanna of the *Acacia-Balanites*-type. Dominant are grasses and aquatics (*Typha,* Cyperaceae). Cerealia are persistent.

The arrows point to the *Casuarina-* and *Eucalyptus* curves.

Three pollen zones could be discriminated on the base of the variation *Typha*-Gramineae for the one and for the other on the base of the diversity of floristic elements:

PZ I. 65–40 cm: The aquatics have high values against the low values of grasses. Arboreal pollen shows a relatively high diversity including some Sudanian/Sahelian elements (*Guiera, Khaya*, Combretaceae).

PZ II. 40–23 cm: The part of the aquatics is reduced by rising values of grasses. The diversity of arboreal and non-arboreal elements is reduced too.

PZ III. 23–0 cm: There is a rise of the aquatics against reduced values of grasses. Trees and shrubs recover but do not reach to the diversity of PZ I.

#### *6.3.4 Charcoal*

The charcoal record, which mainly consists of grass coal, depicts the general presence of fire in the region as it is. It still today comprises flaming of the reeds in order to get place for new fields and also slash and burn on the dune slopes. The sharp rise in PZ III represents an accelerated burning for new fields after the end of the drought period.

### 6.4 Time frame

The nature of the sediments will not allow a radiocarbon dating. However, the presence of *Eucalyptus* and *Casuarina* (see the arrows in the diagram) shows that the sequence is not older than the beginning of the twentieth century. The colonial authorities of Nigeria planted both tree species as ornamental or afforestation elements as well as roadside-trees [84, 85]. Their pollen takes part in the long distance transport. Thus, in combination with sediment modes and the fact that PZ II is apparently contemporaneous to the desiccation of the lake during the drought of the 1970s up to the beginning of the 1980s, the base of the core might be deposited during the l920 years.

The only comparable record reaching to the present time is that of Oursi in Burkina Faso [86], which shows a similar open vegetation due to extensive agriculture and animal breeding. However, the record of grass coal stands unique also compared to the upper parts of the Manga lake records [87–89]. But tiese lakes did not provide suitable sediments to follow them up to the present. This record is the first to discriminate between the two main elements in the charred material (grasses and trees) – at least for the Sahel.

## 7. Regeneration. A confusion of concepts, different observations and realitiy

The discussion about regeneration is controversial. There is the position of a definite or long-time degradation, which Miehe et al. [90] explain by a short time of resilience and then a declining to a lower ecological equilibrium. This corroborates the conclusion from the Guidimouni-record as presented above. Hahn and Kusserow [91] and Kusserow [92] report a severe degradation of the Sahel from remote sensing over a long period and also report the algae crusts on silty/clayey sediments as indicators/results of a definite decline of savannas. However, she states that sandy environments will much faster regenerate. Thus, it is necessary to differentiate between the types of environments and also to take the periods of observation into consideration.

### 7.1 The first steps of regeneration

An example is given by the investigations in the Guidimouni depression (see above). Field work and observations on the lake-cores structure revealed the general regeneration potential of soil surfaces on sandy and clayey sediments. The upper centimetres of the dune tops and their middle slopes expose fine layers of blown sand which are covered and fixed by bacteria/algae films together with their gelly formations (biofilms). These biocrusts are the first stages of reorganisation and they represent a general phenomenon in its bimodal feature: deposition of a mineral layer which is afterwards settled and covered by bacteria and algae. This represents a general phenomenon of soil surface organisation: that of film like OPS/PSO (pellicular surface organisation) in the sense of Pomel [92], see also [93–95]. Thus, it is obvious that even under intensive exploitation, the tendency of regenera-tion of vegetation and soil still exists.

The general mode of sandy crust formation is explained by **Figure 18** (above). The upper series represent the regular repetition of coarse and fine sand layers mainly fixed by cohesion as it is the case for sand layers in the desert described above for the achabs. Anyhow, the normal development is that of biocrusts as represented in the middle series. The ever present spores of cyanobacteria and algae germ rapidly and create a biofilm of jellies and thus stabilise surfaces. The algae belong to the *Nostoc-* and *Lyngbya*-realm with a nodular and chain-formed appearence [91]. When covered by another sand layer the bacteria/algae will rapidly form another cover. Even though these crusts are fragile and may crack easily, they enable a further succession of grasses and herbs as a next step. Seeds may be blown in cracks and they may germ and exploit the nutrition reserves of the biocrusts. Finally, these crusts may be seen as functional types representing the early stages of soil development [96, 97].

On silty/clayey stones or sediments, the situation is different as **Figure 15** (below) may show. The surface of these fine grained rocks or shallow soils are often covered by thick cyanobacteria/algae layers which may reach to several mm thickness (cf. [91]). They are coherent and impede an implantation of seeds. When covered by dust or fine sand, they easily reform. Finally, their smooth surface is water repellent and for longer periods, they may be an obstacle to a colonisation of grasses or herbs – not to speak about trees.

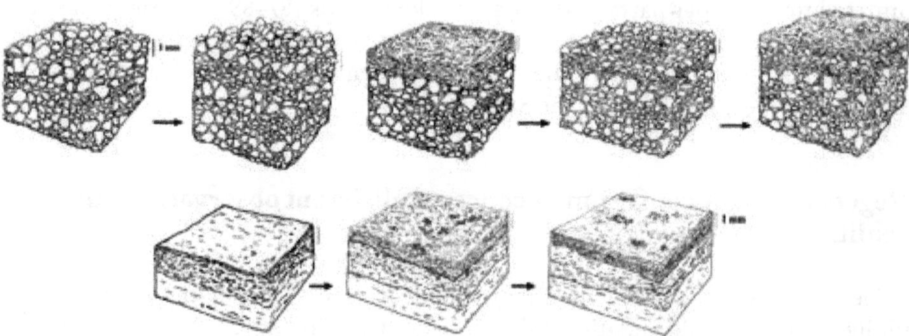

**Figure 18.**
*The first steps of regeneration and biocrusts in sandy (above) and clayey (below) sediments (from [69], modified).*

In contrast to them, the rapid regeneration mode of the cyanobacteria/algae crusts on sandy soils is successfully exploited in the dune rehabilitation of Northern China [98, 99]. The crusts are collected, crushed and afterwards sprayed over loose sand surfaces or dunes, which gives a good example of working with natural succession.

## 7.2 Regeneration in the landscape scale

If one regards the philosophy, performance and success of the various projects which are active or planned in the Sahel, we have to differentiate between the large scale technical ones and those, which are adapted to the conditions of the population.

### *7.2.1 Gourma – guide – great green wall. The limits of regeneration*

Among the large-scale projects, we have the extended dune fixation by fencing and tree plantation [102] or the transcontinental 'Great Green Wall' [17, 18] still based on the idea of an extending desert (see **Figures 8, 19, 20**). The second type is the creation of large natural reserves or national parks in the Sahara and the Sahel. They are initiated or proposed for auto-regeneration of vegetation and wildlife – following mostly the WWF-philosophy see [14, 16]. Very often their aim is to protect emblematic animals, which are supposed to act as key stone organisms (see for both **Figure 1**).

The opposite is the creation of pasture-rotation systems to exploit the limited resources but also guarantee their regeneration. Finally, they are the counterparts of the old shifting/fallow cultivation, which by now in the Sahel is only rarely carried out. Several examples will illustrate these projects.

### *7.2.2 The rotation pasture system 'Gourma'*

The northern part of the Gourma region (Mali) from the Niger-bow to the mountains of Hombori (17°-15°N) is a perfect example of Middle and Northern

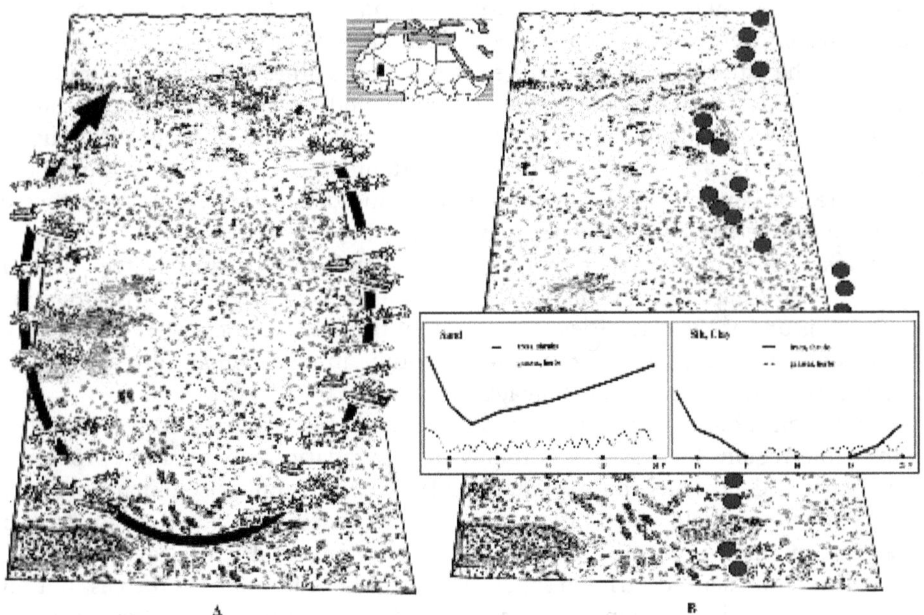


**Figure 19.**
*The pasture-rotation system of Gourma/Mali during the 1970s and 1980s and the long time observation project of regeneration up to 2017. Graphs on regeneration without scale [8, 100]. Drawing Schulz.*

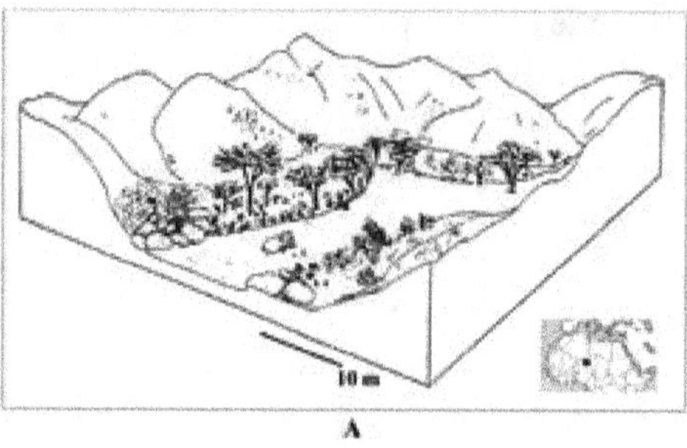

**Figure 20.**
*The pasture rotation project 'Guide' in the Central Air Mts. [43, 101]. (A) The Wadi vegetation in the innerpart of the ring structure. (B) The regeneration of trees and shrubs after the second year of closure. See also the dense herb cover under the* Acacia*-umbrellas. Drawing Schulz.*

Sahel-savannas. They range from a Combretaceae-savanna of tiger bush in the South to the *Acacia-Commiphora-Balanites*-savannas of the North. East of Tombouctou there even existed a real *Acacia*-forest, which was formerlyx exploited for the steam boat lines to Bamako [103–105]. The savannas represented for long time a rich pas-ture, which could not be exploited [100] as only a few wells or water points existed. Thus, human impact was restricted to the river-banks and the northern savannas, where the cattle keepers had constructed a series of hafirs (rain fed pools). The other regions were the oxbows of river Niger in the West and the agriculture areas near the mountains in the South. Instead there was an intensive elephant pasture creatred by the greatest herds of West Africa [106]. Probably, these savannas may be regarded as the gullivers of elephant impact [107], giving a good example for the reports of the voyagers of the nineteenth century (see above). Only a few wells were constructed in the first half of the twentieth century. But the aureoles of overgrazing developed rapidly around them. With the drought of the 1960s, the Sahel started to degrade. However, in the Gourma, it went differently. Together with the local cattle keepers and authorities, the geologist R. Reichelt from CILSS developed a rotation system of new wells and pasture. It was based on the opening and closing of wells depending on the state of the pastures around, which were regularly controlled. When degradation started, the wells were closed and the cattle keepers were obliged to proceed to other wells and pastures. This system worked from the end of the 1960s on and revented the desertification phenomena, which had hit the regions around. But in 1984 – at the peak of the drought, a great number of cattle keepers from then North of Niger River invaded the region. They were not familiar with the rotation system and did not respect it. After severe quarrels of the herders, the rotation system collapsed.

Anyway, for long years, it represented a sustainable pasture system, which saved the Gourma region from the desertification as it occured in the regions around. It is one of the curiosities in science that these experiences were completely forgot-ten and were not taken into account in the whole discussion on desertification and regeneration management.

From about 1984, a long time observation project (see **Figure 19B**) was installed in the same region. Its goal was to follow the degradation-regeneration processes under various conditions and exploitations [8]. It was a multidisciplinary proj-ect mainly based on field observation and remote sensing. It could evaluate the regeneration chances of the different savanna systems and it well demonstrated that regeneration started early on sandy substrates both for herbs and trees, but on clayey sediments degradation continued even after protection. Here, the regeneration started only after a long period, which corroborated the experiences from other regions (see above). But the general insecurity of the regions forced the colleagues to abandon the project in 2014 [4].

### *7.2.3 The pasture rotation project 'Guide' in the central Air Mts*

The extreme degradation of herb and tree pastures in the Air Mts. on the one hand and the octroyance of the Reserve Naturelle de l'Air et Ténéré (see **Figure 6**, [101, 108]) with the exclusion of the herders from traditional pasture areas on the other initiated the planning of a new regeneration concept together with the local authorities of the Timia village in the central Air Mts. [108]. **Figure 8** explains the general situation. It shows the two granite ring structures of Agalak and Aroyan and the upper part of the Wadi Anou Mekkerene, one of the greatest of the Air Mts. And it also depicts the altitudinal change in the Air Mts. from the mountain savannas to the middle stretches of the wadi Anou Mekkerene heading to the West. Within 4 or 5 years, a rotation system, which functioned on the closure of pastures for several years, aimed to assure the regeneration of grasses, herbs and trees. At the same time, a sustainable exploitation system of the pastures should impede a new degeneration by overgrazing or other forms of over-exploitation. The first of these closures was the mountain pasture 'Guide' southeast of Timia (see **Figures 6, 8**). It is situated in the Aroyan-granite ring structure, which could easily be closed in 1986 for 4 years. This area showed the typical transitions from the contracted desert vegetation to the mountain savannas of the Sahelian type. The soil cover of vegetation did not exceed 10% but could rise to 70% under the umbrella of Acacias. The first years showed an enormous growth rate of trees as well for the seedlings-saplings as for branches and twigs −30 cm – for *Acacia* and *Maerua*. Apparently, trees could profit from the good rainy season and from the reduced concurrence of herbs, which suffered from the preceding drought. In 1990, a first two-days-opening was organised for fruit collection and grass cut-ting. The enclosure was mapped for vegetation, a floristic inventory was organised and also some demilunes/half moon sand accumulations and stone lines were constructed in order to collect rain water [101, 109]. Within 4 years, the develop-ment of tree and grass pastures was as astonishing high, and also people from other villages had planned to initiate comparable systems. After the controlled opening in 1991, a second pasture was closed for regeneration. For the long run, the village council discussed the models of an interdiction of pastures but with controlled collection of grass and fruits or controlled pasture. In the mountain savannas, the protection and controlled collection of medicinal herbs was an attractive point too. Anyway, a permanent following up of vegetation development was planned for the future. The Guide-project evidenced the chances of local and accepted regenera-tion initiatives and it could have been a model for other regions. Unfortunately, as for the Gourma project, the rebellion and the successive insecurity put a premature end to this success.

### *7.2.4 Think big! Bridging Sahara and West Africa*

#### *7.2.4.1 National parks or natural reserves*

The creation of extended reserve areas or national parks have been generated by the ideas of an auto-regeneration through excluding further human exploitation or through the protection of emblematic animals as key stone organisms.

For the Sahara, the three National parks or natural reserves of Air-Ténéré (see **Figures 1, 6**), Termit-Tin- Toumma and Wadi Rime-Wadi Achmed in Niger and Chad should protect huge ecosystems and also support regeneration of vegetation and wildlife (see **Figure 1**). These are the greatest protection areas in the Sahara and in Africa as a whole and were supposed bridge the areas of endangered key stone animals [15, 16, 110]. Moreover, there was already a survey on the chances to establish a system of monetary exploitation of ecosystem services [111]. However, these initiatives often disturbed the traditional pasture systems, and due to the insufficient involvement of the local populations, it led to various problems and frictions. Anyhow, the sense of these protections and reserve areas was not really communicated to and accepted by the concerned populations. Thus in the 1990s, with the beginning of the rebellions in Mali, Niger and Chad, these projects were no longer accepted by and the state could no longer maintain them. Today, most of them gained a status of 'being endangered' or 'in suspense' [112]. At present, the natural reserve Air-Ténéré continues in a certain cooperation with the local popula-tion in order to manage resources [113].

#### 7.2.4.2 The 'Great Green Wall'

This is the continental flagship of the protection-regeneration projects and follows still the philosophy of expanding ecosystems and the combat against them (see **Figure 1**). The project was created by the African Union in 2007 [17, 114] as a 7800 km belt from Senegal to Djibouti. Fifteen kilometres wide, it should work as protection against wind and erosion. Afforestation should provide nutrients to the soil and also ameliorate pasture by foliage and shadow too. Finally, the tradition of agroforestry (parks see above) was taken as a model (see also [115]). Research on amelioration of soil and plant fertility is an important part such as investigation on the symbiosis of bacteria/fungi and acacias. Anyway, as for the other smaller or greater projects, this initiative came to an intermediate (?) end caused by the general insecurity in the concerned areas. But the research in the various institutes of the partner states continues in the hope to reactivate and readjust this flagship. However, it already serves for the governmental propaganda. The presidency of Niger claimed to have plated millions of trees in order to reduce soil erosion and to fix dunes [116].

### 7.3 'Small scale' as a chance!?

Several projects and activities concentrate on the regeneration and amelioration of degraded soils in order to restore the soil cover and to assure food production [117, 118]. They are mostly organised on personal or village level and so they are participative. These activities have to be seen on the background of a general extensification of agriculture, parallel to the intensification, e.g., irrigation cultures at favoured places [119]. Most Sahelian farmers are still subsistence-oriented. This means that they mainly crop to nurture their families rather than to produce market products. The steadily increasing population with growth rates of about 3%

**Figure21.**
*Regeneration and food security measurements in Niger. (A) Tassa /Zai-cultivation on the Ader-plateau, Central Niger. (B) Reduced weeding in S-Niger. (C) Intensive irrigation for vegetable production at Niamey, Niger. Drawing Schulz.*

per annum leads either to an expansion of cropped surfaces to marginal land or– where the population density is already high – to decreasing cropland per family. Several examples illustrate these activities (see **Figure 21**).

*7.3.1 The restoration on heavily degraded soils*

The 'tassa' or 'zai' culture (**Figure 21A**) is an old cultivation system of degraded soils [120]. It is based on dug in holes, 10–40 cm in diameter and 10–25 cm in deep in a distance of about 1 m. These holes can store rain and run off and thus support the regeneration of spontaneous vegetation. They may be filled with leaves or compost in order to attract the termites.

Experiments showed the possibilities of 640 kg–800 kg/ha yields of millet. The dug in holes must be renovated each year. As the financial component is quite low and as it is based on personal or village activity, 'Tassa' is the most appropriate and widely accepted cultivation system.

*7.3.2 The amelioration of crop planting by preservation or planting of trees*

In the Haoussa region around Maradi in Niger, average farm size has reached meanwhile about 2 ha. For the simple reason of survival, intensification of cropping is mandatory.

However, a number of obstacles exist that hinder the application of innovations. Among these are traditions, low educational level, low investment capacity and the need for risk management. The latter aspect means that farmers are risk averse and are not – in contrast to the normal economical theory– yield or income maximisers. First of all, the family members need to survive.

So the question is: how does innovation needs to be alike to be acceptable for farmers. The answer is manyfold: the innovation needs to be simple, affordable, relying on local resources, risk reducing, functioning under multiple weather scenarios and it cannot contradict local customs. There are not many innovations that fulfil these criteria, in particular if we want to address the 'regreen-ing' of the Sahel. We can approach the 'regreening' from two angles. One is the re-establishment of ligneous vegetation, and the other is increasing the crop biomass production. At the first glance, these are contradictory objectives. Is this really so? In order to answer this question, we will discuss different options in the following.

*7.3.2.1 Windbreaks (or agro-forestry in a more general sense)*

Heavy convective storms are a regular phenomenon in the Sahel. They lead to erosion on open surfaces at the beginning of the rainy season and homogenisation of soil surface properties through redistribution of particulate matter [121, 122]. The saltating sand grains damage the young seedlings and can lead to crop loss at an early vegetative state. Therefore, it is reasonable to think of windbreaks as a solution to the problem. A lot of research has been done in this respect [118, 123]. However, we hardly see any adoption of this technology by farmers.What are the problems? It begins with legal problems. Planting a tree means to express a claim on property. This is delicate in societies were the land is distributed according to local traditions. Second, planting trees in a hedgerow means an investment that is hardly affordable for a single farmer. A third argument for rejection is the workload for making the trees survive after planting and for pruning in order to reduce competition with the neighbouring crop later. And the competition for land, water, light and nutrients is the fourth argument to set this technology aside. In conclusion, hedgerows are a typical innovation typical for scientists and based on on-station results, thus neglecting the constraints of the rural populations.

Are there more simple and adoptable solutions? One is, i.e., called farmer-managed natural regeneration [124]. It uses the regeneration of ligneous species by re-sprouting from rootstocks. Already Wezel et al. [125] could show in the 1990s that the minimum yield of pearl millet increased with the number of small bushes in the field. This is achieved through the reduction of the negative wind erosion effects and the increase of the organic matter stock that is the major provider of the major limiting nutrient phosphorus. As side effect, fire wood is provided. In contrast to hedgerow planting, with this technology, the only input to be provided is low: i.e., only pruning. The disadvantage is that it is only possible in non-mechanised agriculture. And, the woody species composition is hardly foreseeable. Studies in the Maradi area in Niger have shown that in densely populated areas, all still existing woody species are under use and that their distribution is depending on the distance to settlements (**Figure 22**).

Close to settlements, old *Faidherbia albida* trees dominated are protected, since they deliver high quality animal fodder and do not compete during the rainy season with the crop due to the leaf cover developing in the off-season. Farther away from the settlements, *Piliostigma reticulata* and *Combretum glutinosum* dominate are mainly used as fire wood resources. Also crops differ with distance to the settlements, cash crops like cowpea grown more closely to the settlements on the more fertile sites. Reasons are protection against theft and higher expected

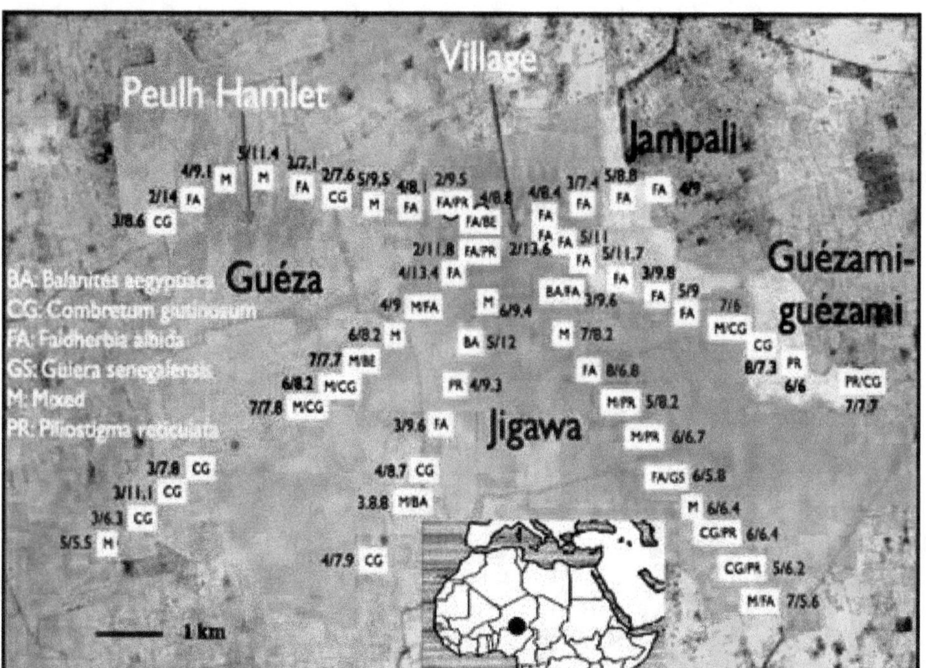


**Figure 22.**
*A survey on tree vegetation in the* Faidherbia*-park zones around a village in southern Niger. Dominance (>40% counts) of woody species >1 m crown diameter (white boxes) and number of woody species detected and their average height (number/heigth m) on transects with increasing distance to the village Warzou, central southern Niger (Jigawa, Jampali, Guezami-guezami and Gueza being local soil names indicating increasing clay content in this order). Survey and assembly by Herrmann 2017.*

yields due to nutrient concentration closer to the settlements. Another variable explaining crop diversity is soil conditions, *Sorghum* preferentially being cropped on the more loamy sites.

#### *7.3.2.2 Partial weeding as wind erosion barrier and intermediate nutrient stock*

If one wants to reduce wind and water erosion effects on cropping, the simple technique of partial weeding is an option (**Figure 21B**). Under Sahelian conditions, sowing and weeding are the most time-consuming agricultural actions. Labour shortage during these periods limits agricultural performance, since crop surfaces fall out of the scheme. Partial weeding, i.e., stripwise weeding in the sowing lines or circular weeding around the sowing pockets, reduces the workload for the first weeding by about 50%. The herbs and grasses left standing then act as a semi-natural erosion barrier. In addition, this vegetation component stores nutrients that were otherwise leached. In this way, the weeds can be used as an intermittent nutrient reservoir that can be managed, and nutrients are provided to the crop when needed by a timely second weeding.

### *7.3.3 Varieties*

The Sahel is the genetic center for the major staple crop pearl millet that is mainly planted on the sandy sites. Many different land races exist that have been developed by local communities by mass selection over generations. These local communities have a quite determined idea about what a variety must provide with regard to pest and drought resistance, taste, and yield, just to name a few aspects. Independent development of so-called 'improved varieties' has repeatedly failed, simply due to the fact that breeders were not aware of the mandatory properties for different communities, and they did breeding on-station under conditions that are not comparable to the farm environment. Therefore, the future agricultural research needs to be more participatory and include the farmers perspective already at the state of objective definition. Then, higher biomass yielding varieties can be developed.

### *7.3.4 Seedballs*

Under the Sahelian conditions, dry sowing before the rainy season is an option if fields are too far from the settlements, if the rainy season starts very later or for women, when they are not able to sow at the time due to the obligation to help her husbands on their fields first. However, dry sowing imposes the risk of seed loss through predation or early droughts. In order to assure a timely establishment of the pearl millet crop, the seedball technology was developed [126]. It uses local resources like sand, loam, seeds and a little bit of fertiliser (NPK or wood ash) to form small balls of about 2 cm diameter. Seedballs have shown to increase biomass and yield by about 30% under all kinds of conditions in sandy low fertility soils. The only constraint is the labour required for seedball production. However, this can be accomplished during the dry season when opportunity costs are low.

### *7.3.5 Microdosing*

The sandy soils of the northern Sahel are characterised by a low chemical fertility, phosphoruns and nitrogen being the main limiting nutrients for cereal crops. The soils are so poor that even the smallest amounts of nutrient addition can boost the yield. Based on this knowledge, micro-dosing as fertiliser strategy has been developed [127, 128]. Micro-dosing means a placed fertilisation (in contrast to broadcast application) into the sowing pocket at sowing or early in the season, where the nutrients are needed most. Only 2 kg of phosporus are able to double the yield on the poorest sites. Micro-dosing at sowing supports the early establishment of the plant. Once the crop is established and crop loss ha not to be expected, further fertilisation can be done without the risk of investment loss.

However, for the poorest farmers in remote areas, even market access to fertiliser is limited. They can rely on wood ash as local fertiliser, since cooking is done with firewood. Wood ash provides soluble phosphours, potassium, calcium and other micro-nutrients. It can be considered as a complex fertiliser, since it stems from plants. Consequently, it provides most nutrients needed by plants. two grams of wood ash placed into the sowing pocket but at little distance to pearl millet seeds has proven to be effective in increasing yield on poor sites. For legumes, this local fertiliser is applied shortly before flowering.

### *7.3.6 OGA*

OGA is fermented human urine that is used as liquid fertiliser. It is an autochthonous innovation developed by the farmer organisation Fuma Gaskiya in the Maradi area of Niger taking Asian practices as example. It mainly contains nitrogen and potassium as fertilising compounds and has shown to consistently increase pearl millet biomass and grain yield. It is a resource that is locally available for free. It is placed application makes it efficient in annihilating the nitrogen constraint of crop production. Combined with wood ash application (as source for soluble phosphorus), two local resources can be used to fight the notorious soil deficiency with respect to these nutrients. In addition, it is reported by farmers that the smell of OGA is effective to chase off harmful insects.

### *7.3.7 Biological insect control*

The head miner became a major during the Sahelian droughts of the 1970s. Pesticide control is out of reach for subsistence farmers. In consequence, a biological control mechanism using the parasitoid wasp *Habrobracon hebetor* was developed. The parasitoid can potentially be produced locally. However, there is still no agro-enterprise that has taken up this innovation. Perhaps, production is too sophisticated and potential price levels or too high for application by subsistence farmers.

### *7.3.8 The diversification or the counter-season production*

Food security shall be enlarged by intensified and irrigated vegetable production. It constitutes by now a widely accepted activity, wherever the bases are given (**Figure 21C**). It ranges from the vegetable and fruit production in the vicinity of towns or to intensive onion production for export [83, 129]. It can be run on as personal activity or as a collective one.

Thus, these small-scale projects proved chances on the personal of village level to earn its own living and to build sustainable base for villages. They fulfil the demand for participativity and local decision on the projects. Moreover, they are less endangered by the overall insecurity and they may develop their systems by own experiences, and guaranteeing thus a long performance, independently from external pressures.

### 7.4 Finally, the 'Greening'

After all there is an augmentation in the plant cover. It is evident too in the southern Sahara and the northern Sahel as well as in the Park region of the southern Sahel, from where it was taken by [130, 131] as a sign of a principal 'regreening'. But there is still degradation of ecosystems parallel to that recovery in some regions [7, 12].

### 7.5 In the long run – future prospects

Finally, the green future of the Sahelian areas needs a landscape approach where the different stakeholders jointly act in a way it takes into account that the multiple angles of natural and socio-economic environment. Short-term action by decision makers who want to see short-term results and who are driven by the dogma of novelty – in particular in science – will not lead to a sound outcome. In contrast, the basics need to be understood, more participatory action is needed, and long-term development concepts need to be supported. Agriculture has to and is able to support the landscape productivity and thus 'greening'. No sophisticated approaches are needed, but the insight that subsistence oriented agriculture needs innovations that are simple, affordable and based on local resources. In a long-term, a re-integration of crop and livestock production is inevitable to partly close the nutrient cycle.

The decade long experience of our colleagues from university of Abdou Moumouni university of Niamey [132, 133] came to the general conclusion on regen-eration possibilities of degraded landscapes (see **Figure 21**). Damage and degradation of *Acacia-albida*-parks and Combretaceae-savannas in the Southeast of Niger (stages 1 and 2) diminished the resources for the local population in such a dimension that an intervention was necessary. The classical stonewalls on the slopes alone provoked runnels climbing up the slope and aggravated the situation (stage 3). Thus, it was necessary to intervene at all points and for a long period in order to stop further linear erosion and to allow the auto-regeneration of vegetation and soil (stage 4). Especially on silty-clayey grounds, it will take time to collect sufficient organic material on the surface to allow an implantation of grasses and herbs as further stages of succession. Mulching, however, turned out to be successful to attract ants of termites to transport

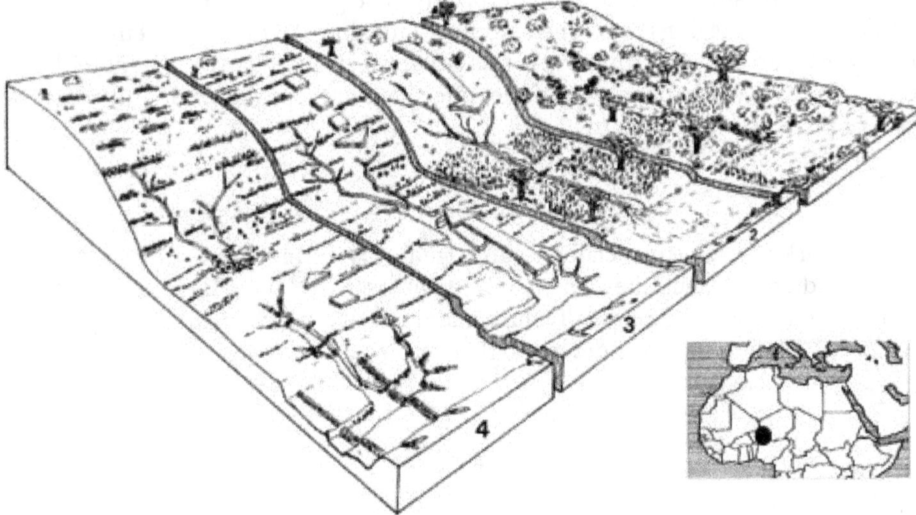


**Figure 23.**
*Experiences with the regeneration of an overexploited* Faidherbia-*park Southwest of Niamey [69, 132], modified.*

fines from the deeper parts of the sediments up to the surface (see also [134]). And finally, also follows the ideas of the different 'Tassa' initiatives (see above).

Supported by strong farmer organisations, farmers can make significant progress indepently from the national political situation. Agricultual research should not only focus on single management measures but also adopt a farming systems approach, where combined innovations are researched always under the paradigm of adoptability taking the farmers' view into account [135, 136].

In the long-term, a part of the population needs to gain its living from activities outside agriculture. The pre-requisites to reach this goal are infrastructure and education. The latter should begin in rural areas with agriculture becoming a regular subject in grammar schools (**Figure 23**).

## 8. Conclusion

Field observation revealed the clear partition of the Sahara in three main landscape types: The Semidesert, the Desert and the (Saharan) Savanna. Thus, the divide between desert and savanna occurs within the Saharan realm. Historical reports and sediment records reveal a stable southern boundary of the desert in the secular scale. Apparently, the boundaries of the desert are the rare climatic ones on the continent. Most savannas South of it are cultural landscapes – including 'elephantscapes' – as preserved in the Gourma/Mali. The degradation-desertification of the last 80 years resulted in a decline to a lower ecological equilibrium. However, the first steps of regeneration are always visible. Their further development, however, depends largely on the type of environments and on human interference. Several projects and initiatives evidenced a principal chance of regeneration or at least preservation. They also showed that small scale projects have a better chance to be accepted and to be continued by the local population. Il became clear that any initiative must be based on the participation of the respective population and must be conceived for a long time. We still do not know how many years or decades the different ecosystems need to fully recover – or if they will remain on a lower level of ecological equilibrium. We should consider the whole discussion and the various activities that take place on the background of a rapidly increasing demography. And finally, the situation changed completely. The general insecurity for the civil population in the regions concerned stopped most initiatives or set them in a state of 'suspense' or 'endangeredness'. As this situation exists already since more than a decade and as it will probably continue, one should accept the latter and adjust all kinds of plans and initiatives to it.

### Acknowledgements

Fieldwork was financially supported by Deutscher Akademischer Austauschdienst, Deutsche Forschungsgemeinschaft and Université Abdou Moumouni de Niamey. J.Merkt, T.Musch, and F. Neagra gave valuable suggestions for the manuscript. We are indepted to them all.

## Author details

Erhard Schulz[1*], Aboubacar Adamou[2], Sani Ibrahim[2], Issa Ousseini[2] and Ludger Herrmann[3]

1 Institut für Geographie und Geologie, Universität Würzburg, Germany

2 Département de Géographie, Université Abdou Moumouni de Niamey, Niger

3 Institut für Bodenkunde und Standortslehre, Universität Hohenheim, Germany

*Address all correspondence to: erhard.schulz@mail.uni-wuerzburg.de

## References

[1] Brandt M, Grau T, Mbow C, Samimi C. Modelling soil and woody vegetation in the Senegal Sahel in the context of environmental change. Landscape. 2014;**3**:770-790

[2] Brandt M, Grau T, Mbow C, Samimi C. Woody vegetation die off and regeneration in response to rainfall variability in the West African Sahel. Remote Sensing. 2017;**9**:39. DOI: 10.3390/rs9010039. 21 p

[3] Brito JC, Godinho R, Martines Feira F, Pleguezuelos JM, Rebelo H, Vale GG, et al. Unravelling biodiversity, evolution and threats to conservation in the Sahara-Sahel. Biological Research. 2014;**89**:215-231

[4] Dandel C, Kergoot L, Hiernaux P, Mogin F, Grippa M, Tucker CJ. Regreening Sahel: 30 years of remote sensing data and field observations (Mali, Niger). Remote Sensing of Environment. 2014;**140**:350-364

[5] Gonzales P. Desertitication and a shift of formations in the West African Sahel. Climate Research. 2001;**17**:217-228

[6] Hellden U. Desertification-time for an reassessment. Ambio. 1991;**20**:372-383

[7] Herrmann SM, Tappan GG. Vegetation impoverishment despite greening: A case study from Central Senegal. Journal of Arid Environments. 2013;**90**:55-66

[8] Hiernaux P, Diarra L, Soumagel N, Lavenu F, Tracol Y, Diawara M. Sahelian rangeland response to changes in rainfall over two decades in the Gourma region, Mali. Journal of Hydrology. 2009;**371**(1-2):114-127

[9] Hutchinson CF, Herrmann SM, Mankonen T, Weber J. Introduction: The greening of the Sahel. Journal of Arid Environments. 2005;**63**:535-537

[10] Karlson M, Oswald M. Remote sensing of vegetation in the Soudano-Sahelain zone: A literature review from 1975-2104. Journal of Arid Environments. 2016;**124**:257-269

[11] Olson DM, Eklung L. Aido: A recent trend greening oft he Sahara-trends patterns and potential causes. Journal of Arid Environments. 2005;**63**:556-566

[12] Ousseini I. A green Sahel: Perceptions, facts and perspectives. In: Symposium Documentation Green Sahel, GIZ, Division 45 Agriculture, Fisheries and Food; 2010. pp. 14-20

[13] Schulz E, Hagedorn H. Die Wüste, wächst sie denn wirklich? Die Geowissenschaften;**12**:204-210

[14] Boulanodji E, Harouna A. Réserves sahélo-sahariennes du Niger et du Tchad. Noé; 2018. 3 p

[15] Giazzi F. La Réserve Naturelle Nationale de l'Air et de Ténéré. Analyse descriptice. Gland, MH/E-WWF-UICN; 1996. 712 p

[16] Newby JE, Dulieu D, Lebrun JP. Avant –Projet de classement d'une aire protegée dans lÂir et le Ténéré (Republique du Niger). Niamey: UICN, WWF; 1982. 122 p

[17] Dia A, Duponnais R. Le projet majeur africain de la Grand Muraille Verte. In: Concepts et mises en oeuvre. Mayenne: IRD; 2010. 44p

[18] Escadafal R. Le projet africain de Grande Muraille Verte. Quels conseils les scientifiques peuvent-ils apporter? Montpellier: Dossier d'actualité; 2011. CSF-désertification.org/grande-muraille–verte. 45 p

[19] Stebbing EP. The enchroaching Sahara: The threat to the West African colonies. The Geographical Journal. 1935;**95**:500-551

[20] Aubreville A, Paterson JR, Collier ES, Brynmor J, Dundas J, Mathes J, et al. Rapport de la mission forestière anglo-francaise Nigeria-Niger (décembre 1936-février 1937). Revue Bois et Forets des Tropiques. 1973;**148**:3-26

[21] Tucker CJ, Dregne H, Newcomb WW. Expansion and contraction of the Sahara desert from 1980 to 1990. Science. 1991;**253**:299-301

[22] Thomas V, Nigaru S. 20th century climate change over. Africa: Seasonal hydroclimate trends and Sahara desert expansion. Journal of Climate. 2018;**31**:9

[23] Chevalier A. La végétation de la région de Tombouctou. Actes du IIIme Congres international de botanique; 1900. pp. 248-276

[24] Médail F, Quezel P. Biogéographie de la flore du Sahara. Une biodiversité en situation extrème. Marseille: IRD; 2018. 366 p

[25] Barry JP, Celles JC, Musso J. Le problème des divisions bioclimatiques et floristiques du Sahara. Note V: du Sahara au Sahel. Un essai de définition de cette marche africaine aux alentours. Ecologia Mediterranea. 1986;**XII**:187-235

[26] Reichelt R, Faure H, Maley J. Die Entwicklung des Klimas im randtropischen Sahara-Sahelbereich während des Jungquartärs- ein Beitrag zur angewandten Klimakunde. Petermanns geographische Mitteilungen. 1992;**136**(2-3):69-79

[27] Schulz E, Abichou A, Adamou A, Ballouche A, Ousseini I. The desert in the Sahara. Transitions and boundaries. In: Baumhauer R, Runge J, editors. Holocene Palaeoenvironmental History of the Central Sahara. Palaeoecology of Africa. Vol. 29. 2009. pp. 63-89

[28] Cole M. The Savannas. Biogeography and Geobotany. London: Academic Press; 1978. 418 p

[29] Le Houerou HN. La végétation de la Tunisie steppique. Tunis: Annals of the National Institute of Agronomic Research; 1969. 142 p

[30] Walter H. Die Vegetation der Erde. In: Ökophysiologischer Betrachtung. Vol. 2. Jena: VEB G.Fischer; 1964

[31] Hagedorn H. Formen und Bilder der Wüste am Beispiel der Sahara. Natur und Museum. 1985;**115**:210-230

[32] Rognon P. Biographie d'un Desert. Paris: Plon; 1989. 348 p

[33] Bornkamm R, Kehl H. Pflanzengeographische Zonen in der Marmarika. Flora. 1985;**176**:141-151

[34] Kehl H, Bornkamm R. Landscape ecology and vegetation units of the western desert of Egypt. Catena Supplement. 1993;**26**:155-176

[35] Stahr K, Bornkamm R, Gauer J, Kehl H. Veränderungen von Böden und Vegetation am Übergang von Halbwüste zur Volwüste zwischen Mittelmeehr und Quattara Depression in Ägypten. Geoökodynamik. 1985;**6**:99-120

[36] Stokker O. Steppe-Wüste-Savanne. Veröff. Geobot. Inst ETH Zürich (Rübel). 1962;**37**:234-243

[37] Stokker O. The water-photosynthesis syndrome and the geographical plant sistribution in the Saharan desert. In: Lange O, Kappen L, Schultze ED, editors. Ecological Studies, Analysis and Synthesis, 11. 1976. pp. 506-521

[38] Blanco J, Genin D, Carriere SM. The influence of Saharan agro pastoralism on the structure and dynamic of acacia stands. Agriculture, Ecosystems and Environments. 2015;**213**:21-31

[39] Jaouadi W, Hamroudi L, Khouja ML. Phénologie de *Acacia tortilis* ssp. *raddiana* dans le Parcc N atgional de Bou Hedma en Tunisie, effet du site sur les phénophases de l'espèce. Boir et Forets des Tropiques. 2012;**312**(2):31-41

[40] Noumi Z, Chaieb M. Dynamics of *Acacia tortilis* (Forsk) Hayne subsp. radd*iana* (Savi) Brenan in arid zones of Tunisia, Acta Botanica Gallica. Botany Letters. 2012;**159**:121-125

[41] Sghari A. A propos de la presence d'une steppe tropicale au Jebel Bouhedma en Tunisie présaharienne. Approche géomorphologique. Quaternaire. 2009;**202**(20/2):255-264

[42] Van Collie F, Delaplac EK, Gabriels D, De Smet K, Ouessa M, Belgacem AO, et al. Monotemporal assessment of the population strucure of *Acacia tortilis* (Forssk.) Hayne ssp.*raddiana* (Savi) Brenan in Bou Hedam National Park, Tunisia: A terrestrial and reomote sensing approach. Journal of Arid Environments. 2016;**129**:80-92

[43] Schulz E, Adamou A. Die Vegetation des Aïr-Gebirges in Nord-Niger und ihre traditionelle Nutzung. Giessener Beiträge zur Entwicklungsforschung, I. 1998;**17**:75-86

[44] Monod T. Modes "contracté et diffus" de végetation saharienne. In: Cloudsley-Thompson JL, editor. Biology of Desert, 3. London; 1954. pp. 5-44

[45] Voss F, Krall P. Principaux biotopes du criquet pelerin dans le Nord du Tilemsi (Mali). Berlin: GTZ/Institute of Geography, Technical University of Berlin; 1994a

[46] Voss F, Krall P. Principaux biotopes du criquet pelerin dans l'Adrar des Iforas (Mali). Berlin: Gtz/Institute of Geography, Technical University of Berlin; 1994b

[47] Schulz E, Adamou A. Die Grenzen der "neolithischen Revolution". Gab es einen frühen Ackerbau in der Sahara? Würzburger Geographische Arbeiten. 1997;**92**:71-95

[48] Quezel P. La végétation du Sahara duTchad à la Mauretanie. Geobotanica selecta. Stuttgart: G. Fischer; 1965. 333 p

[49] Morel A. Les hauts massifs de l'Air (Niger) et leurs piemonts. Etude géomorphologique. Paris; 1984. 404 p

[50] Barry JP. La frontière méridionale du Sahara entre l'Adrar des Iforas et Tombouctou. Ecologia Mediterranea. 1982;**7**:3

[51] Barry JP, Jaquen X, Musso J, Riser J. Le problème des divisions bioclimatiques au Sahara. Note VI: entre Sahel et Sahara. L'Adrar mauretanien. Approches bio-géographiques et géomorphologiques. Ecologia Mediterranea. 1987;**XIII**(1/2):131-142

[52] Schulz E. The southern margin of the Sahara in the Republic of Chad. Vegetation, soil, and present pollen rain. Zentralblatt für Geologie und Paläontologie. 1999;**I**:483-496

[53] Aktar-Schuster M. Degradations prozesse und Desertifikation im semiariden randtropischen Gebiet der Butana/Rep. Sudan. In: Göttinger Beiträge Land- und Forstwirtschaft in den Tropen und Subtropen, 105. 1995. pp. 1-165

[54] Dubief J. Le climat du Sahara. Trav. de l'Inst. Rech. Sahar. Alger 312, 1956. 1963. 275 p

[55] White F. The Vegetation of Africa. Paris: Natural Resources Research, Paris: UNESCO; 1983. 356 p

[56] Anthelme F, Waziri Mato M, Maley J. Elevation an local refuge ensure persistance of mountain refuge vegetation in the Nigerien Sahara. Journal of Arid Environments. 2008;**72**:2232-2242

[57] Linder HP, DeKlerk MM, BornJ, Burgess ND, Fieldsa J, Rahbeck C. The partition of Africa. Statistically defined biogeographical regions in Sub Saharan Africa. Journal of Biogeography, 39-1189-1205; 2012

[58] Barbour KM, Oguntoyinbo JS, Onyemelukwe JOC, Nwafor JC. Nigeria in Maps. London: Hoder and Stroughton; 1982. 148 p

[59] Letourzey R. Végetation. In: Laclavère G, editor. Atlas of the United Republic of the Cameroons. Paris: Editions J.A; 1980. pp. 20-24

[60] Krings T. Kulturbaumparke in den Agrarlandschaften Westafrikas – eine Form autochtoner Agroforstwirtschaft. Die Erde. 1991;**122**(2):117-129

[61] Seignobos C. Matières grasses et civilisations agraires (Tchad et Cameroun). Annales de l'universite de Tchad. Ser Lettres, Langues vivantes et schiences humains Vol spec. 39-119; 1979

[62] Seignobos C. Strategies de survie dans les économies des Razzias (Ronier, Ficus, et Tubercules sauvages). Annales de l'université de Tchad. Ser Lettres, Langues vivantes et sciences humaines Vol spec. 1-37; 1979

[63] Seignobos C. Des mondes oubliés. IRD Editions: Marseille; 2017. 310 p

[64] Ballouche A, Dolidon H. Forets claire et savane ouest-africaines: Dynamique et évolution des systèmes complexes a l'interface nature/societe. In: Taabni M, editor. La fôret: enyeux comparés de formes a l'approproation de gestion et exploitation dans les politiques environnementales et de contexte d'urbanisation géneralisé. Poitiers Maisons des Sciences d l'homme et de la societé. Université de Poitiers; 2003. pp. 56-70

[65] Dolidon H. L'espace des feux en Afrique del'Ouest [thesis]. Caen: Université de Caen; 2005. 345 p

[66] Pomel S, Pomel-Rigeaud, Schulz E. Les indicateurs anthropogènes de la végétation et des sols de quelques savanes de l'Afrique de l'Quest. In: Maire R, Pomel S, Salomon HN, editors. Enregistrateurs et indicateurs de l'evolution de l'environnement en zone tropicale. Bordeaux: Presses Universitaires de Bordeaux; 1994. pp. 173-200

[67] Monod T. The Sahel zone north of the equator. In: Evenary M, Nov-Meier J, editors. Hot Deserts and Arid Shrublands. Oxford; 1986. pp. 203-242

[68] Le Houerou HN. The Grazing Land Ecosystem of the African Sahel, Ecol Studies 75. Berlin: Springer; 1989. 282 p

[69] Ibrahim S, Schulz E. At the sources of fear. Zentralblatt für Geologie und Paläontologie. 2017;**1**(2):11-25

[70] Nicholson SE, Dezfuli AK, Klotter D. A two-century precipitation data set for. The continent of Africa. American Meteorological Society. 2012;**93**:1219-1231

[71] Denham D. Mission to the Niger. The Bornu Mission. Cambridge: Reprint. 1822-1825. 325 p

[72] Barth H. Reisen und Entdeckungen in Nord- und Central-Afrika in den Jahren 1848 bis 1855. Vol. 4. Gotha: Justus Perthes; 1857. 1858

[73] Nachtigal G. Sahara und Sudan. Ergebnisse sechsjähriger Reisen in Afrika. Vol. 2. Parey: Weidmann; 1879-1881

[74] Rohlfs G. Reise vom Mittelmeer nach dem Tschadsee und zum Golf von Guinea. Vol. 2. Leipzig: Brockhaus; 1974

[75] Vischer H. Across the Sahara from Tripoli to Bornu. London; 1910. 308 p

[76] Gardi R. Ténéré. Bern: Benteli; 1978. 288 p

[77] Rabe K, Kilian, N. Georg Schweinfurth. Sammlung botanischr Zeichnungen im BGBM. www.bgbm.org/Schweinfurth (15.4.2019)

[78] Bouché P, Doamba B, Sissoko B, Boujju S. The elephants in Gourma, Maili: Status and threats to their preservation. Pachyderm. 2008;**45**(1):47-56

[79] Canney S. Les élephants du Gourma. Une synthese des connaisances, des récherches et des récommendations. Wild Foundation. 2007. summary-fr_19dec07_with–map.pdf

[80] Dublin HJ, Sinclair ARE, McGald J. Elephants and fire as causes of multiple stable states in the Serengeti-Mara (Tanzania) woodlands. Journal of Animal Ecology. 1990;**59**:1147-1164

[81] Guldemond RAR. The influence of savannah elephants on vegetation: A case study in the Tecah Elephant Park, South Africa [thesis]. Pretoria: Pretoria University; 2006. 163 p

[82] IUCN, SOS. Combating a new elephant poacing threat in the Gourma region of Mali. 2012. 3 p

[83] Ibrahim S. Evolution des paysages dunaires fixés par la végétation au Niger. Schriftenreihe Junges Afrikazentrum, 2016;**4**:171

[84] Buffe J. La plantation des *Casuarina equisetifolia* (Filao) dans le sud du Dahomey. Revue Bois et Forets des Tropiques. 1962;**84**:13-20

[85] Louppe D, Depommier D. Expansion, research and development of the Eucalypts in Africa. Wood production, livelihood and environmental issues: an unlikely reconciliation? Bujumbura: FAO/MEEATOU workshop "Eucalyptus in East Africa"; 2010. 9 p

[86] Ballouche A, Neumann K. A new contribution to the Holocene vegetation history of the West African Sahel: Pollen from Oursi, Burkina Faso and charcoal of three sites in northern Nigeria. Vegetation History and Archaeobotany. 1995;**4**:31-39

[87] Salzmann U, Waller M. The Holocene vegetational history of the Nigerian Sahel based on multiple pollen profiles. Review of Palaeobotany and Palynology. 1998;**100**:39-72

[88] Waller M, Salzmann U. Holocene vegetation changes in the Sahelian zone of NE Nigeria: The detection of anthropogenic activity. Palaeoecology of Africa. 1999:85-102

[89] Waller M, Street-Perrott A, Wang H. Holocene vegetation history of the Sahel: Pollen, sedimentological and geochemical data from Jikariya lake, northeastern Nigeria. Journal of Biogeography. 2007;**34**(9):1575-1590

[90] Miehe S, Kluge J, Von Wehrden H, Retzer V. Long term degradation of Sahelian rangeland detected by 27 years of field study in Senegal. Journal of Applied Ecology. 2010;**47**:692-700

[91] Hahn A, Kusserow H. Spatial and temporal distribution of algae in soil crusts in the Sahel of W-Africa: Preliminary results. Willdenowia. 1989;**28**:227-223

[92] Kusserow H. Desertification, resilience and regreening on the African Sahel. A matter of observation? Earth System Dynamics. 2017;**8**:1141-1170

[93] Pomel S. La mémoire des sols. Bordeaux: Presses Universitaires de Bordeaux; 2008. 343 p

[94] Abichou A. Les changements de paysages du basin-versant de l'oued Tataouine-Fessi (sud-est tunisien): Etude multiscalaire et micromorphologie des remplissages des sebkhas et études des étâts de surface. PHD – Université Michel de Montaigne – Bordeaux 3, France; 2002

[95] Belnap J, Büdel B. Biological soil crust as soil stabilisators. In: Weber B, Büdel B, Belnap J, editors. Biological Soil Crusts. An Organising Principal in Drylands. Ecological Studies 226. Switzerland: Springer; 2016

[96] Buis E, Veldkamp A, Boeken B, van Bremen N. Controls of plant functional surface cover types along a precipitation gradient in the Negev Desert of Israel. Journal of Arid Environments. 2009;**73**(1):82-90

[97] Prasse R, Bornkamm R. Effect of microbiotic soil surface crusts on emergence of vascular plants. Plant Ecology. 2000;**150**:65-75

[98] Guan P, Zhang X, Yu J, Cheng Y, Li Q, Andriuzzi WS, et al. Soil microbial food web channels associated with biological soil crusting and desertification renaturation: The carbon flow from microbes to nematodes. Soil Biology and Biochemistry. 2018;**116**:82-90

[99] Liu Y, Li X, Jia R, Lei H, Gao Y. Effects of biological soil crusts on soil nematode communities following dune stabilisation in the Tengger desert, northern China. Applied Soil Ecology. 2011;**49**:118-124

[100] Reichelt R. L'Hydraulique Pastorale et la Desertification au Sahel des Nomades en Afrique de l'Ouest – Réalités et Perspectives. Geologisches Jahrbuch, C. 1989;**52**:3-32

[101] Spittler G. Dürren, Krieg und Hungerkrisen bei den Kel Ewey (1900-1985). Studien zur Kulturkunde 89. Wiesbaden: F. Steiner; 1989. 199 p

[102] Hallard J. Le barrage vert algérien est un exemple de lute contre la désertification des territoires. ISIAS, Arbres Forets Agroécologie Climat, 2. www.isias.lautre,net/spip.php/article547 15.4.1028; 2016

[103] Gallais J. Pasteurs et paysans du Gourma. Mém. Bordeaux: Centre d'Etudes de Geographic Tropicale CEGET; 1975. 209 p

[104] Catella AM. Modifications de la végétation de Tombouctou depuis dix siècles. Ecologia Mediterranea. 1988;**XIV**(1-2):185-197

[105] Becker LC. Seein g green in Mali's woods: Colonial legacy, forest use and local control. Annals of the Association of American Geographers. 2001;**91**(3):504-526

[106] Canney S. Les élephants du Gourma. Une synthèse des connaissances, des récherches et des recommendations. Wild Foundation. 2007. summary-fr_19dec07_withmap.pdf

[107] Morrison TA, Holdo RM, Anderson TH. Elephant damage and fire or rainfall explains mortality of overstorey trees in Serenget. Journal of Ecology. 2015;**104**:409-418

[108] Rèpublique du Niger: Réserve Naturelle Nationale de l'Air et du Ténéré. Sanctuaire des Addax. Journal Officiel 4. 1988;**15**(3):4

[109] Maas I. Weiderotation in Timia, Niger. Stuttgart: Bericht über fünf Jahre Erfahrung; 1991. 57 p

[110] Magrin G, van Vliet G. La réserve du Termit Tin-Toumma et l'exploitation petrolière au Niger: etat de lieux et piste d'action. Niamey: CIRAD, NOE; 2014. 34 p

[111] Zonon A, Hervé C, Behnke R. A preliminary asssessment of the economic values of the goods and services provided by dryland ecosystems of the Air and Ténéré. The World Conservation Union; 2007. 36 p

[112] UICN/PACO. Parcs et réserves du Niger: evaluation de l'efficacitdé de gestion des aires protégées. Ougadougou, BF: UiCN/PACO; 2010. 78 p

[113] Koudemoukpo B, Nignon P. Final Report of the Terminal Evaluation of the Niger COGERAT Project PIMS 2294. Geneva: UNDP; 2014. 82 p

[114] République du Niger: Grand Muralle Verte. Niamey Minist de l' Hydrauliques et de l'environnement. 109 p

[115] Bayala J, Kalingare A, Tchoundjev Z, Sonclair F, Garrily D. Conservation Agriculture with Trees in the West African Sahel – A Review. Nairobi: World Agriforestry Centre Occas; 2006. Papers 14, 1. 57 p

[116] Presidence de la République du Niger: Niger. Renaissance au sommet. Jeune Afrique. 2019;**3039**:7

[117] DFG. Atlas of Natural and Agronomic Resources of Niger and Benin. Bonn: CDRom; 2000

[118] Roose E, Dugue D, Rodrigues L. Une nouvelle strategie de lutte antei-érosive appliquée a l'amenagement du terroirs en zone sudano-sahelienne en Burkina-Faso. Bois et Forets des Tropiques. 1992;**233**:49-63

[119] Garraud S, Mahamane L. Evolution des pratiques d'adaptation des communautés agropastorales de la zone de Tillabery Nord et de Tahoua au Niger dans un context de changement climatiques. Secheresse. 2012;**23**:24-730

[120] Bouzou JM, Nomao DL. Le "Tassa" une technique de conservation des eaux et des sols bien adaptée aux conditions physiques et socio-économiques des glacis des régions seimiarides (Niger). Revue de Géographie Alpine. 2004;**1**:61-70

[121] Herrmann L. Staubdeposition auf Böden West-Afrikas - Eigenschaften und Herkunftsgebiete der Stäube und ihr Einfluß auf Boden- und Standortseigenschaften. Hohenheimer Bodenkundliche Hefte 36; 1996

[122] Bielders CL, Alvey S, Cronyn N. Wind erosion - the perspective of grass-root communities in the Sahel. Land Degradation and Development. 2001;**12**:57-70

[123] Michels K, Lamers J, Bürkert A. Effects of windbreak species and mulching on wind erosion and millet yield in the Sahel. Experimental Agriculture. 1998;**34**(4):449-464

[124] Reij C, Garrity D. Scaling up farmer-managed natural regeneration in Africa to restore degraded landscapes. Biotropica. 2016;**48**(6):834-843

[125] Wezel A. Scattered shrubs in pearl millet fields in semiarid Niger: Effect on millet production. Agroforestry Systems. 2000;**48**:219-228

[126] Nwankwo CI, Mühlena J, Biegert K, Butzer D, Neumann G, Herrmann L. Physical and chemical optimisation of the seedball technology addressing pearl millet under Sahelian conditions. Journal of Agriculture and Rural Development in the Tropics and Subtropics. 2018;**119**(2):67-79

[127] Buerkert A, Bationo A, Piepho H-P. Efficient phosphorus application strategies for increased crop production in sub-Saharan West Africa. Field Crops Research. 2001;**72**:1-15

[128] Hayashi K, Abdoulaye T, Gerard B, Bationo A. Evaluation of application timing in fertilizer microdosing technology on millet production in Niger, West Africa. Nutrient Cycling in Agroecosystems. 2018;**80**:257-265

[129] Tardiati M, Robbiota G, Rafiou SM. The onion sector of West Africa. Comparative study of Niger and Benin. Cahiers Agriculture. 2013;**22**(2):112-123

[130] Lawarnou M, Abdoulaye M, Reij C. Etude de la régéneration naturelle assisté dans la région de Zinder (Niger): une première exploration d'un phénomèn spectaculaire. Washington DC: Internat. Resources group for the US Agency for International Development; 2006

[131] Reij C. Agroenvironmental transformation in the Sahel. Another kind of "Green Revolution" Washington, D.C. IFPI Discussion paper; 2009;**00914**:54

[132] Bender H, Ousseini I. La protection des bas fonds au Sahel: Transfert de connaissances. ETH Zürich et Université de Niamey; 2000. 133 p

[133] Ousseini I, Karimou A. Amenagement de dix (10) sousbassins versants au moyen de seuils de crue dans le département de Tahoua. Analyse des impacts sur l'environnement. Cooperation Nigero-allemande LUCOP; 2007. 48 p

[134] Larwanou M, Saadou M. Biodiversity of ligneous species in semiarid zones of southwestern Niger according to anthropogene and natural factors. Agriculture, Ecosystems and Environment. 2005;**105**:267-271

[135] Schulz E. Indicateurs de l'influence antropique sur la vegétation actuelle et passée. In: Maire R, Pomel S, Salomon JN, editors. Enregistreurs et indicateurs de l'evolution de l'environnement en zone tropicale. Espaces tropicaux 13. Bordeaux: PUB; 1994. pp. 129-142

[136] Casenave A, Valentin C. Les etâts de surface de la zone sahelienne. Paris: ORSTOM; 1989. 229 p

# 7

# Eastern Poison Ivy (*Toxicodendron radicans* L.): A Bioindicator of Natural and Anthropogenic Stress in Fields and Forests

*Dean G. Fitzgerald, David R. Wade and Patrick Fox*

**Abstract**

This chapter considers herbaceous and woody plants near the 1800s era hydrocarbon extraction areas (HEAs) in Wiikwemkoong Unceded Territory (WUT) on Manitoulin Island, Lake Huron, Ontario, Canada. Plant community assessment used patterns of diversity and distribution at five field and six forest sites to assess the response of the plants to HEAs. These sites receive brine episodically from HEAs and natural seeps over the Collingwood Oil Shale Formation. This brine contains high concentrations of chloride and sodium along with total dissolved solids that exceed 100,000 mg/L. Exposure to brine is identified as the causative factor shaping plant distribution, survival, size, leaf bleaching (i.e., chlorosis), and dead branches on woody stems. These sites demonstrate an ecotone of disturbance defined by transition from natural plant community to dominance by eastern poison ivy (EPI, *Toxicodendron radicans* L.). Disturbed sites within brine drainage areas are dominated by EPI, reflecting tolerance to elevated salinity due to rhizome growth strategy. Evaluation of the plant communities and EPI allowed for preparation of a framework that can be used to guide interpretation of response of plants to drainage of brine from HEAs and natural sources beyond WUT.

**Keywords:** brine, community responses, hydrocarbon, *Toxicodendron radicans*, Manitoulin Island, Canada

## 1. Introduction

Abiotic and biotic factors can act synergistically to influence growth, longevity, and distribution of plants [1]. Abiotic factors include disturbance regimes, light intensity, availability of water, and microelemental concentrations, among other factors [1, 2]. Biotic factors include competition, predation, and propagule pres-sure, among other factors [2, 3]. When plants respond to abiotic and biotic factors, the resulting growth, longevity, and distribution can be used to understand the dominant factor(s) affecting success in a habitat. This understanding arises when the abiotic and biotic factors are known, and then analyses can be completed to identify the factor(s) responsible for observed growth, longevity, and distribution. In these settings, it is often feasible to identify one or more plant species as bioindicators, to represent patterns [1–5]. This approach uses existing knowledge on ecological, physiological, and ontological requirements of bioindicator species. Hence, the status of bioindicator species in a habitat can be used as a surrogate to identify the dominant abiotic and biotic factors shaping plants within an area of interest [5, 6].

Studies of plants and plant communities exposed to brine from oil and gas wells, referred to herein as hydrocarbon extraction areas (HEAs), demonstrated the short- and long-term consequences of this type of episodic and/or persistent disturbance [7–12]. Historically, brine was allowed to drain away from HEAs and was then observed to kill all exposed plants [7, 8]. Best practices now involve the capture of brine for safe disposal [7]. Brine from HEAs varies from locale to locale but always contains high concentrations of elements, like Cl >50,000 mg/L, Na >25,000 mg/L, Ca > 10,000 mg/L, Mg >1000 mg/L, $SO_4^{-2}$ > 500 mg/L, and Fe > 200 mg/L with total dissolved solids (TDS) > 100,000 mg/L [7, 13, 14]. When brine drains to adjacent plant communities, most species will show a short-term response involving the leaves turning white, indicating the loss of chlorophyll, referred to as chlorosis, with leaf drop soon thereafter [7, 15, 16]. The process of chlorosis is attributed to the loss of ionic balance in the roots and leaves of the plant, attributable to the high concentrations of elements such as Cl and Na in the brine [12, 15].

A detailed study of the response of plants to long-term exposure of brine was completed by government scientists in former oak (*Quercus* sp.)-dominated forest of Oklahoma, USA [7]. This Oklahoma study documented how herbaceous and woody vegetations were completely absent in areas that received brine run-off during the past, while trees downslope were short and demonstrated dead branches. In contrast, herbaceous and woody vegetation upslope and adjacent to the brine-exposed areas showed no evidence of stress [1]. The authors attributed the loss of ground vegetation and the short height and dead branches of the trees to long-term exposure to brine from HEAs. Another study that documented the response of plants to brine exposure was completed in the Allegheny State Forest in Pennsylvania, USA [15, 16]. This forest patch was dominated by trees such as eastern hemlock (*Tsuga canadensis* L. Carrière), red maple (*Acer rubrum* L.), American beech (*Fagus grandifolia* Ehrh.), northern red oak (*Quercus rubra* L.), and yellow birch (*Betula alleghaniensis* Britton). This forest was exposed to brine from a leaking impoundment over a period of 3 years. The path followed by the brine resulted in the death of all ground vegetation during the first season and all trees within 2 years. Walters and Auchmoody ([15], p. 124) stated: "The swiftness and completeness of the kill attests to the extremely toxic nature of the spilled brine. Ground cover was eliminated immediately, and trees showed visual symptoms of stress during the first growing season...." The last plants to die were the old growth (>300 years old) eastern hemlock and American beech in the drainage area. The extended survival of these large and old trees was attributed to the deep roots providing some tolerance to the brine, but they still died. It was also reported that within 2 years, plants returned to the brine-disturbed forest areas, with typical pioneer species, including ferns, as first to appear, followed by previously evident woody species [16]. Studies that document the response of plants to brine exposure during short- and long-term periods represent an opportunity for learning about species and community response patterns to this type of disturbance [17, 18]. Schindler [18] suggested such severe types of disturbance are useful for learning about responses of species and communities to a disturbance but do not necessar-ily represent the key variables to help elucidate exact response patterns. Schindler [18] also noted that disturbance regimes can provide a basis to resolve cause-effect relationships and are instructive if the response patterns are indeed outside of the typical normal range. Schindler [18] also suggested that understanding the exact responses of species and communities to disturbance can lead to the development of a predictive framework of responses for low-level disturbance. Using this basis, additional studies of plants associated with HEAs and brine are justified, to resolve growth, longevity, and distribution of plants in these areas, as a basis to refine future rehabilitation activities.

This chapter reports how the distribution of eastern poison ivy (EPI; *Toxicodendron radicans* L.) has been used as a diagnostic indicator to locate lost HEAs that include oil and gas wells in fields and forests. These lost HEAs exist on Wiikwemkoong Unceded Territory (WUT) #26, an area that extends along the entire east shoreline of Manitoulin Island, Lake Huron, Ontario, Canada (**Figure 1**). Portions of Manitoulin Island and WUT are located over the Collingwood Oil Shale Formation (Ordovician origin) containing oil and gas deposits within the porous limestone of the Trenton Formation [19, 20]. Since 2014, members of WUT have been working with Premier Environmental Services Inc. (Premier), to rediscover lost HEAs using an approach that integrates oral traditional knowledge (TK) with plant ecology and chemistry-based analyses [21]. This discovery process has been refined since 2014, with the integration of TK and science to help understand the distribution of EPI along with the status of the plant communities in proximity to candidate HEAs. Direct experience at WUT allowed for the refinement of this understanding of how EPI represents a bioindicator species to represent the responses of other herbaceous and woody plant species in these habitats at 50+ HEAs. Documentation of the response of EPI and plant communities associated with HEAs was achieved with detailed studies of five field and six forest sites, including a groundwater seep in a field and natural hydrocarbon seep in a forest. This representation of the responses of plants to local habitat features provides the basis for learning about the key environmental factors shaping the growth, longevity, and distribution of plants in areas with HEAs and natural seeps.

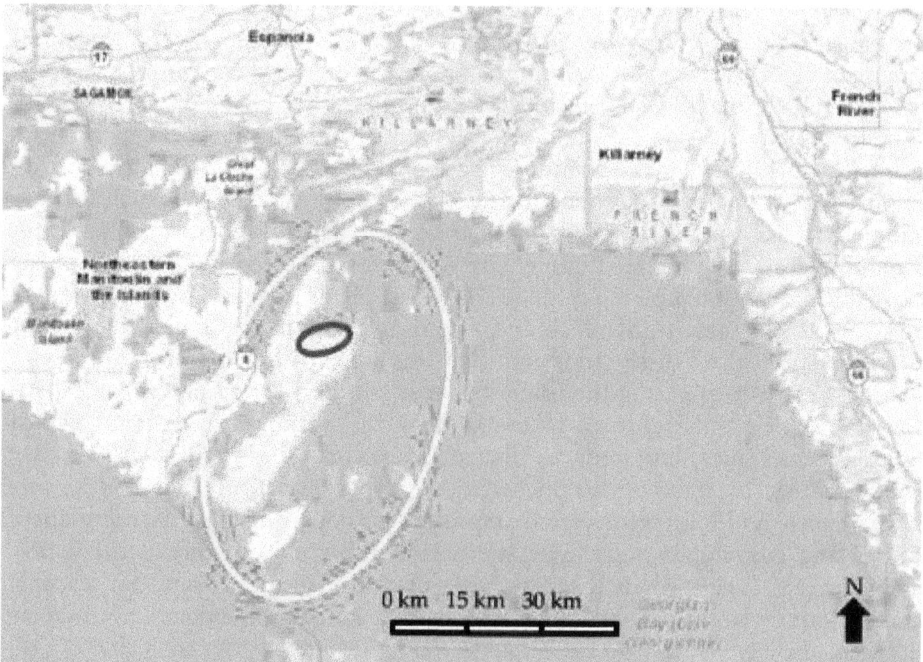


**Figure 1.**
*View of Manitoulin Island, Lake Huron, Ontario. The approximate boundaries (less adjacent lands) of WUT are noted within the yellow oval, along the entire eastern end of the island. The general area of the five field and six forest sites are within the brown oval.*

## 2. Taxonomy and physiological basis for study

Eastern poison ivy is a member of the Anacardiaceae family (sumac-cashew) and distributed across Eastern North America, whereas western poison ivy (*Toxicodendron rydbergii* (Small ex Rydb.) Greene) is distributed across Western North America [22, 23]. These two species are morphologically variable (e.g., leaf color, size, shape), and they can hybridize within overlapping habitats. Eastern poison ivy shows varied growth forms, with the most common as a woody rhizome that grows underground in all directions, with no nodes but roots and leaves at varied intervals. This variable rooting leads to the establishment of patches over large areas and varying types of soil. An alternate form of EPI demonstrates a woody vine that will climb trees and other hard surfaces. Both growth forms are identified as shade-intolerant and are often found in habitats with direct sunlight such as the edge of forests, along roadways, disturbed areas, fields, and wetlands. Both forms produce the toxin urushiol as a defense mechanism, and these toxins often cause contact dermatitis in two of three humans following exposure to as little as 1 nanogram [24]. These toxins are released from the roots and leaves to the air as oil droplets [24]. On Manitoulin Island, EPI is most commonly found along edges of forests and farm fields, roadways, and other disturbed areas [25]; EPI is considered a rare species in natural forests with closed canopies and wetland areas on the island [25], as this species prefers direct sunlight [22, 23].

All species within the *Toxicodendron* genus show some tolerance to soil salinity, and this provides the plants with the ability to maintain viable populations within disturbed habitats [22]. Kuester et al. [26] reported that elevated tolerance to soil salinity was a common characteristic of successful weed species in North America. The United States Department of Agriculture (USDA) Plants Database (http://plants.usda.gov/charinfo) provides a standard definition to represent how plants tolerate soil salinity. Such representation of plant tolerance to soil salinity reflects studies of plant growth performance that classified species among four tolerance categories (zero, low, medium, high), to represent range of responses. These USDA responses include zero tolerance to soil solutions with electrical conductivity of 0–2 dS/m, low tolerance to 2.1–4.0 dS/m, medium tolerance to 4.1–8.0 dS/m, and high tolerance to >8.0 dS/m. Plants are defined to tolerate salinity if there is zero to slight reduction in growth (<10%). Eastern poison ivy demonstrates an ability to tolerate elevated concentrations of soil salinity, often within the medium tolerance category [22, 23].

Such reports of tolerance of EPI to soil salinity include a range of responses, depending on the locale of the study [22, 23]. It is likely this range of responses is a direct response to rhizome growth forms with roots and stems extending over large areas. This pattern indicates that the rhizome allows the plant to tolerate a wide range of soil salinities, as well as other forms of local disturbance. The presence of rhizome growth form was identified as the causative factor for rapid reestablishment of EPI after flooding destroyed surface stems and leaves [27, 28]. The observation of tolerance of EPI to environmental disturbance, due to ecological and physiological adaptations, justifies the consideration of EPI as a candidate bioindicator species. The use of EPI as a bioindicator species identifies that the phenological response patterns of the species to different environmental, ecological, and physiological factors is understood. This chapter demonstrates the current understanding of the responses of EPI and plant communities near HEAs and two natural seeps and represents an illustrative example of plant phenology in field and forest ecosystems.

## 3. The setting

At WUT, oral history combined with TK led to the initial documentation of old and un-abandoned HEAs that were observed to be harming native vegetation as well as fouling agricultural fields. Specifically, resident observations included woody vegetation downslope of old HEAs that were dead or showed dead branches and short height compared with specimens in adjacent habitats that were live and taller. Other observations included livestock that shun hay cut from a field with an HEA releasing an oil-water slurry downslope through the field. These observations led to the abandonment of these HEAs. After these HEAs were abandoned, Premier was invited to participate in the discovery of lost HEAs, and this recent work represents the basis for this study.

It is prudent to briefly review the history of how HEAs were established at WUT during the 1800s, reflecting oral history, government reports, scientific articles, photographs, and letters written by Jesuit missionaries This history also reveals that an extremely large forest fire occurred across Manitoulin Island during 1865, with areas on WUT having nearly all woody stems burned, and this land described afterward as fully cleared, whereas some wet areas and shorelines were spared [25, 29]. This regional fire disturbance at WUT likely aided development of HEAs, through enhancement of access following the loss of dense forest tracts. A second regional disturbance within the WUT forest has been the recent loss of ash (*Fraxinus* spp.) trees due to invasion by emerald ash borer (EAB, *Agrilus planipennis* ). This beetle from Asia invaded southern Canada during the early 2000s and now has spread extensively within North America [30]. At WUT, most ash have perished during the last decade, including black ash (*Fraxinus nigra* Marshall), green ash (*F. pennsylvanica* Marshall), and white ash (*F. americana* L.) due to the EAB infestation.

Initial advancement of HEAs at WUT occurred during the summer of 1865 while other portions of the island burned [31, 32]. The first HEAs were planned to focus on oil and followed the McDougall Treaty involving WUT and Upper Canada, signed during 1862; it is noteworthy to identify that the WUT community did not fully embrace this treaty as only two community members actually signed the document [32, 33]. After this treaty, staff from Milwaukee Petroleum Company arrived at WUT on May 1 of 1865 to negotiate access. After access was tentatively granted by WUT, John Ward, an experienced pioneer in oil well drilling from Oil Springs, Ontario, arrived about a month later. Mr. Ward focused the drilling activity on limestone outcrops found in close proximity to natural oil seeps evident along Smith's Bay shoreline, an area directly accessible by boat. This initial extraction activity yielded a "schooner load" of oil that was shipped south with a request for 500 wooden barrels to be shipped north. However, the oil extraction ended during autumn in 1865 when the Wiikwemkoong community members asked the oil men to leave on short notice; their equipment was reportedly left behind [31, 33].

Representatives from the oil industry returned to WUT during the early 1880s and proceeded to develop a large number of HEAs despite opposition from the community [33]. This development was approved by the Government of Canada, as oil well licenses facilitated access to WUT for lease areas along the recently described Trenton Formation. When these 1880s HEAs were prepared, the initial disturbance involved road construction, as land away from the shoreline was targeted. After roads were prepared, HEAs were constructed in agricultural fields and forest settings. In agricultural fields, site preparation was a relatively simple activity, while forest settings required additional effort. It is probable that some of these forest areas used for HEAs in the 1880s were burned during 1865 and included younger trees. After a site was cleared, then a wooden oil rig was built around the extraction point. A wood barrel was then usually placed next to the rig, to collect the oil-water slurry; natural gas wells also had a collection barrel. These barrels allowed for the separation of oil and water via gravity, but it is not clear if it was necessary to separate water from the natural gas. Abandonment studies to date have revealed collection barrels in agricultural fields which were placed on grade, while in the forests, they were buried. Collection barrels documented at WUT in fields had dimensions with a width of 3–4 m and height of 3–4 m, while barrels abandoned during 2016 in forests had widths of 3–4 and depth up to 6 m [21]. This design implies the larger barrels needed to be placed below grade, to support the weight and facilitate gravity separation of the oil and water slurry [21].

All wells were drilled by hand, with help from horses. Abandonment stud-ies led by WUT from 2014 to 2017 years revealed the depth of the drilled holes ranging from 105 to 130 m for oil sites and up to 337 m for natural gas [21]. The presence of porous limestone of the Trenton Formation across WUT provides a simple explanation why the HEAs involved this range of depths. These depths also reveal that the wells could have been more shallow, as it is now known that these depths penetrated the hydrocarbon-bearing strata and then entered the deep groundwater strata [34]. It was reported [34] that drilling of hydrocarbon wells in Ontario during the 1800s and 1900s included the systemic problem of over-drilling oil strata followed by penetration of deep groundwater. When an oil well is over-drilled, it is often associated with initial high yields of an oil-water slurry; however, the longevity of the well is shortened due to the high volumes of water that are interacting with the oil at depth. In the case of the natural gas wells at WUT, it appears they did not produce excessive quantities of water, based on abandonment activities [21].

Hydrocarbon extraction at WUT ended in 1905, after 20+ years, when the men were evicted and the structures burned [21]. Pipes at HEAs were often cut below grade. Then, some pipe holes were filled with soil and rock, and graded, to reduce risk from falling in to these hazards. This history identifies that all known HEAs were burned, indicating a common disturbance history for the forest and field settings. This 1905 burning followed the 1865 forest fire, representing key disturbance events shaping the plant communities associated with most HEAs at WUT. After this period, many HEAs near or within forests experienced regeneration, although the species composition of trees differs from historical forest composition (described below). In contrast, HEAs located in agricultural fields were often maintained as fields, through regular cutting of the herbaceous vegetation. Studies [15] at WUT documented that for seven HEAs, the soil concentrations of hydrocarbons were elevated in close proximity to HEAs, but these concentrations rapidly declined with distance from the HEAs. This elevation of hydrocarbons at an HEA was attributed to extraction while the lower concentrations attributed to bacterial degradation of hydrocarbons downslope. These low concentrations of soil hydrocarbons downslope from HEAs provide additional evidence that it is brine causing disturbance to the plant communities [21].

This study considers the diversity and distribution of plants and EPI associated with HEAs located in forest and field settings at WUT, as described in **Table 1**. These plant associations reflect past field surveys that focused on the common plants [25] near HEAs. Initial observations at HEAs revealed that the plant community was less diverse and these plants were distributed in what was documented as a predictable manner in proximity to HEAs and this led to the preparation of this study [21]. Hence, this study does not focus on the morphological or phenological aspects of plant specimens near or around HEAs. However, aspects of morphology and phenology are considered while interpreting the diversity and distribution of plants in forest and field settings associated with HEAs.

| Site (current habitat) | Plant association with EPI | HEA |
|---|---|---|
| A (forest) | Ostrich fern (*Matteuccia struthiopteris* L.), balsam poplar (*Populus balsamifera* L.) | Wood crib |
| B (forest) | Marsh horsetail (*Equisetum palustre* L.), balsam poplar | Metal pipe |
| C (forest) | Ostrich fern, balsam poplar | Metal pipe |
| D (forest) | Smooth Solomon's seal (*Polygonatum biflorum* Walter), balsam poplar | Metal pipe |
| E (forest) | Common lady fern (*Athyrium filix-femina* L.), balsam poplar | Wood crib |
| F (forest) | Common lady fern, balsam poplar | Wood crib |
| G (field) | Red osier dogwood (*Cornus sericea* L.) | Metal pipe |
| H (field) | Timothy *Phleum pratense* (L.), Virginia strawberry (*Fragaria virginiana* Duchesne) | Metal pipe |
| I (field) | Timothy, Virginia strawberry | Metal pipe |
| J (roadside ditch) | Raspberry (*Rubus arcticus* (L.), Virginia strawberry | Metal pipe |
| K (field) | Virginia strawberry | Natural seep |

*At each site, the main plant association with EPI and HEA is noted along with the type of HEA. Plant identifications follow the standard guide for Manitoulin Island [25].*

**Table 1.**
*Summary of forest and field sites considered in this study.*

## 4. Chemistry of water from HEAs and seeps

Studies of oil and gas wells generally, as well as for those located across the Trenton Formation, revealed brine is readily evident and the chemistry reflects local limestone composition [19, 20, 35]. Hence, wells within a short distance can show variations in concentrations of key elements. General features of brine within the Trenton Formation describe this water as having a near neutral pH (5.5–7.0) with elevated concentrations of elements that can be described generally as follows: Cl >100,000 mg/L, Na >50,000 mg/L, Ca > 20,000 mg/L, Sr. > 5000 mg/L, Mg >2000 mg/L, K > 2000 mg/L, Ba >2000 mg/L, Bo >1000 mg/L, $SO_4^{-2}$ > 500 mg/L, and Fe > 300 mg/L [19, 35]. This general composition of brine represents concentrations of elements that are potentially harmful to aquatic and terrestrial life and why this water is regulated under Ontario's *Oil, Gas and Salt Resources Act*. This need to control the release of brine can be illustrated by considering the Cl surface water quality guideline in Canada to protect aquatic life during short-term exposure is 640 mg/L and long-term exposure is 120 mg/L [36] while brine may contain Cl at >100,000 mg/L. Thus, water from HEAs can be regarded as hazardous to plants and wildlife, even in small quantities. Water samples collected by WUT confirmed the presence of brine at HEAs.

## 5. Results

Integration of available information from WUT, scientific literature, and field inspections allowed for the identification of a list of plants associated with HEAs in field and forest habitats (**Tables 2–4**). The first forest habitat type is dominated by balsam poplar (*Populus balsamifera* L.) and the second forest habitat dominated by balsam fir (*Abies balsamea* L.) and eastern white cedar (*Thuja occidentalis* L.). When the plant associations with HEAs were identified, they were intended to represent spatial patterns concerning the general plant community found in each area, as

| Plants found in poplar forest settings within 30 m of HEA | Plants within 5 m of HEA | | | | Plants within 1 m of HEA | | | |
|---|---|---|---|---|---|---|---|---|
| | A | B | C | D | A | B | C | D |
| Apple (*Malus domestica* Borkh.) | X | X | X | X | | | | |
| Arctic sweet coltsfoot (*Petasites frigidus* L.) | | | | | | | | |
| Balsam poplar (*Populus balsamifera* L.) | X | X | X | X | | X | X | X |
| Balsam fir (*Abies balsamea* L.) | | | X | X | | | X | |
| Bebb's sedge (*Carex bebbii* Olney ex Fernald) | | | | | | | | |
| Black ash (*Fraxinus nigra* Marshall) | X | X | X | X | | | | |
| Black cherry (*Prunus serotina* Ehrh.) | | | | | | | | |
| Black medic (*Medicago lupulina* L.) | X | | | X | | | | |
| Black spruce (*Picea mariana* Mill.) | | | | X | | | | |
| Blue-joint (*Calamagrostis canadensis* Michx.) | | | | | | | | |
| Bur oak (*Quercus macrocarpa* Michx.) | | | | | | | | |
| Canada goldenrod (*Solidago canadensis* L.) | X | X | X | X | | | | |
| Canada thistle (*Cirsium arvense* L.) | | | | | | | | |
| Cinnamon fern (*Osmunda cinnamomea* L.) | | | | | | | | |
| Chestnut sedge (*Carex castanea* Wahlenb.) | | | X | | | | | |
| Columbine (*Aquilegia canadensis* L.) | | | X | | | | | |
| Common buckthorn (*Rhamnus cathartica* L.) | | | | X | | | | |
| Common dandelion (*Taraxacum officinale* Ledeb.) | X | X | X | X | | | | |
| Common ragweed (*Ambrosia artemisiifolia* L.) | | | | | | | | |
| Common evening primrose (*Oenothera biennis* L.) | | | | | | | | |
| Common lady fern (*Athyrium filix-femina* L.) | | | | | | | | |
| Common milkweed (*Asclepias syriaca* L.) | | | | | | | | |
| Common mullein (*Verbascum thapsus* L.) | | | | | | | | |
| Common raspberry (*Rubus arcticus* L.) | X | X | X | X | | | | |
| Common yarrow (*Achillea millefolium* L.) | | | | | | | | |
| Daisy fleabane (*Erigeron annuus* L.) | | | | | | | | |
| Eastern bracken fern (*Pteridium aquilinum* L.) | | | | | | | | |
| Eastern cottonwood (*Populus deltoides* W. Bartram ex Marshall) | | | | | | | | |
| Eastern poison ivy (*Toxicodendron radicans* L.) | X | | | | X | X | X | X |
| Eastern white cedar (*Thuja occidentalis* L.) | | | | | | | | |
| False Solomon's seal (*Maianthemum racemosum* L.) | | | | | | | | |
| Green ash (*Fraxinus pennsylvanica* Marshall) | | | | | | | | |
| Ground juniper (*Juniperus communis* L.) | | | | X | | | | |
| Golden sedge (*Carex aurea* Nutt.) | X | X | | | | | | |
| Heal-all (*Prunella vulgaris* L.) | | | | | | | | |
| Lesser burdock (*Arctium minus* Bernh.) | | | | | | | | |
| Longroot smartweed (*Polygonum coccineum* Muhl. ex Willd.) | | | | | | | | |
| Mapleleaf viburnum (*Viburnum acerifolium* L.) | X | X | X | | | | | |
| Marginal wood fern (*Dryopteris marginalis* L.) | | | | X | | | | |
| Marsh horsetail (*Equisetum palustre* L.) | | X | X | | | X | X | |
| Multiflora rose (*Rosa multiflora* Thunb.) | | | | | | | | |

| Plants found in poplar forest settings within 30 m of HEA | Plants within 5 m of HEA | | | | Plants within 1 m of HEA | | | |
|---|---|---|---|---|---|---|---|---|
| | A | B | C | D | A | B | C | D |
| New England aster (*Symphyotrichum novae-angliae* L.) | | | X | X | | | | |
| Northern bugleweed (*Lycopus uniflorus* Michx.) | | | | | | | | |
| Northern red oak (*Quercus rubra* L.) | | | | | | | | |
| Orange daylily (*Hemerocallis fulva* L.) | X | | | | | | | |
| Ostrich fern (*Matteuccia struthiopteris* L.) | X | | | | X | | | |
| Quaking aspen (*Populus tremuloides* Michx.) | | | | | | | | |
| Red maple (*Acer rubrum* L.) | | | X | | | | X | |
| Red osier dogwood (*Cornus sericea* L.) | | | | | | | | |
| Riverbank grape (*Vitis riparia* Michx.) | X | X | X | X | | | X | |
| Rough bedstraw (*Galium asprellum* Michx.) | | | | | | | | |
| Sensitive fern (*Onoclea sensibilis* L.) | | | | | | | | |
| Silver maple (*Acer saccharinum* L.) | | | | | | | | |
| Smooth Solomon's seal (*Polygonatum biflorum* Walter) | | | | X | | | | |
| Spotted jewelweed (*Impatiens capensis* Meerb.) | X | X | X | X | | | | |
| Spotted joe-pye weed (*Eutrochium maculatum* L.) | | | | | | | | |
| Staghorn sumac (*Rhus typhina* L.) | | | | | | | | |
| St. John's wort (*Hypericum perforatum* L.) | | | | | | | | |
| Stinging nettle (*Urtica dioica* L.) | | X | | X | | | | |
| Sweet white clover (*Melilotus officinalis* L.) | | | | | | | | |
| Sugar maple (*Acer saccharum* Marshall) | | | | | | | | |
| Tall buttercup (*Ranunculus acris* L.) | | | | X | | | | |
| Tamarack (*Larix laricina* Du Roi) | | | | | | | | |
| Virginia strawberry (*Fragaria virginiana* Duchesne) | X | X | X | X | | | X | |
| White ash (*Fraxinus americana* L.) | | | | X | | | | |
| White birch (*Betula papyrifera* Marshall) | | | X | X | | | X | |
| White oak (*Quercus alba* L.) | | | | | | | | |
| White spruce (*Picea glauca* Moench) | | | | | | | | |
| White trillium (*Trillium grandiflorum* Michx.) | | | | | | | | |
| Wild carrot (*Daucus carota* L.) | X | X | X | X | | | | |
| Yellow daylily (*Hemerocallis lilioasphodelus* L.) | | | | | | | | |
| Yellow hawkweed (*Hieracium piloselloides* Vill.) | | | X | X | | | | |
| Yellow sedge (*Carex flava* L.) | | | | | | | | |

*These settings included site A with wood crib, site B with pipe that was bubbling natural gas, site C with pipe draining water to long drainage channel, and site D representing a dry pipe in a forest.*

**Table 2.**
*Representation of common plant species found at different distances from HEAs in forests dominated by balsam poplar [25].*

well as the plant species tolerant of these habitats, within ~5 m and within 1 m of an HEA. These community associations show how diverse plant communities become depauperate with increased proximity to the HEAs and how EPI is consistently the most common species within 1 m of HEAs.

Using Ontario's Ecological Land Classification (ELC) strategy [37] and the lists of common plants associated with the HEAs (**Tables 2–4**), the ecosites associated

| Plants found in white cedar-balsam Fir forest within 100 m of HEA | Plants within 5 m of HEA | | Plants within 1 m of HEA | |
|---|---|---|---|---|
| | E | F | E | F |
| American basswood (*Tilia americana* L.) | | | | |
| Apple (*Malus domestica* L.) | | | | |
| Arctic sweet coltsfoot (Pe*tasites frigidus* L.) | | | | |
| Balsam poplar (*Populus balsamifera* L.) | X | X | X | X |
| Balsam fir (*Abies balsamea* L.) | X | X | X | X |
| Bebb's sedge (*Carex bebbii* Olney ex Fernald) | | | | |
| Bebb's willow (*Salix bebbiana* Sarg.) | | | | |
| Black ash (*Fraxinus nigra* Marshall) | X | X | | |
| Black cherry (*Prunus serotina* Ehrh.) | | | | |
| Black medic (*Medicago lupulina* L.) | | | | |
| Black raspberry (*Rubus occidentalis* L.) | | | | |
| Black spruce (*Picea mariana* Mill.) | | | | |
| Blue-joint (*Calamagrostis canadensis* Michx.) | | | | |
| Canada goldenrod (*Solidago canadensis* L.) | X | | | |
| Canada thistle (*Cirsium arvense* L.) | | | | |
| Cinnamon fern (*Osmunda cinnamomea* L.) | | | | |
| Chestnut sedge (*Carex castanea* Wahlenb.) | | | | |
| Columbine (*Aquilegia canadensis* L.) | | | | |
| Common buckthorn (*Rhamnus cathartica* L.) | | | | |
| Common dandelion (*Taraxacum officinale* Ledeb.) | X | X | | |
| Common ragweed (*Ambrosia artemisiifolia* L.) | | | | |
| Common evening primrose (*Oenothera biennis* L.) | X | | | |
| Common lady fern (*Athyrium filix-femina* L.) | X | X | | |
| Common milkweed (*Asclepias syriaca* L.) | | | | |
| Common mullein (*Verbascum thapsus* L.) | | | | |
| Common raspberry (*Rubus arcticus* L.) | X | X | | |
| Common yarrow (*Achillea millefolium* L.) | | | | |
| Daisy fleabane (*Erigeron annuus* L.) | | | | |
| Eastern bracken fern (*Pteridium aquilinum* L.) | X | X | X | X |
| Eastern cottonwood (*Populus deltoides* W. Bartram ex Marshall) | | | | |
| Eastern poison ivy (*Toxicodendron radicans* L.) | | | X | X |
| Eastern white cedar (*Thuja occidentalis* L.) | X | X | | |
| False Solomon's seal (*Maianthemum racemosum* L.) | X | X | | |
| Green ash (*Fraxinus pennsylvanica* Marshall) | | | | |
| Ground juniper (*Juniperus communis* L.) | X | X | | |
| Golden sedge (*Carex aurea* Nutt.) | | | | |
| Heal-all (*Prunella vulgaris* L.) | | | | |
| Hoary willow (*Salix candida* Flügge ex Willd.) | | | | |
| Large-toothed aspen (*Populus grandidentata* Michaux) | | | | |
| Lesser burdock (*Arctium minus* Bernh.) | | | | |

| Plants found in white cedar-balsam Fir forest within 100 m of HEA | Plants within 5 m of HEA | | Plants within 1 m of HEA | |
|---|---|---|---|---|
| | E | F | E | F |
| Longroot smartweed (*Polygonum coccineum* Muhl. ex Willd.) | | | | |
| Northern maidenhair fern (*Adiantum pedatum* L.) | | | | |
| Mapleleaf viburnum (*Viburnum acerifolium* L.) | X | X | | X |
| Marginal wood fern (*Dryopteris marginalis* L.) | | | | |
| Marsh horsetail (*Equisetum palustre* L.) | | | | |
| Multiflora rose (*Rosa multiflora* Thunb.) | | | | |
| New England aster (*Symphyotrichum novae-angliae* L.) | | | | |
| Northern bugleweed (*Lycopus uniflorus* Michx.) | | | | |
| Northern maidenhair (*Adiantum pedatum* L.) | X | X | | |
| Northern red oak (*Quercus rubra* L.) | | | | |
| Orange daylily (*Hemerocallis fulva* L.) | | | | |
| Ostrich fern (*Matteuccia struthiopteris* L.) | | | | |
| Pussy willow (*Salix discolor* Muhlenb.) | | | | |
| Quaking aspen (*Populus tremuloides* Michx.) | | | | |
| Red maple (*Acer rubrum* L.) | X | X | X | X |
| Red osier dogwood (*Cornus sericea* L.) | | | | |
| Red spruce (*Picea rubens* Sarg.) | | | | |
| Riverbank grape (*Vitis riparia* Michx.) | X | X | | |
| Rough bedstraw (*Galium asprellum* Michx.) | | | | |
| Sensitive fern (*Onoclea sensibilis* L.) | | | | |
| Shining willow (*Salix lucida* Muhlenb.) | | | | |
| Silver maple (*Acer saccharinum* L.) | | | | |
| Smooth Solomon's seal (*Polygonatum biflorum* Walter) | | | | |
| Spotted jewelweed (*Impatiens capensis* Meerb.) | X | X | | |
| Spotted joe-pye weed (*Eutrochium maculatum* L.) | | | | |
| St. John's wort (*Hypericum perforatum* L.) | | | | |
| Stinging nettle (*Urtica dioica* L.) | | | | |
| Sweet white clover (*Melilotus officinalis* L.) | | | | |
| Sugar maple (*Acer saccharum* Marshall) | X | | | |
| Tall buttercup (*Ranunculus acris* L.) | | | | |
| Tamarack (*Larix laricina* Du Roi) | | | | |
| Virginia strawberry (*Fragaria virginiana* Duchesne) | X | X | X | |
| Western bracken fern (*Pteridium aquilinum* L. Kuhn) | | | | |
| White ash (*Fraxinus americana* L.) | | | | |
| White birch (*Betula papyrifera* Marshall) | X | X | X | |
| White spruce (*Picea glauca* Moench) | | | | |
| White trillium (*Trillium grandiflorum* Michx.) | | | | |
| Wild carrot (*Daucus carota* L.) | | | | |
| Yellow birch (*Betula alleghaniensis* Britton) | X | X | | |
| Yellow daylily (*Hemerocallis lilioasphodelus* L.) | | | | |

| Plants found in white cedar-balsam Fir forest within 100 m of HEA | Plants within 5 m of HEA | | Plants within 1 m of HEA | |
|---|---|---|---|---|
| | E | F | E | F |
| Yellow hawkweed (*Hieracium piloselloides* Vill.) | | | | |
| Yellow sedge (*Carex flava* L.) | | | | |

*These settings included site E with wood crib and site F with wood crib.*

**Table 3.**
*Representation of common plant species found at different distances from HEAs in forests dominated by eastern white cedar-balsam fir forest along the shoreline of Cape Smith [25].*

| Plants found in field settings at WUT with 100 m of HEA | Plants within 5 m of HEA | | | | | Plants within 1 m of HEA | | | | |
|---|---|---|---|---|---|---|---|---|---|---|
| | H | I | J | K | L | H | I | J | K | L |
| Apple (*Malus domestica* L.) | X | X | X | X | | | | | | |
| Black raspberry (*Rubus occidentalis* L.) | | X | X | | | | | | | |
| Bedstraw (*Galium aparine* L.) | | X | X | | | | | | | |
| Black locust (*Robinia pseudoacacia* L.) | X | | | | | X | | | | |
| Calico aster (*Symphyotrichum lateriflorum* L.) | | X | X | | X | | | | | |
| Canada anemone (*Anemone canadensis* L.) | | X | | | | | | | | |
| Canada goldenrod (*Solidago canadensis* L.) | X | X | X | X | X | | | | | |
| Columbine (*Aquilegia canadensis* L.) | | X | | | X | | | | | |
| Common buckthorn (*Rhamnus cathartica* L.) | | | | | | | | | | |
| Common dandelion (*Taraxacum officinale* Ledeb.) | X | X | X | X | X | | X | | | |
| Common ragweed (*Ambrosia artemisiifolia* L.) | X | X | X | X | X | | X | | | |
| Common evening primrose (*Oenothera biennis* L.) | | X | X | | X | | X | | | |
| Common milkweed (*Asclepias syriaca* L.) | | X | X | | X | | | | | |
| Common mullein (*Verbascum thapsus* L.) | | | | | | | | | | |
| Common raspberry (*Rubus arcticus* L.) | | X | X | X | | | X | | | |
| Common yarrow (*Achillea millefolium* L.) | | X | X | | | | | | | |
| Crawe's sedge (*Carex crawei* Dewey) | | X | | | | | | | | |
| Daisy fleabane (*Erigeron annuus* L.) | | X | X | | | | | | | |
| Eastern poison ivy (*Toxicodendron radicans* L.) | | | | X | | X | X | X | X | X |
| Ebony sedge (*Carex eburnea* Boott) | | X | X | | X | | | | | |
| Glossy buckthorn (*Rhamnus frangula* Mill.) | | | | | | | | | | |
| Hairy goldenrod (*Solidago hispida* Muhl. ex Willd.) | | X | | | | | | | | |
| Heal-all (*Prunella vulgaris* L.) | | | X | | X | | | | | |
| Houghton's goldenrod (*Oligoneuron houghtonii* Torr. & A. Gray ex A. Gray) | | | | | | | | | | |
| Lesser burdock (*Arctium minus* Bernh.) | | X | X | X | | | | | | |
| Little bluestem (*Schizachyrium scoparium* Michx.) | X | X | X | | X | | | | | |
| Little green sedge (*Carex viridula* Michx.) | | | | | | | | | | |
| Multiflora rose (*Rosa multiflora* Thunb.) | X | | | X | | X | | | | |
| New England aster (*Symphyotrichum novae-angliae* L.) | X | X | X | X | X | | | | | |

| Plants found in field settings at WUT with 100 m of HEA | Plants within 5 m of HEA | | | | | Plants within 1 m of HEA | | | | |
|---|---|---|---|---|---|---|---|---|---|---|
| | H | I | J | K | L | H | I | J | K | L |
| Ohio goldenrod (*Oligoneuron ohioense* Frank ex Riddell) | | X | | | | | | | | |
| Orange daylily (*Hemerocallis fulva* L.) | | X | | X | | | | | | |
| Poverty oatgrass (*Danthonia spicata* L.) | X | X | X | | | | | | | |
| Prairie smoke (*Geum triflorum* Pursh) | | X | X | | | | | | | |
| Red osier dogwood (*Cornus sericea* L.) | X | | | | | X | | | X | |
| Riverbank grape (*Vitis riparia* Michx.) | X | X | X | X | X | X | | | | |
| Shrubby cinquefoil (*Dasiphora fruticosa* L.) | X | X | X | | | | | | | |
| Smooth blue aster (*Symphyotrichum laeve* L.) | | | | | | | | | | |
| St. John's wort (*Hypericum perforatum* L.) | | X | X | | | | | | | |
| Staghorn sumac (*Rhus typhina* L.) | | | | | | | | | | |
| Stinging nettle (*Urtica dioica* L.) | | X | X | | | | | | | |
| Sweet white clover (*Melilotus officinalis* L.) | X | | | | | | | | | |
| Tall buttercup (*Ranunculus acris* L.) | | X | X | | X | | | | | |
| Timothy (*Phleum pratense* L.) | X | X | X | X | X | | X | | | |
| Virginia strawberry (*Fragaria virginiana* Duchesne) | X | X | X | X | X | X | X | | | |
| Upland white goldenrod (*Oligoneuron album* Nutt.) | | | | | | | | | | |
| White clover (*Trifolium repens* L.) | X | X | X | X | X | | | | | |
| White snakeroot (*Ageratina altissima* L.) | | | | | | | | | | |
| Wild bergamot (*Monarda fistulosa* L.) | | X | X | | | | | | | |
| Wild carrot (*Daucus carota* L.) | | X | X | X | | | X | | | |
| Woolly panic grass (*Dichanthelium acuminatum* Sw.) | | | | | | | | | | |
| Yellow daylily (*Hemerocallis lilioasphodelus* L.) | | X | | X | | | | | | |
| Yellow hawkweed (*Hieracium piloselloides* Vill.) | X | X | X | | X | | | | | |

*These settings included site H with a pipe actively draining oil brine, site I with capped pipe, site J with pipe with no cap and no brine, site K with pipe in roadside ditch, and site L as patch of EPI in field representing a groundwater seep.*

**Table 4.**
*Representation of common plant species found at different distances from HEAs in field settings [25].*

with HEAs at WUT were identified. This included forest dominated by balsam poplar reflecting past forest clearing efforts during the hydrocarbon extraction period of the late 1800s. These forest areas now represent Fresh-Moist Poplar Deciduous Forest (FOD8-1) with balsam poplar as the dominant species. Other tree species include conifers such as eastern white cedar, balsam fir, white spruce (*Picea glauca* Moench), black spruce (*P. mariana* Mill.), and red spruce (*P. rubens* Sarg.). Deciduous species include American basswood (*Tilia americana* L.), American elm (*Ulmus Americana* L.), black cherry (*Prunus serotina* Ehrh.), red maple, red oak, staghorn sumac (*Rhus typhina* L.), sugar maple (*A. saccharum* Marshall), tamarack (*Larix laricina* Du Roi), white birch (*Betula papyrifera* Marshall), and white oak (*Q. alba* L.); ash species included black ash predominantly in wetlands along with green ash and white ash in well-drained areas. Nearly all ash species have died recently due to infestation by EAB. Most American elms were also dead at WUT, due to past exposure to Dutch elm disease. In contrast, bur oak (*Q. macrocarpa* Michx.) is a tree only periodically found in these forests, as it has been previously reported to prefer soil conditions on limestone alvars along areas with recent fire history on Manitoulin Island [25, 29].

Using ELC [37], the shoreline forest areas of Smith's Bay were identified as Fresh-Moist White Cedar-Balsam Fir Coniferous Forest (FOC4-3). That is, the dominant tree species are eastern white cedar and balsam fir along with smaller coverage of species such as balsam poplar, white birch, red maple, and yellow birch; the herbaceous ground cover is depauperate and sparse, likely due to the closed canopy. Oral history revealed these shoreline areas did not burn during the 1865 fire and were dominated at the time by eastern hemlock and yellow birch in poorly drained areas, while eastern white pine (*Pinus strobus* L.) and red pine (*Pinus resi-nosa* Aiton) dominated the well-drained soils. Due to forestry, eastern hemlock, red pine, and white pine are now essentially absent from the shoreline of Smith's Bay with HEAs. Small areas of remnant eastern hemlock pine forests still occur at WUT but were not targeted for HEAs, as they are in remote areas with very limited access.

The ELC [37] interpretation identified the hay fields as cultural meadow (CUM) due to the long history of regular cutting. Although these fields have been cut on a regular basis for more than 100 years and show very few trees, the plant community includes a diverse array of other herbaceous species likely from seed dispersal, such as prairie smoke (*Geum triflorum* Pursh) along with typical grasses such as timothy (*Phleum pratense* L.).

Each area around the former oil well pipes and wood cribs in forest and field settings show depressions in the soil that vary from 5 to 30 cm in depth. This observed pattern of soil subsidence suggests compaction during HEA installation and past hydrocarbon extraction activities. Recent abandonment activities revealed that these areas demonstrate compacted soil to depths ~ 5 m below grade [21]. Most metal pipes reported in this study have been abandoned and had well depths >110 m, while the wood cribs were variable in construction, and some abandoned with well depths >110 m [21].

At forest site A, a rotted wooden crib (~1.2 × ~1.2 m) was found and inferred to be an HEA with no well in the structure. In this area, the dominant plants (>75% coverage) were ostrich fern (*Matteuccia struthiopteris* L.) along with a few other herbaceous species with the overstory dominated by balsam poplar; other trees in close proximity included balsam fir, red maple, and red osier dogwood. The ostrich fern was replaced by nearly complete cover (>95%) of EPI within 3–4 m of the crib (**Figure 2**). It is noteworthy to identify that specimens of EPI showed mostly bright green leaves and tall plants within 1 m of the crib, while the few specimens of ostrich fern that remained showed leaves with chlorosis (**Figure 3**). In addition, the ostrich ferns closest to the crib were ~¼–½ of the height of the specimens a few meters away from the crib; some EPI adjacent to the crib also showed chlorosis and stunting (**Figure 3**). This reduction of height of ostrich fern was attributed to the HEA. At this site, the balsam poplar was absent within 2–3 m of the crib. The balsam poplar in proximity to the crib demonstrated dead branches on the side facing the crib (**Figure 4**). For this crib, the soil was completely devoid of plants and covered by detritus and leaf matter (**Figure 5**). This pattern of herbaceous and woody plant dieback with robust EPI around HEAs was consistently evident across all forest sites.

Forest site B was in proximity to a metal pipe that contained water with bubbling natural gas. The dominant herbaceous plant in the area was marsh horsetail (*Equisetum palustre*) with the overstory dominated by balsam poplar; other trees were balsam fir, red maple, and red osier dogwood. At site B, the marsh horsetail was essentially absent within about 2–3 m of the pipe with EPI as the dominant (>90%) plant in this area. The balsam poplar was also absent within 2–3 m of the

**Figure 2.**
*Views from August 2, 2016, show site A in the forest dominated by ostrich fern as groundcover that transitions to EPI in close proximity to a rotted wood crib from the 1800s. The view in the figure shows how EPI within 5 m of the wood crib was mostly healthy, except for some small EPI specimen with partial chlorosis represented by yellow leaves (yellow arrow). However, the ostrich fern demonstrates white and green fronds, also indicative of partial chlorosis (white arrow). Also, some ostrich fern shows stunted size and chlorosis (blue arrow).*

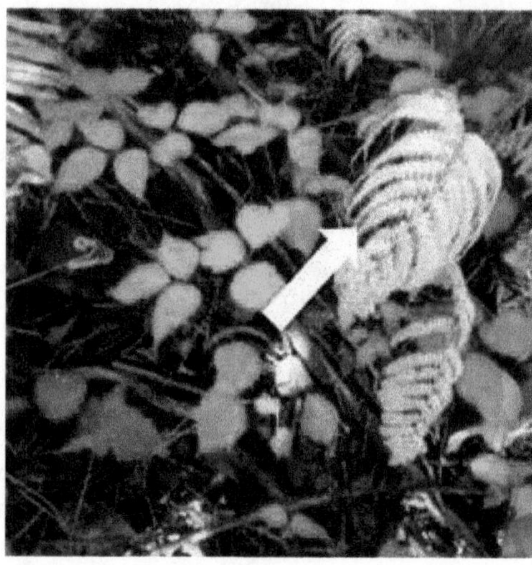

**Figure 3.**
*Views from August 2, 2016, show site A in the forest dominated by ostrich fern as groundcover that transitions to EPI in close proximity to a rotted wood crib from the 1800s. The view in Figure 2 shows how EPI within 5 m of the wood crib were mostly healthy, except some small EPI specimen with partial chlorosis represented by yellow leaves. However, the ostrich fern demonstrates white and green fronds also indicative of partial chlorosis (white arrow). Also, some ostrich fern shows stunted size and chlorosis. Figure shows the area within 1.5 m of the crib where the density of all plants declines with total chlorosis evident on ostrich fern (white arrow).*

pipe, and balsam poplar in proximity to the pipe demonstrated dead branches on the side of the tree facing the pipe, in a similar manner as observed at site A. At 1 m from the pipe, coverage was about 95% EPI. At 30 cm from the pipe, the soil was bare and devoid of plants.

Forest site C also demonstrated extensive marsh horsetail and ostrich fern in an area that receives water from a pipe that was initially observed to be dry. At this site, a channel extends from the pipe, and EPI is the dominant plant (>90%) along the edge of the channel, for a distance of at least 50 m. That is, the soil of the channel

**Figure 4.**
*Views from August 2, 2016, show the groundcover and tree overstory at the rotted wood crib at site A. Figure shows a complete absence of live plants within the former wood crib.*

**Figure 5.**
*Views from August 2, 2016, show the groundcover and tree overstory at the rotted wood crib at site A. Figure shows dieback of the balsam poplar branches within canopy (yellow arrows) over the former wood crib. Dieback in these overstory branches indicates root death.*

demonstrates bare earth along with patches of EPI and a few specimens of other herbaceous species. The overstory trees in this area are codominant balsam poplar and balsam fir; other trees in the area included red maple, red oak, and eastern white cedar away from the channel. Most balsam poplars along the channel were dead; live balsam poplar was set back from the channel and shows dead branches on the side of the channel. Also the balsam fir close to the channel is short, with a height of $<2$ m, while balsam fir upslope was up to ~15 m tall.

At forest site C on September 9, 2017, the channel was observed to contain water that appears to have originated from the pipe. This pipe did not generate water until after rain storms from the previous 2 weeks. On this date, red maple saplings growing close to the channel had green leaves and yellow leaves on the same plant. Other red maple found close to the channel was shorter than specimens away from

**Figure 6.**
*View of site C with iron-stained water that originates from a pipe on September 9, 2017. Figure 6 shows an area that is in close proximity to the pipe, with chlorosis evident in red maple sapling (yellow arrow) and EPI (white arrow).*

**Figure 7.**
*View of site C with iron-stained water that originates from a pipe on September 9, 2017. Figure shows an area in very close proximity to the pipe, with very low density of plants with those evident showing severely stunted size.*

the channel (**Figure 6**). For the EPI in this area, it showed a gradient of responses to exposure to channel, based on distance to the water. Eastern poison ivy specimens that were in direct contact with the water had all three leaves showing red color, while other specimens on the channel above the water line had leaves that were red on the water side and yellow-green leaves upslope of the water. The EPI away from the channel demonstrated less yellow and red color in the leaves with green color evident. In contrast, specimens of EPI found in a short distance upslope had bright green leaves with no discoloration (**Figure 6**). The area closest to the pipe had a few very small plants evident (**Figure 7**). It is inferred the direct contact with water is triggering the rapid onset of chlorosis in leaves. Stunted plant height is attributed to exposure of soil to brine in the past. These phenological responses to brine exposure suggest a mechanism to explain the paucity of plant species, reduced height, and low areal coverage.

Observations at forest site D revealed smooth Solomon's seal as the most common species in close proximity to a pipe that contained no water for a period of 6 months prior to abandonment [21]. A second common plant in the area was nonnative dandelion (*Taraxacum officinale* Ledeb.), likely due to the close proximity to an adjacent abandoned field with extensive dandelion. The overstory trees in this area were codominant balsam poplar and black ash (all dead) with some balsam fir, red maple, red oak, and white spruce. In contrast, the balsam poplar were extensive in the area but absent within 2–3 m of the pipe. The balsam poplar in proximity to the pipe demonstrated dead branches on the side facing the pipe, as observed elsewhere. At 2.0 m from the pipe, coverage was >95% EPI with bare earth and essentially zero plants within 0.5 m of the pipe.

Site E was within the shoreline forest along Smith's Bay, Lake Huron, and contained a wooden crib that measured 1.2 m by 1.2 m and contained water (**Figure 8**). The surrounding forest was dominated by eastern white cedar, while balsam poplar and balsam fir were the most common trees in close proximity to the crib. Following the removal of the wood crib, the area was scanned using electromagnetic induction (EMI: GSSI, Model: Profiler EMP-400). The EMI scanning involves transmission of an electrical field to create a primary magnetic field in the ground. This induced current also then generates a secondary magnetic field in the ground. Both magnetic fields are then quantified as a map of the conductivity of the earth. Soil features such as reinforced concrete impair the performance of the scans, whereas it is well suited for raw soils in the forests. The EMI scans of this site identified no pipes or other metal infrastructure at depth and groundwater close to surface [21]. Abandonment of site F was completed and revealed an oak barrel structure evident below the wood crib that extended to about 9 m depth (**Figure 9**). Below the barrel, an oil well was identified within an eastern white cedar wooden box, and the well was abandoned with cement via pipe to a depth of 130 m ([21]; **Figure 9**).

The herbaceous plant community in proximity to site E was depauperate and sparse within 5 m of the wood crib, likely attributable to the low light levels from the closed forest canopy and heavy leaf litter. Some common lady fern (*Athyrium filix-femina* L.) were evident within 3–5 m of the crib along with EPI. Very few plants were evident in close proximity to the crib, except for some EPI. Balsam pop-lar growing over the wood crib had dead branches on the side of the crib and live branches on all other sides. The balsam fir that was downslope of the wood crib was short with a maximum height of <2 m, while other balsam fir upslope and adjacent areas had heights of >15 m. No eastern white cedar was evident directly downslope of the wood crib in the drainage path but was upslope.

**Figure 8.**
*Views of site E with wood crib within the shoreline forest of Smith's Bay. Figure is on November 10, 2015, at the time it was found, after it was burned during 1905.*

**Figure 9.**
*Views of site E with wood crib within the shoreline forest of Smith's Bay. Figure is from November 1, 2016, and shows the excavation in progress with the discovery of a barrel at ~5 m below grade but still 4 m above the well box [21].*

**Figure 10.**
*View of site F near the shoreline of Smith's Bay, Lake Huron, on June 29, 2016. Figure shows how the wood crib exists over a natural hydrocarbon seep, about 20 m upslope from Smith's Bay.*

Site F was approximately 100 m west of site E within the shoreline forest of Smith's Bay, Lake Huron, and was about 20 m upslope from the shoreline. Site F included a wood crib that measured 1.8 by 1.8 m and contained water (**Figure 10**). After the wood crib was removed, the area was scanned using EMI, similar to site E. This EMI scanning identified no pipes or other metal infrastructure at depth but suggested that groundwater was close to the surface [21]. During abandonment, no metal well pipe was found, but extensive volumes of groundwater were observed within 1 m below grade. The overstory trees in the area were eastern white cedar; however, no mature specimens were evident at the HEA, as it was dominated by

**Figure 11.**
*View of site F near the shoreline of Smith's Bay, Lake Huron, on June 29, 2016. Figure shows the path where oil-water seepage drains to the shoreline that lacks live vegetation, with white residue along the path.*

balsam poplar and balsam fir directly around the wood crib. Balsam poplar growing over the wood crib has dead branches on the side of the crib and live branches on all other sides. The balsam fir that was downslope of the wood crib was short with a maximum height of <2 m, while other balsam fir upslope and adjacent areas had heights of >15 m. No eastern white cedar was evident directly downslope of the wood crib but was evident upslope with a range of heights (<1 m to 15+ m). The area directly downslope of the wood crib lacked live vegetation within an area about 1 m wide (**Figure 11**).

The herbaceous plant community in proximity to site F was depauperate and sparse within 5 m of the wood crib, similar to site E (**Figure 10**). This pattern of low density and diversity of herbaceous species was also likely partially attributable to the low light levels due to the nearly closed forest canopy and heavy leaf litter. Some mapleleaf viburnum (*Viburnum acerifolium* L.) and common lady fern were also evident within 3–5 m of the crib along with abundant EPI. Very few plants were evident in close proximity to the crib, except a few EPI. There was also a channel that drained water from the crib to Smith's Bay that was 1.0 m wide (**Figure 11**). This drainage channel lacked herbaceous species, while EPI was one of the only species periodically evident along the edge. Also balsam fir trees were evident along the channel and were all short (<2 m); no eastern white cedar were evident along the channel. Specimens of balsam fir away from this channel were all tall (10–15 m). Also rocks downslope from the seep show white particulate residue in the path to the water (**Figure 12**). Also, rocks in Smith's Bay also showed this white particulate and lacked submerged aquatic vegetation (SAV; **Figure 13**). It is prudent to note that SAV was absent only along this area of the shoreline (Figure 13).

Abandonment activity at site G initially involved the excavation of the soils around the wood crib. Soils on the upslope side of the excavated pit appeared to be fine textured with little visual or olfactory evidence of hydrocarbons, while the soils on the downslope were coarse and contained free petroleum product mixed with water. A short time after the excavator removed the surficial soil, groundwater with brown hydrocarbons filled the excavation to approximately 2.0 m below grade. A vacuum truck was used to ensure this oil-water slurry did not drain to the shoreline. Follow-up excavations and dewatering determined clay was extensive at depth below the excavation with no pipe or infrastructure evident. The initial

**Figure 12.**
*Views of site F. Figure shows the seepage path from site F to the lake on April 29, 2016, with white foam arising from the wooden crib.*

**Figure 13.**
*Views of site F. Figure from October 12, 2016, shows the shoreline of Smith's Bay, Lake Huron, with a gap in the vegetation (black arrow) and a person with orange vest standing in front of the wood crib in the background. Inspections of the rocks in the water demonstrated the presence of white particulate residue from the seep and zero SAV in this area, while adjacent areas had SAV [21].*

determination was this site represented a natural hydrocarbon seep. A soil test pit survey was conducted to map the hydrocarbon distribution below grade, and chemical analyses tracked the plume 75 m upslope from the shoreline, where the survey ended. The results from this tracking exercise led to the final determination that this was a natural seep, not a lost HEA, as no evidence of infrastructure was found in the area upslope [21].

Field site H included a pipe that was actively draining an oil-water slurry downslope through a hay field to a wetland-forest complex (**Figure 14**). Since the farmer was not cutting the vegetation around the pipe, black locust (*Robinia*

**Figure 14.**
*View of site H within a hay field on August 8, 2014. Figure shows the oil-water slurry runoff from a pipe. The vegetation near the pipe demonstrated evidence of leaf chlorosis and stunted size, including red osier dogwood that was upslope and had branches over the pipe (blue arrow), black locust (black arrow), EPI (yellow arrow), and multiflora rose (red arrow).*

*pseudoacacia* L.) and red osier dogwood were evident in the area, along with a few small green ash that had not yet been attacked by the EAB. The herbaceous community upslope of the pipe was typical of the field, with dominance by timothy grass. In contrast, the herbaceous community within 1 m of the pipe was dominated (>75%) by EPI. The EPI specimens located about 1 m from the pipe showed bright green leaves, while specimens in contact with the oil-water slurry showed yellow leaves. The soil within 30 cm of the pipe lacked plants; no plants were evident within the path immediately downslope used by the oil-water slurry. Branches of red osier dogwood in close proximity to the pipe included discoloration of the leaves, suggestive of chlorosis. Black locust leaves also included discoloration. The EPI found in association with the oil-water slurry also demonstrated small size and discoloration of the leaves compared with specimens 1 m away from the oil-water slurry. A few specimens of Virginia strawberry (*Fragaria virginiana* Duchesne) and multiflora rose (*Rosa multiflora* Thunb.) were also evident on the edge of the oil-water slurry and demonstrate discoloration of leaves as well as stunted size. The EPI dominates the vegetation along the path followed by the oil-water slurry downslope, whereas the oil-water slurry is not evident on surface at 15 m from the pipe. Even though the oil-water slurry is not evident at the surface, the EPI follows a path to the edge of the wetland forest, approximately 40 m downslope, and was rare on the edges of the path. When the trees in the wetland forest were inspected, it revealed that those trees in the path of the oil-water slurry were shorter and show a high frequency of dead branches compared with the areas on either side despite no other differences in land use or drainage (**Figure 15**). Inspection of the hay field downslope of the pipe identified EPI was the dominant plant (>50% areal coverage) right through the field to the edge of the woodland. Such disparities in height and survival of varied plants downslope of the pipe imply the oil-water slurry is the causative factor responsible for the local plant responses to exposure to the slurry.

Field site I included a pipe that was capped because it was observed to release large quantities of natural gas in the past by the landowner. The herbaceous plant community within 5 m of the pipe was diverse and essentially identical to the adjacent hay field, including Canada goldenrod, common evening primrose

**Figure 15.**
*View of site H within a hay field on August 8, 2014. Figure shows the forest-wetland downslope of the pipe. The trees in the area were shorter (stunted) and also showed dead stems relative to adjacent areas, and EPI was evident downslope but absent from adjacent field areas.*

(*Oenothera biennis* L.), tall buttercup (*Ranunculus acris* L.), and yellow hawkweed (*Hieracium piloselloides* Vill.). However, within 1 m of the pipe, EPI was dominant (>90%). At 30 cm from the pipe, the soil was bare. This pattern of reduced plant diversity and elevated EPI implies that groundwater likely rises periodically around the base of the capped pipe.

Field site J was essentially identical to site I except the pipe lacked a cap. This pipe was observed to contain no water during the 6 months prior to abandonment. The herbaceous plant community within 5 m of the pipe was somewhat similar to the adjacent hay field. However, within 2–3 m of the pipe, EPI was dominant (>90%). At 30 cm from the pipe, the soil was essentially devoid of plants.

Field site K was within a roadside ditch, and the pipe was within 10 m of a residential driveway. This pipe contained no water, but the resident stated it would release water after spring snow melt. This ditch had an array of species evident within 5 m of the pipe, including staghorn sumac; American elm along with specimens of common raspberry (*Rubus idaeus* L.), riverbank grape, Virginia strawberry, and multiflora rose. The plant community was diverse within 1 m of the pipe. At 0.5 m from the pipe, EPI was dominant (>90%). There was bare earth below the EPI vines growing near the pipe.

Field site L was within a hay field. The site is defined by an abrupt ecotone between the field plant community with transition to dominance by EPI (>95%) with a few Virginia strawberry. This patch of EPI measured approximately 3 × 3 m. This patch demonstrated healthy EPI with no bare earth. Detailed inspections were completed within this patch. Initially, the area was inspected using a hand-held metal detector, to assess possible presence of a metal pipe below grade. This inspection led to no observations, suggesting no presence of metal. Then the area was scanned using EMI, similar to the studies at sites E and F. The EMI scans of this site identified no pipes or other infrastructure at depth but suggested groundwater was close to surface [21]. After the EMI scans were completed, the patch of EPI was excavated, to further inspect the area for oil-stained soil to a depth of 1 m, to complete the search for a lost HEA, with none found [21]. After, the excavation filled with water, and the EPI patch was attributed to a groundwater seep.

Information on the herbaceous and woody plant communities along with the distri-bution of EPI observed within forests and fields at sites A to L at WUT allowed for the resolution of the spatial responses of plants to the presence of HEAs, natural hydro-carbon seep, and groundwater seep (**Table 5**). Species that show growth morphology via rhizome or vine seem to show an ability to occur within 5 m of HEAs, such as EPI, riverbank grape, and Virginia strawberry. Since these patterns were resolved across

| **Response of plants and EPI relative to distance from HEA or natural groundwater seep (as approximate % EPI coverage in an area)** | | | | |
|---|---|---|---|---|
| **Setting** | **Upslope** | **5–10 m from HEA or seep** | **1–5 m from HEA or seep** | **0 m from HEA or seep** |
| Field: HEA pipe with cap and herbaceous plants | No response with diverse plant community (<5%) | Diverse plant community (< 5%) | A few plant species evident, some EPI with most species absent (>90%) | No plants |
| Field: HEA and EPI | EPI absent or rare (<5%) | EPI absent or rare (<5%) | EPI dominant (>90%) | No EPI; no plants |
| Field: groundwater seep and EPI | EPI absent or rare (<5%) | EPI absent or rare (<5%) | EPI dominant (>95%) | EPI dominant (>95%) |
| Forest: HEA and herbaceous plants | No response with diverse community (<5%) | Loss of diversity, shorter plants within drainage path (25–95%) | A few plant specimens with EPI dominant (>95%) | No plants |
| Forest: hydrocarbon seep and herbaceous plants | No response with diverse community (<5%) | No plants in drainage path of seepage to Lake Huron | No plants in path of seepage; EPI evident on edge of drainage | No plants |
| Forest: HEA and trees | No response with diverse community (<5%) | Sensitive trees absent; tree branches on side of HEA dead; other trees stunted in height | Trees dead; other trees stunted with dead branches near HEAs with dominant EPI (>90%) | No trees |
| Forest: hydrocarbon seep and trees | No response with diverse community (>5%) | Sensitive trees absent from path of groundwater. Tree branches dead along drainage path. Other trees stunted in height | Sensitive trees absent from path of groundwater. Tree branches dead along path. Other trees stunted in height | No trees |
| Forest: HEA and EPI | EPI absent or rare (<5%) | EPI dominant, usually >50% of specimens | EPI dominant (>95%) | No EPI and no plants |
| Forest: hydrocarbon seep and EPI | EPI absent or rare (<5%) | EPI dominant (>95%) along edge of path of groundwater; no EPI in path | EPI dominant (>95%) along edge of path of groundwater; no EPI in path | No EPI |

*Findings from this study demonstrate the plant communities respond to the HEA and seeps based on water drainage, attributable to brine that contains elevated concentrations of a suite of elements.*

**Table 5.**
*Summary of response patterns for vegetation found in association with hydrocarbon extraction area, groundwater seep in a field, and hydrocarbon seep in a forest setting, represented as approximate % coverage of EPI in an area.*

forest and field sites, it provides justification for the use of EPI as an indicator species of disturbance. This resolution of general response patterns of plants to this type of disturbance suggests the response patterns could also be used to predict the sites of HEAs, hydrocarbon seeps, or groundwater seeps in areas with oil shale formations.

## 6. Discussion

Wiikwemkoong Unceded Territory history identifies the community-associated HEAs with disturbed land, and this led to the expulsion of oil men during 1905 [21, 31]. Such identification of disturbance was attributed to the construction of infrastructure like roads as well as clearing of forest in addition to using valuable farmland for HEAs. Community members also described the presence of large barrels used to separate oil from water in fields and forests [21]. The oil men would direct the oil-water slurry through pipes to these barrels. When the decision was made to evict the oil men, the HEAs were burned, and this common disturbance history contributed to the responses of herbaceous and woody plants described in this study. Oral history also identifies that these burned areas were considered scars on Mother Earth that were very slow to recover. However, this recovery of the HEAs was inferred to be slower than expected and resulted in different plants at these sites compared with adjacent areas [21]. Since 109 years passed from the time the HEAs were burned until the initial assessments during 2014, this represents a unique opportunity for learning, to understand how the herbaceous and woody plants responded to this common disturbance history [17]. This understanding arises from unreplicated activity associated with large-scale disturbance of fields and forests at WUT due to the development of HEAs. Such activities represent an opportunity for learning that is consistent to Carpenter's [17] recommendation to consider ecological settings to quantify interactions involving species and document responses of plants and animals to large-scale environmental perturbations. Carpenter [17] recommended that such perturbations, if studied, often provide the chance to document nonrandom change; a reasonable way to interpret such change is with causal inference. In an earlier study [18], a similar idea was expressed about severe perturbations representing a chance to document response of plants and animals but provided the caveat that such events possibly generate unique responses that are difficult to quantify and apply to other settings. Using this documentation of the response of herbaceous and woody plants near HEAs, it pointed to the use of EPI as a bioindicator of the plant community responses to brine as a way to find HEAs. At WUT, it is useful to use EPI as a bioindicator, as it is the only species consistently found within 1 m of HEAs. Other species, such as Virginia strawberry, also show distributions near HEAs, likely attributable to growth morphology via vine. Due to the wide distribution and salinity tolerance of EPI, it is probable this species could be evaluated for use to resolve the response of plants to HEAs beyond WUT.

This study reports the observed responses of herbaceous and woody species to episodic exposure to brine in a setting with a common disturbance history. These observations at WUT identified a pattern of reduced diversity of herbaceous and woody species as well as increased coverage of EPI in proximity to drainage from HEAs and natural seeps in fields, and FOD8-1 and FOC4-3 woodlands. These plant communities demonstrate a spatial response to water that originates from HEAs and natural seeps with rapid transition from diverse plant community to dominance by EPI over short distances despite similar exposure to sunlight and a common disturbance history. The identification of this response pattern involving a reduction of plant diversity and dieback of overstory tree branches was directly attributable to brine found near HEAs and seepage in fields and forests, representing an ecotone that defines the zone of disturbance. Observations from site C during September 2017 demonstrated that chlorosis will start within a few days of the arrival of brine; at this site, the brine is able to travel an extended distance from the source, due to the topography, causing stress and death to plants all along the drainage path. Site G demonstrated the response of herbaceous and woody species that reflects natural seepage of hydrocarbons and brine with similar patterns of species selection leading to nearly full coverage of the area by EPI. In contrast, the capped well at site I showed evidence of very little disturbance at 5 m and the focal presence of EPI directly around the pipe. Species such as Virginia strawberry and multiflora rose that grow with vines appear alongside EPI. Long-term responses include herbaceous plant communities essentially absent in close proximity to HEAs and natural seeps as well as an overstory tree canopy dominated by species somewhat tolerant to salt with dead branches near the source of brine.

Other observations from HEAs identified additional plant associations within these areas. Specifically, all HEAs include a few very mature common buckthorn (*Rhamnus cathartica* L.) and/or glossy buckthorn (*Rhamnus frangula* Mill.). It is inferred the seeds of these nonindigenous buckthorn were accidentally introduced to WUT during the 1800s via transport on equipment used for HEAs [21]. Inspections always reveal that buckthorn has spread extensively outward, in a radial pattern, from each of the HEAs. Hence, when buckthorn is found at WUT, surveys will follow these nonindigenous trees, to see if they lead to large specimens in prox-imity to an HEA. Another consistent observation is that one apple tree (often *Malus domestica* Borkh.) was always planted upslope from each of the HEAs. Further, in a field with numerous HEAs represented as metal pipes, one apple tree is located upslope and within 10 m of each pipe [21]. Oral history from WUT revealed the oil men regarded an apple tree as good luck and a source of food during the autumn season. At WUT, the presence of a large (>30 cm DBH) apple tree in close associa-tion with buckthorn is now used as preliminary indicators of possible presence of HEAs in an area.

Studies at WUT [21] documented soil at HEAs containing elevated hydrocarbon concentrations, especially for soil in contact with wood from the former facilities, whereas soil concentrations of hydrocarbons rapidly declined with distance from HEAs, attributed to bacterial degradation of the hydrocarbons over time [21]. These low concentrations of soil hydrocarbons downslope from HEAs provide additional evidence that it is brine from the HEAs that plants are responding to in these areas. The response pattern of the plants, including EPI, to brine at WUT forms the framework, as represented in **Table 5**, as a guide to find lost HEAs as well as understand influence of brine from natural seeps on plant communities.

A literature review documented a range of responses of herbaceous and woody species to brine from HEAs, but none was found for water arising directly from oil shale formations. The study in Oklahoma, USA, for plants downslope of HEAs attributed responses to the brine from the HEAs [6]. A separate study reported that herbaceous and woody plants recovered initially from a brine leak at an impoundment within 1 year after the leak was stopped with no soil treatments; plant community recovery was attributed to frequent rain washing the brine away [15]. At the site of this brine leak [16], the plant community 4 years later was populated by a diversity of herbaceous and woody species, attributed to the seed bank and live adjacent vegetation [15, 16]. Another study reported treatments for bare soils within 1 m deep former reserve pits used to store drill cuttings, drilling fluid, and brine, as preparation for replanting [12]. This bare earth was disked, mulch added, irrigated, and then seeded with local range grasses as well as planted seedlings. Success of seed germination and seedling survival was higher on mulch-treated soils than untreated soils. Over time, the mulch-treated soils showed declines in the seeded and planted species due to colonization by other local species. This mulch treatment led to the identification that establishment of salt-tolerant plants in the reserve pits following treatment can act as the basis for future plant establishment and reclamation; without treatment, seeding, and/or planting of seedlings, the pits require extended periods before plants naturally establish on these soils. Another study recommended the use of local, native plant species for seeding on land disturbed by hydrocarbon extraction activities, with emphasis on species adapted to the soils associated with the area that often show elevated tolerance to soil salinity [38]. Such seeding recommendations included the use of local sources, due to likely adaptations to regional climate and soils, as the process of reestablishing vegeta-tion on sites disturbed by HEAs is challenging due to the chemistry as well as other physical factors, such as soil compaction [3, 6, 32]. As a means to address challenges for seeding and planting on high-salinity soils, computer software was recently presented to guide such activities [39].

Other areas of study are warranted, to resolve the morphological and physiological response of herbaceous and woody species associated with HEAs and natural seeps, based on observations at WUT. For example, anecdotal observations indicated a trend that herbaceous species near HEAs often had fewer and smaller flowers compared with specimens in adjacent areas. Other plants showed differences in height and leaf size in proximity to HEAs. This study did not provide for the opportunity to quantify the range of possible responses of plants to exposure to brine arising from HEAs. If the source of stress on these plants was not known, the differences in flowering, plant height, leaf size, etc. could be documented as a phenological response to regional factors, not local responses to brine exposure. Such opportunities for documentation of response of plants, including separation of cause between stresses in proximity to HEAs, may be useful in the future. The HEAs may also be resulting in genotypic changes to plants, due to long-term episodic exposure to brine. Plants like prairie smoke (*Geum triflorum* Pursh) have been observed periodically at WUT in close proximity to HEAs, whereas the USDA iden-tifies this plant as having no tolerance to soil salinity. It may be interesting to resolve if prairie smoke has developed tolerance to elevated soil salinity at WUT. Responses of aquatic plants to brine exposure represent another understudied topic. At site F, the brine-oil slurry drained downslope to the shoreline of Smith's Bay. In the area of the shoreline, white residue was evident on the rocks on the shoreline and in the water. This area also lacked SAV, while adjacent areas demonstrated SAV among the rocks. These varied themes represent candidate areas for further study, to resolve the range of apparent responses of plants relative to episodic environmental stress from brine arising from HEAs and separate them from phenological responses to regional processes like climate.

This study resolved the response of plant communities to a common disturbance history as well as the episodic release of brine from HEAs and seeps within about 30 $km^2$. The resolution of the response of herbaceous and woody species and EPI to HEAs and natural seeps represents a framework that could be applied elsewhere, to assess the intensity of disturbance as well as to find lost HEAs. The ecotone of disturbance at WUT for HEAs is defined by the brine drainage area within a local area of subsidence where EPI is able to dominate the herbaceous plant community. It is probable EPI is able to achieve dominance in these areas due to growth via rhizome and can ameliorate the extreme chemical concentrations in brine. Another consideration at WUT is the native plant community contains very few species tolerant of elevated soil salinity [25] and led to the opportunity for EPI to dominate areas with HEAs. This study has confirmed the distribution of EPI is useful to represent contemporaneous disturbances to herbaceous and woody species attributable to HEAs and natural seeps in forests and fields and applicable to other areas with HEAs.

## 7. Conclusions

Large-scale disturbance of plant communities at WUT occurred with the 1865 forest fires followed by development of HEAs in forests and fields. These disturbed plant communities at the HEAs were burned in 1905 and were followed by establishment of pioneer plant species such as balsam poplar and eastern white cedar in forests and Canada goldenrod and Virginia strawberry in fields. Areas in close proximity to the HEAs have not developed diverse plant communities during the last 109 years but are dominated by EPI, and this pattern is attributed to the periodic expulsion of brine from HEAs. Such dominance by EPI in these areas demonstrates intolerance of brine by most plants, whereas the EPI flourishes, due in part to some tolerance of salt as well as rhizome growth strategy. With this demonstration of EPI as the dominant species near HEAs, it is a phenological response to local habitats that can be used to rediscover lost HEAs. Studies of natural groundwater seeps located near HEAs were dominated by EPI, and confirmed plant community responses are attributable to exposure to brine. This confirmation that EPI represents a bioindicator for understanding environmental disturbance through the analyses of phenological responses at WUT can likely be applied to other areas with EPI and other salt-tolerant plant species.

## Acknowledgements

Thanks to Jean Pitawanakwat, who provided assistance on many aspects of this study. Dr. Laima Kott, University of Guelph, provided confirmation on the identification of some plants, including the ostrich fern. Ann Rocchi also assisted with some plant identifications. Wiikwemkoong Unceded Territory tribal council provided administrative and financial support for this study. Special thanks to the unpaid members of WUT whom have contributed their time to help find lost HEAs as well as share other resources.

## Abbreviations

| | |
|---|---|
| CUM | cultural meadow |
| EPI | eastern poison ivy |
| FOD8-1 | Fresh-Moist Poplar Deciduous Forest |
| FOC4-3 | Fresh-Moist White Cedar–Balsam Fir Coniferous Forest |
| HEAs | hydrocarbon extraction areas |
| Premier | Premier Environmental Services |
| TDS | total dissolved solids |
| TK | traditional knowledge |
| USDA | United States Department of Agriculture |
| WUT | Wiikwemkoong Unceded Territory |

## Author details

Dean G. Fitzgerald[1*], David R. Wade[1] and Patrick Fox[2]

1 Premier Environmental Services Inc., Cambridge, Canada

2 Wiikwemkoong Unceded Territory, Department of Lands and Resources,
Mide kaaning, Wikwemikong, Canada

*Address all correspondence to: dean@elminc.ca

## References

[1] Larcher W. Physiological Plant Ecology: Ecophysiology and Stress Physiology of Functional Groups. New York: Springer Science; 2003

[2] Munns R. Comparative physiology of salt and water stress. Plant, Cell & Environment. 2002;**25**(2):239-250. DOI: 10.1046/j.0016-8025.2001.00808.x

[3] Blum A. Plant Breeding for Stress Environments. Boca Raton: CRC Press; 1988

[4] Shabala S, Munns R. Salinity stress: Physiological constraints and adaptive mechanisms. In: Shabala S, editor. Plant Stress Physiology. 2nd ed. Boston: Cabi; 2017. pp. 24-64

[5] Xie Y, Wang X, Silander JA. Deciduous forest responses to temperature, precipitation, and drought imply complex climate change impacts. Proceedings of the National Academy of Sciences. 2015;**112**(44):13585-13590. DOI: 10.1073/pnas.1509991112

[6] Atkinson NJ, Jain R, Urwin PE. The response of plants to simultaneous biotic and abiotic stress. In: Mahalingam R, editor. Combined Stresses in Plants. New York: Springer International Publishing; 2015. pp. 181-201

[7] Otton JK, Zielinski RA, Smith BD, Abbott MM, Keeland BD. Environmental impacts of oil production on soil, bedrock, and vegetation at the US geological survey Osage–Skiatook petroleum environmental research site A, Osage County, Oklahoma. Environmental Geosciences. 2005;**12**(2):73-87. DOI: 10.1306/eg.09280404030

[8] Merrill SD, Lang KJ, Doll EC. Contamination of soil with oilfield brine and reclamation with calcium chloride. Soil Science. 1990;**150**(1):469-475

[9] Adams MB, Edwards PJ, Ford WM, Johnson JB, Schuler TM, Thomas-Van Gundy M, et al. Effects of Development of a Natural Gas Well and Associated Pipeline on the Natural and Scientific Resources of the Fernow Experimental Forest. Newtown Square, PA: US Department of Agriculture Forest Service, Northern Research Station. General Technical Report NRS-76; 2011

[10] Eldrige JD, Redente EF, Paschke M. The use of seedbed modifications and wood chips to accelerate restoration of well pad sites in Western Colorado, U.S.A. Restoration Ecology. 2012;**20**:524-531. DOI: 10.1111/j.1526-100X.2011.00783.x

[11] Collins CM, Racine CH, Walsh ME. The physical, chemical, and biological effects of crude oil spills after 15 years on a black spruce forest, interior Alaska. Arctic. 1994;**47**:164-175

[12] McFarland ML, Ueckert DN, Hartmann S. Revegetation of oil well reserve pits in West Texas. Journal of Range Management. 1987;**40**:122-127

[13] Dresel PE, Rose AW. Chemistry and Origin of Oil and Gas Well Brines in Western Pennsylvania. Pennsylvania Geological Survey, Open-File Report OFOG, 10(01.00). Harrisburg, PA; 2010

[14] Johnson RH, Huntley LG. Principles of Oil and Gas Production. 1st ed. New York: Wiley; 1916

[15] Walters RS, Auchmoody LR. Vegetation re-establishment on a hardwood forest site denuded by brine. Landscape and Urban Planning. 1989;**17**:127-133

[16] Auchmoody LR, Walters RS. Revegetation of brine-killed forest site. Soil Science Society of America Journal. 1988;**52**:277-280. DOI: 10.2136/sssaj1988. 03615995005200010049x

[17] Carpenter SR. Large-scale perturbations: Opportunities for innovation. Ecology. 1990;**71**:2038-2043. DOI: 10.2307/1938617

[18] Schindler DW. Detecting ecosystem responses to anthropogenic stress. Canadian Journal of Fisheries and Aquatic Sciences. 1987;**44**(S1):s6-s25. DOI: 10.1139/f87-276

[19] Keith BD, editor. The Trenton Group (Upper Ordovician Series) of Eastern North America. American Association of Petroleum Geological Studies in Geology: Deposition, Diagenesis, and Petroleum. Tulsa, Oklahoma: American Association of Petroleum; 1988. DOI: 10.1306/St29491

[20] Russell DJ, Telford PG. Revisions to the stratigraphy of the upper Orodovician Collingwood beds of Ontario–a potential oil shale. Canadian Journal of Sedimentary Petrology. 1983;**59**:114-126. DOI: 10.1139/e83-170

[21] Premier Environmental Services Inc. (Premier). Mide Kaaning Project Report Wiikwemkoong Unceded Territory. Submitted to Wiikwemkoong Unceded Territory #26, Wikwemikong, ON. 31 March 2017. 657 p

[22] Gillis WT. The systematics and ecology of poison-ivy and the poison-oaks (*Toxicodendron*, Anacardiaceae). Rhodora. 1971;**73**(793):72-159

[23] Innes RJ. *Toxicodendron radicans*, *T. rydbergii*. In: Fire Effects Information System, [Online]. U.S. Department of Agriculture, Forest Service, Rocky Mountain Research Station, Fire Sciences Laboratory (Producer). 2012. Available from: http://www.fs.fed.us/database/feis/ (Accessed: 10 November 2017)

[24] Tanner T. *Rhus* (*toxicodendron*) dermatitis. Journal of Primary Care and Community Health. 2000;**27**:493-501. DOI: 10.1016/S0095-4543(05)70209-8

[25] Morton J, Venn J. The Flora of Manitoulin Island. 3rd ed. Waterloo: Department of Biology, University of Waterloo, Ontario; 2000

[26] Kuester A, Conner JK, Culley T, Baucom RS. How weeds emerge: A taxonomic and trait-based examination using United States data. The New Phytologist. 2014;**202**(3):1055-1068. DOI: 10.1111/nph.12698

[27] Deller AS, Baldassarre GA. Effects of flooding on the forest community in a Greentree reservoir 18 years after flood cessation. Wetlands. 1998;**18**(1):90-99

[28] Noble RE, Murphy PK. Short term effects of prolonged backwater flooding on understory vegetation. Castanea. 1975;**40**(3):228-238

[29] Jones J. Fire history of the bur oak savannas of Sheguiandah township, Manitoulin Island, Ontario. The Michigan Botanist. 2000;**39**:3-15

[30] Herms DA, McCullough DG. Emerald ash borer invasion of North America: History, biology, ecology, impacts, and management. Annual Review of Entomology. 2014;**59**:13-30. DOI: 10.1146/annurev-ento-011613-162051

[31] Assiginack Historical Society (AHS). A Time to Remember: A History of the Municipality of Assiginack. Manitowaning, ON: Manitoulin Printing; 1996

[32] Chamberlain B. Canada Indian Treaties and Surrenders, from 1680 to 1890. Vol. 1. Ottawa: Government of Canada; 1891

[33] Point N. Memoirs of the Jesuit Mission at Wikwemikong, Manitoulin Island, Mid 1800's. Transcribed by Shelly Pearen. Toronto, ON: Translated and Published by William Lonc for The Canadian Institute of Jesuit Studies; 2009

[34] Newton AC. Offshore Exploration for Gas under the Canadian Waters of the Great Lakes. Ontario Department of Mines, Geological Circular. Number 7; 1964

[35] Purser BH, Tucker ME, Zenger DH, editors. Dolomites: A Volume in Honour of Dolomieu. Vol. 67. Wiley Blackwell: Hobokan; 2009

[36] Canadian Council of Ministers of the Environment (CCME). Canadian Water Quality Guidelines: Chloride Ion. Scientific Criteria Document. Winnipeg, MB: Canadian Council of Ministers of the Environment; 2011

[37] Lee, H, Bakowsky, W, Riley, J, Bowles, J, Puddister, M, Uhlig, P, et al. Ecological Land Classification for Southern Ontario: First Approximation and Its Application. Ontario Ministry of Natural Resources, South Central Science Section, Science Development and Transfer Branch. SCSS Field Guide FG-02; 1998

[38] Pessarakli M. Formation of saline and sodic soils and their reclamation. Journal of Environmental Science & Health Part A. 1991;**26**(7):1303-1320

[39] Falk AD, Pawelek KA, Smith FS, Cash V, Schnupp M. Evaluation of locally-adapted native seed sources and impacts of livestock grazing for restoration of historic oil pad sites in South Texas. Ecological Restoration. 2017;**35**:120-126. DOI: 10.3368/er.35.2.120

8

# Phenological Behaviour of Early Spring Flowering Trees

*Herminia García-Mozo*

**Abstract**

This chapter reports the phenological trends (reproductive and vegetative events) of some early spring and late winter flowering trees all around the world and especially Europe: *Corylus avellana* L. (hazel); *Quercus robur* L. (common oak); *Quercus ilex* subsp. *ballota,* (Desf.) Samp. (holm oak); *Betula* spp. (birch); *Salix alba* L. (willow); *Fraxinus angustifolia* Vahl. (ash); and *Morus alba* L. (white mulberry). They are deciduous and perennial trees growing in different climatic areas of Europe. They have anemophilous pollination liberating huge pollen concentrations to the atmosphere. Aerobiological surveys give us reproductive phenological information of these wind-pollinated species. The phenological response to climate during the last years was analysed, including budburst, leaf unfolding, flowering, fruit ripening, fruit harvesting, leaf colour change, and leaf fall. The response of each taxon to climate was different; most of the revised species and sites presented an advance of the early spring phenophases, espe-cially budburst. On the contrary, some studies detected a delay in autumn vegeta-tive phases, especially leaf fall events. The statistical analyses indicated that phenological advances are a consequence of the increasing temperature trend—minimum temperature being one of the most influential factors. The increase of temperature influenced that leaf unfolding and flowering dates showed a general advance expressed by negative correlations with temperature data, whereas the leaf colour change and leaf- fall presented positive correlations due to the delay of the colder temperatures. The phenological revised results can be considered as reliable and valuable bio-indicators of the impact of the recent climate change in the Northern Hemisphere, and especially Central and Southern Europe.

**Keywords:** phenology, anemophilous trees, climate change,
*Corylus avellana* L. (hazel), *Quercus robur* L. (common oak),
*Quercus ilex* subsp. *ballota,* (Desf.) Samp. (holm oak), *Betula* spp. (birch),
*Salix alba* L. (willow), *Fraxinus angustifolia* Vahl. (ash),
*Morus alba* L. (white mulberry)

## 1. Phenology

Phenology is derived from the Greek word phaino, meaning 'to show' or 'to appear'. This science studies the recurring biological events as part of the animal and plant life cycles. These events are the phenological stages or phenological phases. Phenology not only studies the timing but also their relationships with weather and climate [1].

Sprouting and flowering of plants in spring and leaves' colour change in the fall are examples of plant phenological events [2].

Phenology has been used as a proxy for climate and weather through all the human history, particularly in relation with agriculture, but only from the last century has emerged as a science in its own right [1]. In last years it is being recog-nized as an integrative measure of plant responses to the environment changes that can be scaled from a local to a global scale, including climate change. During the last 100 years, the Earth's climate has warmed by approximately 0.6°C. In this last century, two main periods of warming have been detected. The first one was between 1910 and 1945, and the second one from 1976 onward [3]. In this second period, the rate of warming is being doubled than in the first and greater than at any other time during the last 1000 years [3]. The response of the different ecosystems and species is not a global response to a global climate average [4]. To know the regional responses can be more relevant in the context of ecological response to climatic change. In this sense, phenological behaviour data are the more reliable actual bio-indicator of the climate change response. Moreover, sessile life-style characteristic of plants has led them to develop high plasticity phenotypes in order to reach better phenological adaptations to deal with environmental changes [5]. These changes include climate changes that are of critical ecological importance as they affect species competitive ability and net primary productivity. These changes can even prompt ecosystem structure transformations [6]. Therefore, the analysis of trends of spring phenological phases for the past decades could provide important information about changes in climate and the impact on sessile organisms' phenology such us plants and specially trees, with longer lifetimes and shorter capacity of area distribution change.

This study presents a review of recent studies on both vegetative and reproductive field phenological development of different tree species characterized by their foliation or flowering during early spring. The phenological response of different tree species in the North Hemisphere was reviewed: hazel (*Corylus avellana* L.), alder (*Alnus glutinosa* (L.) Gaertn), willow (*Salix alba* L.), birch (*Betula pendula* L.), holm oak (*Quercus ilex* subsp. *ballota,* (Desf.) Samp.) in South Europe and common oak *(Quercus robur* L.) in Central Europe, ash (*Fraxinus angustifolia* Vahl.), and white mulberry (*Morus alba* L.) [7]. All of them are anemophilous species producing high quantities of pollen grains spread to the atmosphere provoking allergy to the sensi-tized population [8]. Their huge quantities of pollen grains are also a phenological bio-indicator detected through aerobiological studies [9], also revised for the present review. Their phenological behaviour during last 40 years and the impact of the cli-mate change on it were analysed. Particularly remarkable is the fact that the revised species are important for aerobiology and allergy studies, and therefore the changes experimented on their phenology have a special interest. This review offers valuable information due to the scarce number of researches studying field phenological data including those from the last quarter of the twentieth century.

## 2. Climate change

Climate change due to human activities has been witnessed for at least the last 100 years and is projected to continue for centuries to come. Climate change involves the whole climate system, including not only our atmosphere but also our hydrosphere, cryosphere, land surface, and biosphere [10].

Greenhouse gases and atmospheric concentrations have exponentially increased since the start of the Industrial Era (1750). Moreover, from this time the $CO_2$ concentrations have increased by 41% mostly due to the global use of oil fuel [3]. The latest measures of the year 2013 of the National Oceanic and Atmospheric Administration reveal that global annual mean atmospheric $CO_2$ concentration was 395.22 parts per million (ppm) [11], an increase of over 100 ppm from the

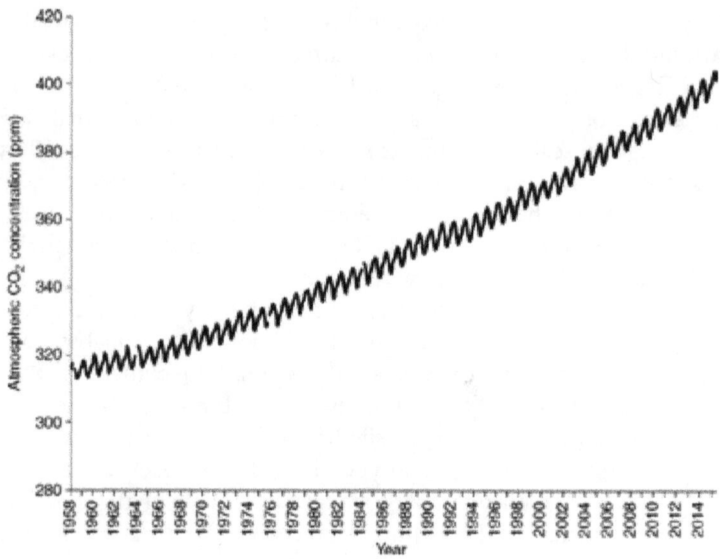


**Figure 1.**
*Monthly mean atmospheric carbon dioxide concentration from March 1958 to July 2015. Source: [11].*

pre-Industrial Era. **Figure 1** shows that the increase in atmospheric CO2 concentration since 1750 has not been linear, being higher in the last 60 years [11].

This increase in the atmospheric concentration of greenhouse gases such as $CO_2$ has led to warm the climate. Between 1880 and 2012, the Earth's average surface temperature warmed by 0.85°C [3]. Most of this warming occurred after 1951 [12]; the warming of the Earth's surface varies over space, the land surfaces tending to warm more than the oceans; and some parts of the Earth's surface temperatures have increased the double than other places. Changes in precipitation have also been observed. Since 1901, precipitation has increased over the mid-latitude land areas of the Northern Hemisphere, especially in intensity, with more frequent heavy rain episodes [12].

By the end of this century (2081–2100), global mean surface temperatures are projected to increase from 0.3°C to 4.8°C depending on the sites. The mean warming over land will be larger than over the ocean, and the Arctic region will warm more rapidly than the global mean [3]. With regard to precipitation, it is projected to increase by the end of this century [13]. However, there will be substantial spatial variation in precipitation changes, with some regions experiencing increases, some decreases, and some no change at all [13]. All these factors and the actual impact on phenology of early spring trees will be reviewed in the present chapter.

## 3. General phenological behaviour of all studied tree taxa

The revised species were selected because of their flowering time in early spring or late winter and because of their strong presence and distribution in Europe. They are anemophilous, deciduous, and perennial trees growing in different climatic areas of Europe: hazel (*Corylus avellana* L.) in Central and South Europe, oak (holm oak (*Quercus ilex* subsp. *ballota,* (Desf.) Samp.)) in Southern Europe, common oak (*Quercus robur* L.) in Central Europe, birch (*Betula* spp.) in Central and North Europe, willow (*Salix alba* L.) in Central and South Europe, ash (*Fraxinus angusti-folia* Vahl.) in South Europe, and white mulberry (*Morus alba* L.) in South Europe [14]. All of them are endemic European species expect for white mulberry [7].

Most of the revised species belong to the Fagales order, divided into the *Betulaceae*, *Salicaceae*, and *Fagaceae* families [14]. On the other hand, *Fraxinus* genus is in the *Oleaceae* family of the Lamiales order, and *Morus* in the *Moraceae* family of the Rosales order.

Fagales order comprises three families: *Betulaceae*, including the genera *Betula* (birch), *Alnus* (alder), and *Corylus* (hazel); *Fagaceae*, including the genera *Quercus* (oak) and *Fagus* (beech); and *Salicaceace* including the genus *Salix* (willow) [14]. These wind-pollinated trees have catkins, which dangle from the branch so that pollen is easily shaken loose in the wind. Interestingly, catkins in deciduous species emerge before the leaves, allowing the pollen to travel further away from the parent without the obstruction of foliage [15].

Birch is the major pollen allergen-producing tree in Northern Europe, although there are high levels of allergenic cross-reactivity between the representative plants of the genera of the order Fagales [16].

As it has been already mentioned, all the revised species are foliating or flowering in early spring in Europe and North America; nevertheless, there are some specific characteristics for each one.

Hazel and alder are the first (December–April) to blossom and to shed pollen in the outdoor air in Europe, followed by birch. This fact joint to an allergenic cross-reactivity between hazel and alder provokes that pollen from these species can act as a primer of allergic sensitization to *Betulaceae* pollen allergens. Consequently, clinical symptoms become more marked during the birch pollen season [17, 18]. In the central Alpine regions, the highest concentrations of *Alnus* pollen are found at the end of May and in early June [17].

In the case of *Betula*, the budburst occurs at March–April depending on the latitude and altitude. In South and Western Europe, the main flowering period usually starts at the end of March, whereas in Central and Eastern Europe, it occurs at early April. In northern areas the flowering season starts from late April to late May depending on the latitude [8]. Pollen values peak 1–3 weeks after the start of the season, so they are recorded in April in South Europe and in May in Northern Europe. Far shorter or longer periods, with yearly alternating low and high pollen production, have been observed in various European regions [17].

On the other hand, the onset of the oak season in spring, shortly before the beech pollen season, which is usually quite mild, can prolong the season in western, central, and eastern Europe [8]. One important characteristic of the oak pollen is the fact that it includes many species. In South Europe perennial species such as holm oak, kermes oak, and cork oak flower through all the spring from March to June [19]. In Central Europe, the pedunculate oak and the sessile oak usually flower in April–May [8].

Mulberry plants are normally dioecious, but they can also be monoecious on different branches of the same plant. The pendulous pistillate (female) and staminate (male) catkins are arranged on spikes and appear in April and May [20].

All the studied species have their main flowering season on early spring; nevertheless the different phenological phases vary among species, sites, and years depending on the bioclimatic characteristics and fluctuations [8].

## 4. Allergenicity

Birch, followed by alder and hazel, has the greatest allergenic potency in this group of allergenic trees. In Central Europe, these tree pollens are the second most common cause of allergic conditions after grass pollen. In the case of birch, the major allergen is Bet v1, and the percentage of subjects with a positivity skin prick test to birch allergens ranges from 5% in the Netherlands to 54% in Zurich (Switzerland) [17, 21].

In recent years, the popularity of *Betula* as an ornamental plant loved by architects has caused a significant increase in allergic sensitization to this allergen [22, 23]. In a large study of cross-sensitization between allergenic plants in adult patients with asthma or rhinitis, it was found that sensitization to birch pollen allergens was frequently associ-ated with other allergens, that it induced mostly nasal symptoms, and that respiratory symptoms started at about 30 years of age [14].

Pollen from the common alder, major pollen allergen Aln g 1, is an important cause of pollen allergy. This pollen has similar physicochemical properties than the pollen of birch, hazel, hornbeam, and oak. The joint presence of these pollen grains in the atmosphere makes difficult to separate out their individual effects [24].

Hazel is well distributed in Europe, and it typically has a flowering occurring from winter to early spring. The major allergen is Cor a 1, cross reactive with Bet v1 [25]. In the case of *Corylus* pollen, a recent study performed in Poland revealed that ~11% of allergy patients had positive skin reactions to *Corylus* pollen allergens, and most of these (94.4%) reacted to pollen allergens from other members of the Betulaceae family—alder or birch [26].

Beech trees are related to oaks. These trees are considered as low allergenic [27]. The European beech sheds much more pollen than the American species, but both have been reported to have minor allergenic importance. Despite the large amounts of pollen grains detected in the European atmosphere, *Quercus* pollen, which is a steno-palynous pollen type for all the genera, does not provoke actual allergy problems [8].

Although willows elicit strong allergic responses from individuals in allergy tests, willows tend to be pollinated more by insects than by wind and therefore present fewer people with the allergenic challenge than other tree types [28]. In fact, the impact of the increase in *Salix* atmospheric pollen upon asthma admissions is insignificant [29, 30].

Mulberry pollen grains cause allergenic symptoms such as rhinitis, conjunctivitis, and asthma [31, 32]. A study from Tucson, Arizona, USA, concluded that it is an important allergen for children raised in a semiarid environment [33]. In other climate areas, Mulberry tree pollen has been revealed as an important aeroallergen. This is the case of the tropical area of Caracas, Venezuela [34], the Mediterranean area [35, 36], and the Atlantic temperate climate of Argentina [37].

The most important allergenic species revised here, birch, alder, and hazel, have their main pollen emission time mostly in early spring although the exact time depends on the response of these trees to climate [8].

## 5. Effects of climate changes on phenology of all tree taxa

Different phenological studies are showing a clear link between anthropogenic climate change, warming winter and spring temperatures, and changes in phenology, especially earlier flowering times and late leaf fall in autumn [8, 10, 17, 18, 21]. This occurs in a wide variety of tree species including the early spring species that are analysed here (**Table 1**).

There is considerable variation in these studies that reflects the time examined and regional differences in temperature, etc.; however, for all tree species examined, flowering is now occurring, on average, approximately 2 weeks earlier than it did relative to the mid-twentieth-century temperature average [38–41].

Some studies have shown the impact of climate change on phenology and pollen and therefore on aeroallergens and allergic diseases describing the influence on the amount, distribution, allergenicity, and pollen season of pollen grains [8, 10, 17, 18, 21]. A global comparative study of the International Phenological Gardens in Europe (covering 69–42°N to 10°W–27°E) of current data compared and early

| Taxa | Country | Time period (a-b) | Start (a) | Start (b) | Difference | Reference |
|---|---|---|---|---|---|---|
| *Fraxinus* | The Netherlands | 1970–1990s | 92 | 88 | −3 | [38] |
| *Betula* | Belgium | 1982–2000 | 102 | 84 | −18* | [39] |
| *Betula* | Finland | 1975–2004 | 130 | 118 | −12* | [40] |
| *Betula* | The Netherlands | 1970–1990s | 106 | 94 | −10* | [38] |
| *Betula* | Switzerland | 1982–2000 | 105 | 85 | −20* | [39] |
| *Quercus* | The Netherlands | 1970–1990s | 135 | 117 | −18*** | [38] |
| *Quercus* | Spain | 1970–1990s | 89 | 78 | −11 | [41] |
| *Corylus* | The Netherlands | 1970–1990s | 84 | 66 | −18** | [38] |
| *Salix* | The Netherlands | 1970–1990s | 82 | 70 | −12* | [38] |

**Table 1.**
*Statistically significant differences in start date between the start and the end of the time are indicated with* $^{*}p < 0.05$, $^{**}p < 0.01$, *and* $^{***}p < 0.001$.

1960s phenological data indicated the advance of spring events, such as flowering (+6 days), whereas autumn phenophases have been delayed by 4.8 days [42].

Speaking about early spring species, in the case of *Corylus avellana*, an ear-lier flowering onset was observed at 80% of the studied localities of the Iberian Peninsula, earlier fruit ripening at all sampling sites, and earlier fruit harvesting at 75% of them [43]. *Salix alba* presented a trend towards earlier budburst and earlier leaf unfolding at 67% of the studied Iberian localities. In the case of autumn phases, delay in leaf fall at all sampling sites [40]. Holm oak is suffering a strong advance in the flowering start, as it was previously indicated in the Iberian Peninsula [41]. Northern species such as birch, poplar, or willow are also showing the impact of climate change on phenology [38, 39].

As it was demonstrated by [41, 43] among others, the relationship between the phenological observations and weather is so clear for tree species and especially for early spring species. The statistical analyses show that in the 55% of the stud-ied localities of the Iberian Peninsula, the temperature is influencing these trees' phenology. In 58% of the sites affected by temperature, the correlation between phenology and minimum temperatures was negative, which is provoking an advance in phenology. The mean temperature results showed negative correlation in 54% of the sites, although different behaviour was observed depending on species and phenophases.

On average, the length of the growing season in Europe increased by 10–11 days during the last 30 years. Trends in pollen amount over the latter decades of the 1900s increased according to local rises in temperature [8, 44–46]. The increased $CO_2$ concentration can be affecting pollen production as it has been demonstrated in experimental conditions [47, 48]. Regarding the pollen season length, it is also extending especially in late spring and summer flowering species [49]. Moreover, temperature is influencing towards stronger allergenicity in tree pollen [17, 50].

An earlier pollen season starts, and peak is being more pronounced in early spring flowering species [43]. Due to this earlier onset, the seasons are more often interrupted by adverse weather conditions in late winter/early spring [51].

Finally, changes in climate appear to have altered the spatial distribution of pollens. New patterns of atmospheric circulation over Europe might increase the number of long-distance transport episodes of allergenic pollen, increasing the risk of new sensitizations among the allergic population [52]. On the other hand, the temperature increases, and the changes in rainfall regime are provoking the geographical spread of some vegetal species to new areas. In the future the effect of the expected rate of warming (0.5°C per decade) could increase this geographi-cal migration although the effect on pollen distribution is expected to be less pronounced than the effect of changes on land as well as international transport of plant species [53].

## 6. Conclusions

The review made about the recent response of the phenology of different species of anemophilous trees to climate change reveals that, apart from the field phenology data, aerobiological pollen data are a valuable tool to obtain reproductive phenological information of wind-pollinated species.

The response to climate of each studied taxon was different; most of the revised species and sites presented an advance of the early spring phenophases, especially budburst. The statistical analyses of the revised studies indicate that phenological advances are a consequence of the increasing temperature trend—minimum temperature being one of the most influential factors. The increase of temperature influenced that leaf unfolding and flowering dates showed a general advance expressed by negative correlations with temperature data, whereas the leaf colour change and leaf- fall presented positive correlations due to the delay of the colder temperatures.

On the contrary, some studies detected a delay in the autumn vegetative phases, especially on leaf-fall events. Both, leaf colour change and leaf-fall events showed positive correlations with temperature due to the delay of the colder temperatures.

The phenological revised results can be considered as reliable and valuable bio-indicators of the impact of the recent climate change in the Northern Hemisphere and especially in the Central and Southern Europe.

## Acknowledgements

Author wishes to thank for their support to the projects FENOMED, REF. CGL2014-54731-R, funded by the Spanish Ministery of Economy and Competitiveness; and CLIMAQUER, REF. 1260464, European Regional Development Funds (ERDF).

## Author details

Herminia García-Mozo
Departamento de Botánica, Ecología y Fisiología Vegetal, Universidad de Córdoba, Córdoba, España

*Address all correspondence to: bv2gamoh@uco.es

## References

[1] Schwartz MD, editor. Phenology: An Integrative Environmental Science. Netherlands: Springer; 2013

[2] Meier U, Bleiholder H, Buhr L, Feller C, Hack H, Heß M, et al. The BBCH system to coding the phenological growth stages of plants–history and publications. Journal für Kulturpflanzen. 2009;**61**(2):41-52

[3] Stocker TF, Qin D, Plattner GK, Tignor M, Allen SK, Boschung J, et al. IPCC, 2013: Climate Change 2013: The Physical Science Basis. Contribution of Working Group I to the Fifth Assessment Report of the Intergovernmental Panel on Climate Change. United Kingdom: Cambridge University Press; 2013

[4] Walther GR, Post E, Convey P, Menzel A, Parmesan C, Beebee TJ, et al. Ecological responses to recent climate change. Nature. 2002;**416**(6879):389-395

[5] Schlichting CD. The evolution of phenotypic plasticity in plants. Annual Review of Ecology and Systematics. 1986;**17**(1):667-693

[6] Peñuelas J, Filella I, Comas P. Changed plant and animal life cycles from 1952 to 2000 in the Mediterranean region. Global Change Biology. 2002;**8**(6):531-544

[7] De Rigo D, Caudullo G, Houston Durrant T, San-Miguel-Ayanz J. The European atlas of Forest tree species: Modelling, data and information on forest tree species. In: San-Miguel-Ayanz J, de Rigo D, Caudullo G, Houston Durrant T, Mauri A, editors. European Atlas of Forest Tree Species. Luxembourg: Publications Office of the European Union; 2016. p. e01aa69+. Available from: https://w3id. org/mtv/FISE-Comm/v01/e01aa69

[8] D'amato G, Cecchi L, Bonini S, Nunes C, Annesi-Maesano I, Behrendt H, et al. Allergenic pollen and pollen allergy in Europe. Allergy. 2007;**62**(9):976-990

[9] Chuine I, Cambon G, Comtois P. Scaling phenology from the local to the regional level: Advances from species-specific phenological models. Global Change Biology. 2000;**6**(8):943-952

[10] Beggs PJ. Impacts of Climate Change on Aeroallergens and Allergic Diseases. United Kingdom: Cambridge University Press; 2016

[11] NOAA 2015 Annual Global Climate Report. Available from: https://www.ncdc.noaa.gov/news/2015-annual-global-climate-report [Accessed: 15 November 2017]

[12] Hartmann DL, Tank AMK, Rusticucci M, Alexander LV, Brönnimann S, Charabi YAR, et al. Observations: Atmosphere and surface. In: Stocker TF, Qin D, Plattner GK, Tignor M, Allen SK, Boschung J, et al, editors. Climate Change 2013 the Physical Science Basis: Working Group I Contribution to the Fifth Assessment Report of the Intergovernmental Panel on Climate Change. United Kingdom: Cambridge University Press; 2013. pp. 159-254

[13] Collins M, Knutti R, Arblaster J, Dufresne JL, Fichefet T, Friedlingstein P, et al. Long-term climate change: Projections, commitments and irreversibility. In: Stocker TF, Qin D, Plattner GK, Tignor M, Allen SK, Boschung J, Midgley PM, editors. Climate Change 2013: The Physical Science Basis. IPCC Working Group I Contribution to AR5. Eds. IPCC. Cambridge: Cambridge University Press; 2013. pp. 1029-1136

[14] Tutin TG, editor. Flora Europaea. United Kingdom: Cambridge University Press; 1964

[15] Nabors MW, editor. Introduction to Botany. USA: Pearson; 2004

[16] Ebner C, Hirschwehr R, Bauer L, Breiteneder H, Valenta R, Ebner H, et al. Identification of allergens in fruits and vegetables: IgE cross-reactivities with the important birch pollen allergens bet v 1 and bet v 2 (birch profilin). Journal of Allergy and Clinical Immunology. 1995;**95**(5):962-969

[17] Spieksma FTM, Emberlin JC, Hjelmroos M, Jäger S, Leuschner RM. Atmospheric birch (*Betula*) pollen in Europe: Trends and fluctuations in annual quantities and the starting dates of the seasons. Grana. 1995;**34**(1):51-57

[18] D'amato G, Spieksma FTM. Allergenic pollen in Europe. Grana. 1991;**30**(1):67-70

[19] Garcıa-Mozo H, Galán C, Aira MJ, Belmonte J, de la Guardia CD, Fernández D, et al. Modelling start of oak pollen season in different climatic zones in Spain. Agricultural and Forest Meteorology. 2002;**110**(4):247-257

[20] Barbour JR, Read RA, Barnes RL, Morus L. Mulberry. The Woody Plant seed manual. AGRIC. 2008;**727**:728-732

[21] Frei T, Leuschner RM. A change from grass pollen induced allergy to tree pollen induced allergy: 30 years of pollen observation in Switzerland. Aerobiologia. 2000;**16**(3-4):407

[22] Ciprandi G, Comite P, Ferrero F, Bignardi D, Minale P, Voltolini S, et al. Birch allergy and oral allergy syndrome: The practical relevance of serum immunoglobulin E to bet v 1. Allergy and Asthma Proceedings. 2016;**37**(1):43-49

[23] Lavaud F, Fore M, Fontaine JF, Pérotin JM, de Blay F. Birch pollen allergy. Revue des Maladies Respiratoires. 2014;**31**(2):150-161

[24] Niederberger V, Pauli G, Grönlundc H, Fröschla R, Rumpold H, Kraft D, et al. Recombinant birch pollen allergens (rBet v 1 and rBet v 2) contain most of the IgE epitopes present in birch, alder, hornbeam, hazel, and oak pollen: A quantitative IgE inhibition study with sera from different populations. Journal of Allergy and Clinical Immunology. 1998;**102**(4):579-591

[25] Żbikowska-Gotz M, Jóźwiak J, Rędowicz JM, Kuźmiński A, Napiórkowska K, Przybyszewski M, et al. Assessment of cross-reactivity in patients allergic to birch pollen by immunoblotting. Food and Agricultural Immunology. 2013;**24**(4):445-456

[26] Grewling L, Janerowicz D, Nowak M, Polanska A, Jackowiak B, Czarnecka-Operacz M, et al. Clinical relevance of *Corylus* pollen in Poznan, western Poland. Annals of Agricultural and Environmental Medicine. 2014;**21**(1):64-69

[27] Mothes N, Horak F, Valenta R. Transition from a botanical to a molecular classification in tree pollen allergy: Implications for diagnosis and therapy. International Archives of Allergy and Immunology. 2004;**135**(4):357-373

[28] Sofiev M, Bergmann KC, editors. Allergenic Pollen: A Review of the Production, Release, Distribution and Health Impacts. Netherlands: Springer; 2013. p. 159

[29] Eriksson NE, Wihl JA, Arrendal H, Strandhede SO. Tree pollen allergy. II. Sensitization to various tree pollen allergens in Sweden. A multi-Centre study. Allergy. 1984;**39**(8):610-617

[30] Dales RE, Cakmak S, Judek S, Coates F. Tree pollen and hospitalization for asthma in urban Canada. International Archives of Allergy and Immunology. 2008;**146**(3):241-247

[31] Navarro AM, Orta JC, Sanchez MC, Delgado J, Barber D, Lombardero M. Primary sensitization to *Morus alba*. Allergy. 1997;**52**(11):1144-1145

[32] Targow AM. The mulberry tree: A neglected factor in respiratory allergy in Southern California. Annals of Allergy. 1971;**29**(6):318

[33] Tiwari GPK. Impacts of climate variability and change on aeroallergens and their associated health effects: A preliminary review. In: Proceedings of the Symposium on Sustainable Development. 2010. p. 194

[34] Hurtado I, Riegler-Goihman M. Air sampling studies in a tropical area. Four year results. In: Advances in Aerobiology. Birkhäuser: Basel: Springer; 1987. pp. 49-53

[35] Bilisik A, Yenigun A, Bicakci A, Eliacik K, Canitez Y, Malyer H, et al. An observation study of airborne pollen fall in Didim (SW Turkey): Years 2004-2005. Aerobiologia. 2008;**24**(1):61-66

[36] Gonianakis MI, Baritaki MA, Neonakis IK, Gonianakis IM, Kypriotakis Z, Darivianaki E, et al. A 10-year aerobiological study (1994-2003) in the Mediterranean island of Crete, Greece: Trees, aerobiologic data, and botanical and clinical correlations. Allergy and Asthma Proceedings. 2006;**27**(5):371-377

[37] Nitiu DS. Aeropalynologic analysis of La Plata City (Argentina) during a 3-year period. Aerobiologia. 2006;**22**(1):79-87

[38] Van Vliet AJ, Overeem A, De Groot RS , Jacobs AF , Spieksma F. The influence of temperature and climate change on the timing of pollen release in the Netherlands. International Journal of Climatology. 2002;**22**(14):1757-1767

[39] Emberlin J, Detandt M, Gehrig R, Jaeger S, Nolard N, Rantio-Lehtimäki A. Responses in the start of *Betula* (birch) pollen seasons to recent changes in spring temperatures across Europe. International Journal of Biometeorology. 2002;**46**(4):159-170

[40] Yli-Panula E, Fekedulegn DB Green BJ, Ranta H. Analysis of airborne betula pollen in Finland; a 31-year perspective. International Journal of Environmental Research and Public Health. 2009;**6**(6):1706-1723

[41] García-Mozo H, Galán C, Jato V, Belmonte J, de la Guardia CD, Fernández D, et al. *Quercus* pollen season dynamics in the Iberian Peninsula: Response to meteorological parameters and possible consequences of climate change. Annals of Agricultural and Environmental Medicine. 2006;**13**(2):209-212

[42] Huynen M, Menne B. Phenology and human health: Allergic disorders. In: WHO Meeting 2003; 16-17 January 2003; Rome, Italy: Health and Global Environmental Series. EUR/03/5036791. Copenhagen: World Health Organization; 2003. p. 55

[43] Hidalgo-Galvez MD, García-Mozo H, Oteros J, Mestre A, Botey R, Galán C. Phenological behaviour of early spring flowering trees in Spain in response to recent climate changes. Theoretical and Applied Climatology. 2018;**132**(1-2):1-11, 263-273

[44] Corden JM, Millington WM. The long-term trends and seasonal variation of the aeroallergen Alternaria in Derby, UK. Aerobiologia. 2001;**17**(2):127-136

[45] Fitter A, Fitter RSR. Rapid changes in flowering time in British plants. Science. 2002;**296**(5573):1689-1691

[46] García-Mozo H, Oteros JA, Galán C. Impact of land cover changes and climate on the main airborne pollen types in southern Spain. Science of the Total Environment. 2016;**548**:221-228

[47] Wayne P, Foster S, Connolly J, Bazzaz F, Epstein P. Production of allergenic pollen by ragweed (*Ambrosia artemisiifolia* L.) is increased in $CO_2$-enriched atmospheres. Annals of Allergy, Asthma & Immunology. 2002;**88**(3):279-282

[48] Leakey AD, Uribelarrea M, Ainsworth EA, Naidu SL, Rogers A, Ort DR, et al. Photosynthesis, productivity, and yield of maize are not affected by open-air elevation of $CO_2$ concentration in the absence of drought. Plant Physiology. 2006;**140**(2):779-790

[49] García-Mozo H, Mestre A, Galán C. Phenological trends in southern Spain: A response to climate change. Agricultural and Forest Meteorology. 2010;**150**(4):575-580

[50] Ahlholm JU, Helander ML, Savolainen J. Genetic and environmental factors affecting the allergenicity of birch (*Betula pubescens*) pollen. Clinical and Experimental Allergy. 1998;**28**:1384-1388

[51] García-Mozo H, Hidalgo P, Galán C, Gómez-Casero MT, Dominguez E. Catkin frost damage in Mediterranean cork-oak (*Quercus suber* L.). Israel Journal of Plant Sciences. 2001;**49**(1):42-47

[52] Cecchi L, Malaspina TT, Albertini R, Zanca M, Ridolo E, Usberti I, et al. The contribution of long-distance transport to the presence of ambrosia pollen in central northern Italy. Aerobiologia. 2007;**23**(2):145-151

[53] Riotte-Flandrois F, Dechamp C. New legislation from the Politique Agricole commune passed in 1994 and its impact on the spread of ragweed. The public health laws are the responsibility of the mayor, the health department, the general council and the state council. Allergie et Immunologie. 1995;**27**(9):345-346

# 9

# Terroir Zoning: Influence on Grapevine Response (*Vitis vinifera* L.) at Within-vineyard and Between-Vineyard Scale

*Álvaro Martínez and Vicente D. Gómez-Miguel*

**Abstract**

Since ancient times, wines from specific regions have been valued and studies related to *terroir* focus on the elements of the environment that affect wine production. This paper presents the *terroir* variations between vineyards and within the same vineyard, as well as its influence on grape production. A soil zoning is carried out, starting from an aerial photointerpretation (FIA) and studying each soil sector based on its depth analysis (pits). This zoning of the environmental homogeneous units (EHU) is redefined with the normalized difference vegetation index (NDVI), resulting in the proposed *terroir* zoning. The temporal stability of the *terroir* zoning has been tested through the representation of the NDVI during 3 years and the response of the vineyard (yield, vegetative growth, and grape composition) during 4 years. The relationship between the EHUs, soil epipedon particularly, and the response of the vineyard is analyzed from an agglomerative hierarchical cluster-ing (AHC) prior to a principal component analysis (PCA). There is an EHU that is shown to be more vigorous, associated with a material deposition area whose main series of soil is fine-loamy, mixed, mesic, Typic Xerofluvent. This *microterroir* produces grapes with low sugar content, high acidity, and low levels of polyphenolic compounds, including anthocyanins.

**Keywords:** *terroir* variability map, precision viticulture, remote sensing, soil science, vine yield, grape composition

## 1. Introduction

From the Sumerians to our days, including Egyptians, Phoenicians, Greeks, Romans, the Middle Ages, etc., quality wine has been linked to certain regions. This interest in the geographical origin of the wines justifies the *terroir* concept and that its use remains in the general feeling of both popular and market and scientific. The International Organization of Vine and Wine (OIV) in its Resolution VITI 333/2010 [1] collects this idea and defines the *terroir* as follows: "Vitivinicultural *terroir* is a concept which refers to an area in which collective knowledge of the interactions between the identifiable physical and biological environment and applied vitivini-cultural practices develops providing distinctive characteristics for the products originating from this area."

The concept of *terroir* began to be used in the fourteenth century for some production properties of high-quality wines in Côte d'Or, Burgundy [2], being difficult to define the ideal factors that make up the *terroir* due to the interaction that exists between them [3]. This complexity of individual natural factor analysis, soil particularly, is gradually overcome with new tools for detection, management, and data analysis. In any case, although the physical and chemical interactions that affect the vineyard are not known with total accuracy, the dissemination of the *terroir* concept is fostering a better knowledge and use of geology, soil, climate, and wine culture for best wine production [2].

The first scientific studies related to the viticultural environment elements and their interactions are carried out in the last quarter of the twentieth century, being able to consider the doctoral thesis of Professor Morlat [4], one of the pioneering studies on the *terroir* zoning in the modern meaning of the term. The aforementioned work takes place in the middle area of the Loire Valley, and European countries have traditionally given more importance to the environment elements in the wine characterization, thus protecting the origin of the wines. Two examples of this tradition are the current classification of Bordeaux wines, which have been practically unchanged since its creation in 1855, and the classification of port wines that were delimited in 1758 (now that zoning has been expanded) and carried out by "Companhia Geral da Agricultura das Vinhas do Alto Douro" (a company similar to the current Regulatory Councils) at the proposal of the Marquis of Pombal.

The globalization of international wine trade has led to increased production, especially in new countries, of varietal and brand wines, and the adoption of low-input techniques, exerting significant pressure on traditional *terroir* wine producers [5]. Even so, there are many recent scientific publications on the concept of *terroir*, interrelating elements of the environment such as temperature [6], water status [7], light [8], geology [9], soil [10], etc. with the response of the vine.

To study the influence of climate in the vineyard, it is traditional to differentiate between macroclimate, mesoclimate, and microclimate depending on the scale of work. The first refers to the climate of a region and is the main limiting factor for the cultivation of the vine [11], while the mesoclimate is characteristic of a specific topographic and landscape location and affects a set of plants equally in a given geomorphological unit. Finally, the microclimate refers to the vine, surrounds to leaves and clusters and has a great influence in the biological cycle (e.g., it is of great importance in the grape ripening stage), being able to modify through the vineyard management.

Geology and geomorphology allow a synthetic approach adapted to small-scale zonings (≤ 1:50,000), explaining the behavior of the vine only indirectly [12]. The geological or geomorphological maps are useful as a first approximation to the *terroir* zoning, since very different soils can be included in the same map unit, so it is necessary to determine the types of soil [13]. For this reason, many of the approaches to viticultural zoning borrow their approach from pedological cartogra-phy, with some variants [14].

The soil study methodology is specified in the genesis of the soil taxonomic units (STU) and the soil map (or cartographic) units (SMU) during the process of their recognition. The processing of the information generated in the different layers of information by a geographic information system (GIS) results in the quantification of the contents and the possibility of their statistical treatment [15]. This methodol-ogy has been and continues to be used as part of the *terroir* zoning of both small-scale viticultural regions (macrozoning), for example, 1:50,000 or 1:25,000, and in vineyards or sets of smaller vineyards at larger scales (microzonifications) between 1:5000 and 1:10,000.

Depending on the scale used in the zoning of the environment elements, mainly climate and soil, we will talk about macro (below 1:25,000), meso (between 1:25.000 and 1:10,000), or micro *terroir* zoning (above 1: 5000). Once the meso or micro *terroir* zoning has been carried out, the management of the vineyard is susceptible of being executed according to the environment homogeneous unit (EHU) defined by it.

The cartographic delimitation of vineyard sectors or EHUs and its individual-ized management is the basis of precision viticulture (PV), a vineyard management technique that relies mainly on remote sensing [16], to monitor and manage the spatial variability of the vineyard [17–19]. The images obtained by remote sensing are the basis for the creation of maps [20–23], such as the cartographic representation of the normalized difference vegetation index (NDVI) [20]; providing important information, they are very affordable, they facilitate the precision of the limits of the conventional zoning, and they are obtained more quickly than those made from the traditional zoning method [24]. These images are characterized by their temporal, spatial, and spectral resolutions and have been used in meso- and micro-zonifications since the end of the twentieth century [25–27].

In the present work, a methodology of NDVI integration in *terroir* zoning is proposed, redefining the cartographic limits of traditional microzoning. Once these sectors, EHU, or micro *terroirs* are defined, the behavior of the vineyard (yield, vigor, and grape composition) is related to the main factors that characterize them.

## 2. Material and methods

The experimental work is carried out for 4 consecutive years (2012, 2013, 2014, and 2015) in four vineyards (A, B, C, and D) located at an average distance of 2 km from each other, in the municipality of Oyón (Álava). The vineyards are protected by the DOCa Rioja, appellation of origin associated with the Ebro River, and located in the northern third of the Iberian Peninsula.

Regarding the climate of the area where the vineyards are framed, the rainfall and average annual temperature are 459 L $m^{-2}$ and 13.7°C and during the vegetative period (April–October) are 260 L $m^{-2}$ and 18.1°C, respectively. According to the Multicriteria System of Climatic Classification (MSCC) [28], the climate is warm temperate (HI + 1), of cool nights (CI + 1) and moderately dry (DI + 1). Although the dominant wind is from the west, another typical northwest wind (known as *cierzo*) has influence during the grape ripening.

The greater part of the vineyards of the area is grown on sandstone and lutites of Haro's facies (Middle-Upper Miocene) [29]. Some of the soils found on this geology and their associated quaternary system are [30] alfisols (e.g., Calcic Haploxeralf subgroup), entisols (e.g., Typic Xerofluvent subgroup), or inceptisols (e.g., Typic Xerocrept subgroup). For more details of the study area, see [9]. The grape cultivar is Tempranillo, grafted on 41B, and the vines are trained using a single trellis system (bilateral cordon Royat pruning), with 2976 vines/ha, and soil management is by tillage.

A zoning is carried out under viticultural criteria (variety and vine age) in the four vineyards, and in the resulting subplots, a FIA was drawn from digital orthophotographs of 0.25 meters of spectral resolution [31], discriminating sectors (A1, A2, A3 for vineyard A and analogously for the rest of the vineyards) on a scale of 1:2500, that is, on a vineyard scale. In each sector, 12 vines are marked, divided into 2 repetitions of 6 plants. Measurements of vegetative growth and yield are carried out, as well as physical-chemical analysis of the grape on each repetition. For the pedological study, a pit is made next to each of the repetitions of six vines, describing the profile and analyzing the different horizons in the laboratory. In this way, between two and three pits per hectare of vineyard are made, density suitable for very detailed soil zoning [12, 32]. The soil classification proposed by the United States Department of Agriculture has been used [30].

The NDVI is defined as the difference between the radiance value in near infrared and red, divided by their sum [20]. In this work, these radiance values have been obtained from multispectral images captured by the Pléiades satellite (0.5 meters of spatial resolution) on August 25, 2014, and August 19, 2015, and by the SPOT 5 satellite (2.5 meters of spatial resolution) on August 14, 2013. The calcula-tion and graphic representation of the NDVI are carried out pixel by pixel, with the help of the ArcGIS 10.1 software from the Environmental Systems Research Institute (ESRI). The definition of the classes (very low, low, medium, high, and very high) is done according to five quantiles, the first quantile corresponding to the very low class and the fifth quantile to the very high class.

The statistical analysis of the data was carried out through principal component analysis (PCA) and univariate ANOVA, after checking normality and homogeneity of variances of the variables. The significance of these analyses was determined for the probability levels $p < 0.05$ (*), $p < 0.01$ (**), and $p < 0.001$ (***). The means are compared by the Duncan test when there were significant differences in the analysis of variance. The SPSS program, version 15.0 (SPSS Inc. Chicago, Illinois), was used for the ANOVA analyses, and for the rest of the statistical calculations, the XLSTAT 2019.1.2 supplement was used on Microsoft Excel 2007. This complement was also used to perform an agglomerative hierarchical clustering (AHC) reducing the 28 analyzed variables of the epipedon before performing the PCA. In this case, the correlation between variables has been calculated for a significance level alpha = 0.05.

For the geostatistical study of the NDVI distribution, the normalized Moran index (NMI) is used, which measures the spatial autocorrelation allowing to evaluate if the NDVI pattern is clustered, dispersed, or random. For the calculation of this index, as well as the associated z-value, the ArcGIS 10.1 ESRI tool is used.

## 3. Results

**Table 1** and **Figure 1** show that EHUs with high yield and high weight of pruning wood (vigorous EHUs) correspond with low probable alcohol grade, low polyphenolic content, and high acidity. These results were to be expected according to numerous previous studies by other authors [19, 33].

The AHC analysis has allowed grouping the 28 analyzed variables of the surface horizon (results not shown) into 3 homogeneous classes, represented by 1 of its variables. Thus, the fine land (FL); the alpha index (AI), which indicates the exchange capacity of the clay; and the humidity at field capacity (H33C) represent, among other variables, the silt content, the total limestone, and the pH; the content in sand, in clay, and in organic matter and the electrical conductivity and the cation exchange capacity; and the content of coarse elements, the active limestone, and the moisture content of the wilting point, respectively.

Observing the biplot graphics of the PCA (**Figure 1**) carried out for the EHUs (observations) and the characteristics of the grape studied together with the three representative variables of the epipedon, it is possible to differentiate between three groups of EHUs. The first group is formed by C1, C2, D2, and D3, which is characterized by its vigor, high yields, and higher levels of malic acid; the second group is composed of B1, B2, and D1 that could be considered as transitional; and the third group is represented by A1, A2, and A3 with a higher probable alcohol level, anthocyanin content, and total polyphenol index (TPI). Regarding the three elements of

| UHM | Yield | WP | %vol | pH | MA | TPI | ANT |
|---|---|---|---|---|---|---|---|
| A1 | 835 abc | 230 a | 14.45 bc | 3.95 c | 2.3 abc | 54.5 ef | 706 cde |
| A2 | 1082 cd | 407 cd | 14.4 bc | 3.86 ab | 1.96 a | 56.5 f | 734 de |
| A3 | 706 a | 231 a | 14.76 c | 4.09 d | 2.92 cd | 52.5 f | 788 e |
| B1 | 1022 bcd | 390 c | 13.68 a | 3.87 ab | 2.01 ab | 46.8 bcd | 612 abc |
| B2 | 767 ab | 308 b | 13.86 ab | 3.82 ab | 2.17 ab | 48.6 cd | 655 bcd |
| C1 | 2342 g | 559 e | 13.35 a | 3.9 bc | 2.97 cd | 43.6 abc | 562 ab |
| C2 | 2439 g | 584 e | 13.40 a | 3.87 ab | 2.7 bcd | 40.6 a | 530 a |
| D1 | 1275 de | 307 b | 13.45 a | 3.8 a | 2.3 abc | 49.9 de | 673 cd |
| D2 | 1640 f | 690 f | 13.84 ab | 3.81 a | 3.09 d | 43.0 ab | 597 abc |
| D3 | 1506 ef | 461 d | 14.02 ab | 3.83 ab | 2.6 bcd | 46.1 bcd | 637 bcd |
| Sig. | *** | *** | *** | *** | ** | *** | *** |

**Table 1.**
*Results (mean 2012–2015) of each UHM: yield (g), weight of pruning wood (WP, g), probable volumetric alcohol degree (%vol), pH, malic acid (MA, g/l), total polyphenol index, and anthocyanins (ANT, mg/l).*

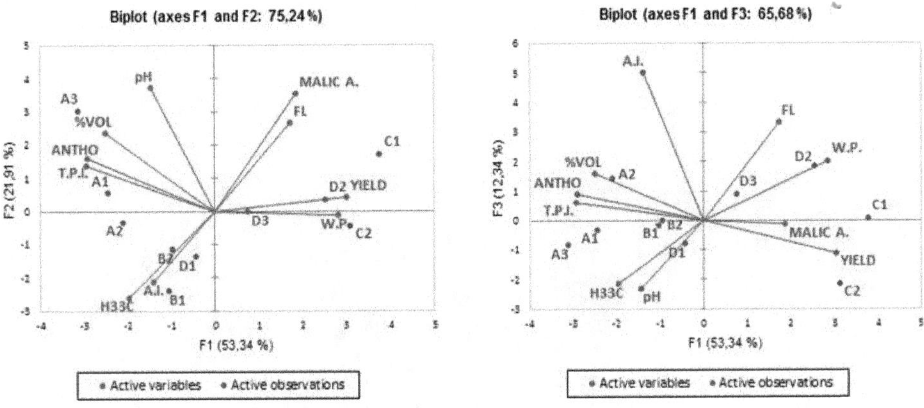


**Figure 1.**
*Biplot graphs of observations (EHU) and variables related to the vineyard (yield, WP, %vol, pH, MA, TPI, and ANT) and to the epipedon (FL, AI, H33C). F1, F2, and F3 are the first three factors of the PCA.*

the soil epipedon and its associated variables according to the AHC, it was found that there are positive correlations between the FL and the content in malic acid and between the H33c and the pH of the grape, as well as negative correlation between the H33c and malic acid (**Figure 1**).

### 3.1 Between-vineyard scale

Depending on the scale used in the class discrimination of the NDVI, differ-ent maps can be obtained, although the values of the distribution are constant. Thus, **Figure 2** presents the graphic representations of the four vineyards with a common classification of values, in contrast to the FIA drawn at the plot scale. On the contrary, in **Figure 3** a classification of the individualized NDVI values has been carried out for each vineyard. Comparing both figures it is observed that in **Figure 3** the NDVI distributions in each vineyard follow a pattern more similar to the FIA; in addition there are greater contrasts, something that facilitates the zoning (**Figures 2A** and **3A** ). In particular, vineyards A and D present a spatial distribution

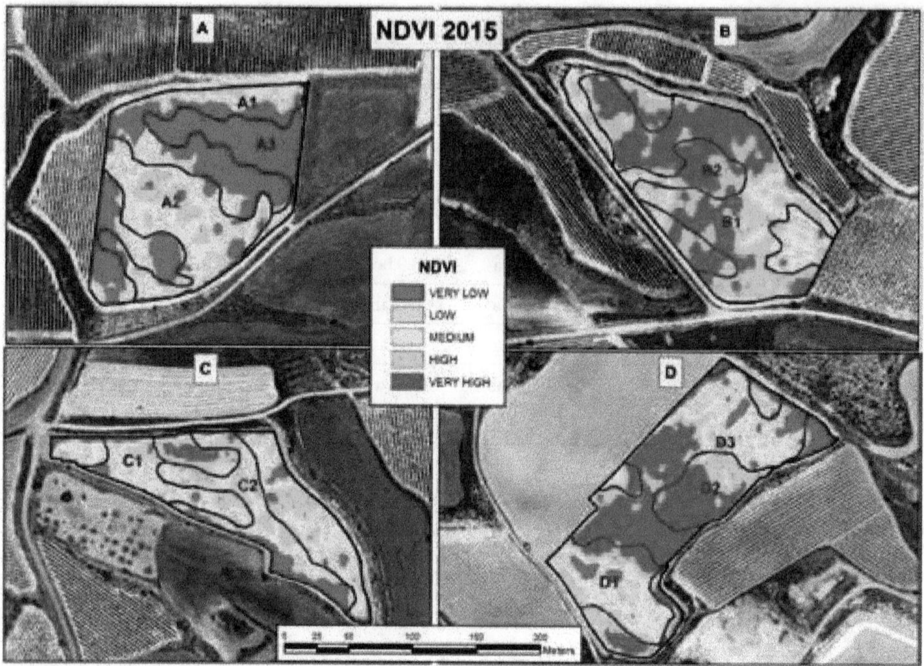


**Figure 2.**
*Spatial distribution of NDVI from Pleiades image. A common classification is carried out for the four vineyards. The labels (A, B, C and D) represent the vineyards of the same name.*

**Figure 3.**
*Spatial distribution of NDVI from Pleiades image. A classification of the individualized NDVI values has been carried out for each vineyard. The labels (A, B, C and D) represent the vineyards of the same name.*

of the NDVI similar to the FIA (**Figure 3**). However, plots B and C seem to have a more chaotic NDVI distribution in relation to the FIA, something that seems to be related to the modification of the terrain before planting. This is also the reason why plot C, which according to **Table 1** includes the most vigorous UHMs in the study, does not include more surface of the high or very high classes of the NDVI map (**Figure 3**).

Considering **Figure 2**, most of vineyard B is included, according to the NDVI, in the very high class, while the results of production and composition of the grape (**Table 1**) do not correspond to this result, agreement with previous studies [16, 19, 34]. Also sector D2 (and part of D3) is included in the very high class (**Figure 2**), but in this case this EHU is characterized by having a high yield and a high prun-ing weight, as well as low sugar levels, pH, IPT and anthocyanins, and the highest content in malic acid; this coincides with the aforementioned works. Again, it seems that the modification of the natural characteristics of the environment prevents the use of the NDVI as a tool for the zoning of the vineyard, at least on this scale of work.

On the contrary, sectors A1 and A3 of vineyard A are included in the very low class (**Figure 2A**). In comparison with the other sectors (**Table 1**), the vines of A1 and A3 have shown the lowest weight of pruning wood, low yields, high levels of probable alcoholic degree, pH, IPT, and total anthocyanins, as well as low levels of malic acid. The other sector of plot A (A2) has been discriminated against by the NDVI since it is not included in the very low class. Sector A2 (**Table 1**) presents higher values of production and weight of pruning wood than sectors A1 and A3. In any case, this discrimination of EHUs is better defined (in relation to the FIA) when making the classification of the vineyard scale distribution (**Figure 3A**).

### 3.2 Intra-vineyard scale

For the study of intra-vineyard variability, we will focus on plot D (**Figure 4**). The difference between the characteristic profiles of each EHU can be observed, appreciating, at first sight, the fluventic character of the D2B profile (**Figure 4B**) and typical of areas of deposition of material or fertile ground. A unique feature of this type of profile is that its agricultural behavior is similar to an addition of slow liberalization fertilizer, which will influence increasing soil fertility and vigor and vineyard production (**Table 1**). Specifically, this profile has a content in organic matter in the epipedon (Ap) of 2.25%, while in the Ab horizon (52–80 cm deep), there is a level of 3.6%. Regarding the NDVI, sector D2 associated with this profile (**Figure 3D**) is included mainly in the high and very high classes. This indicates, as well as the vegetative, grape composition, and pedological results, that it is a UHM that can be characterized as vigorous, in relation to the rest of the UHMs in the vineyard.

In **Figure 4D** a NDVI classification has been carried out independently in each EHU; the result obtained is not technically operational, at least in a vineyard of the size of the studied one (2 ha). In order to refine the zoning delimitation to this scale of work, we could redefine the EHUs with the help of the NDVI distribution from the classification of the vineyard (**Figure 4E**). In this way vineyard D will be zoned in three EHUs:

D1: northeast facing slope whose characteristic profile is a mesic Calcic Haploxeralf with 80 cm effective depth and with an argillic horizon (20–80 cm of depth). It is the least productive EHU of the vineyard, whose characteristics of the grape are high probable alcoholic degree, lower acidity, and high content in total polyphenols and anthocyanins (**Table 1**). Regarding the NDVI, very low, low, and medium classes appear mainly.

D2: material deposition area characterized by fine loamy, mixed, mesic, and Typic Xerofluvent with 145 cm effective depth and darker color of the epipedon. It is expected that it is the freshest sector (and most sensitive to frosts) in relation to the other two sectors of the vineyard. All these factors mean that production is the highest of the three sectors and that the grape composition is the one with the lowest

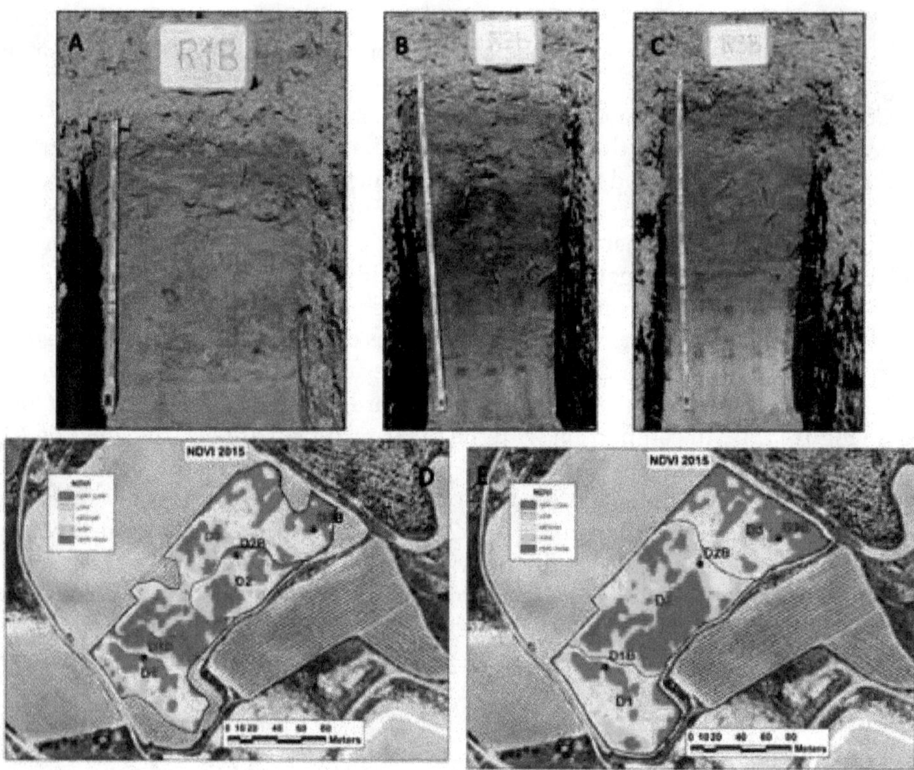


**Figure 4.**
*Characteristic profiles of the EHUs: 4A (sector D1), 4B (sector D2), and 4C (sector D3). NDVI classification carried out independently in each EHU and location of the pits (4D). Definitive* terroir *zoning redefined with the NDVI, with a single classification for the three EHUs (4E).*

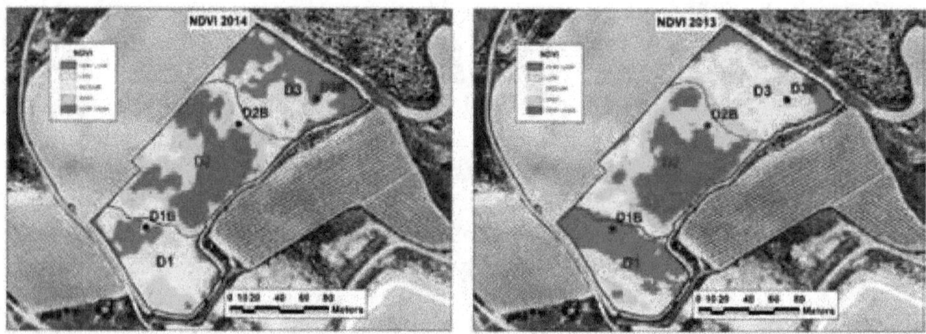


**Figure 5.**
*Spatial distribution of the NDVI in the years 2014 (satellite Pleiades) and 2013 (satellite SPOT 5) and proposed terroir zoning.*

probable alcoholic degree, the highest acidity, and the lowest content of total polyphenols and anthocyanins (**Table 1**). The very high and high classes of the NDVI are the most present in this EHU.

D3: hillside with southwest orientation, whose main soil is fine loamy, mixed, mesic, and Typic Xerothent, with an effective depth of 145 cm. It could be considered as the EHU of intermediate characteristics between the other two.

**Figure 5** shows the graphic representation of the NDVI from other years (2014 and 2013), verifying the temporal stability of the zoning carried out in vineyard

D. It is noteworthy that the image of the year 2013 is of a resolution (2.5 meters) lower than the rest of the years and the distribution pattern of the NDVI is adjusted to the proposed zoning. In order to achieve a harvest as homogeneous as possible, we could recommend practices aimed at reducing vigor in D2 and increasing it in D1, always with the limitations of practical management.

The NMI calculated for the NDVI distribution has a value of 0.998 and a z-score of 397 for the years 2015 and 2014, while in 2013 the value of the NMI is 0.999 and the z-score is 77. For 3 years there is less than 1% likelihood that this clustered pattern could be a result of random chance [35]. In this type of spatial analysis, it seems that the spatial resolution of the starting image does influence, with a lower correlation with lower spatial resolution. In any case, the NDVI distributions have a grouped pattern indicating that there is a link between the NDVI distribution and the landscape or the distribution of the environment elements in space.

## 4. Conclusions

It has been verified that there is similarity in the distribution of the NDVI and the FIA, provided that both cartographies are made at the same scale. In vineyards grown on man-modified soils, it seems that the use of vegetation indexes, such as the NDVI, does not give the expected results, being able to conclude that there is a link between the NDVI and the characteristics of the environment, in particular with those related to the soil and landscape.

In the within-vineyard *terroir* zoning, the EHU associated to a material deposition area and characterized by the main series fine-loamy, mixed, mesic, and Typic Xerofluvent is related to the very high class of the NDVI and at the same time with vigorous properties of the vineyard: high yield, high weight of pruning wood, low probable alcohol grade, high acidity, and low levels of IPT and anthocyanin content. These results have been obtained in comparison with two EHUs associated with hillside and whose main soil series are mesic, Calcic Haploxeralf and fine-loamy, mixed, mesic Typic Xerorthent.

Regarding the NDVI, the interannual stability in the pattern has been demonstrated regardless of the resolution of the image, at least from 0.5 to 2.5 m/pixel.

Epipedon characteristics related to agronomic results have been found. Thus, it was found that there are positive correlations between the FL (and associated variables such as silt content, total limestone, and pH) and content in malic acid and between H33c (and the associated variables such as the content in coarse elements, the active limestone, and the moisture content of the wilting point) and the pH of the grape, as well as the negative correlation between the H33c and malic acid. In the future, it will be interesting to find a methodology that allows to integrate the analytical results, not only of the superficial horizon but also of all the horizons of the soil profile and the vegeto-productive results and grape composition, to be able to relate them to each other as exhaustively as possible.

The importance of the size of the vineyards to find PV applications is noteworthy. In the case of DOCa Rioja, 87% [36] of the vineyards are small (between 0.1 and 2 hectares), making the sectorization of the vineyards technically and economically unviable in order to carry out localized treatments. In any case, it is advisable to project plantations that, as far as possible, facilitate individualized management, particularly in the harvest.

## Author details

Álvaro Martínez* and Vicente D. Gómez-Miguel
Universidad Politécnica de Madrid, Madrid, Spain

*Address all correspondence to: alvaro.martinezh@alumnos.upm.es

## References

[1] OIV. Definición de *terroir* vitivinícola. 2010

[2] Wilson J. Geology and wine 4. The origin and odyssey of terroir. Geoscience Canada. 2001;**28**(3):139-141

[3] van Leeuwen C, Seguin G. The concept of terroir in viticulture. Journal of Wine Research. 2006;**17**(1):1-10. DOI: 10.1080/09571260600633135

[4] Le MR. Terroir viticole: Contribution a l'etude de sa caracterisation et de son influence sur les vins, application aux vignobles rouges de moyenne Vallée de la Loire [thesis]. Universite de Bordeaux II; 1989

[5] Clingeleffer P. Terroir: The application of an old concept in modern viticulture. In: Alfen NKV, editor. Encyclopedia of Agriculture and Food Systems. Oxford: Academic Press; 2014. pp. 277-288. DOI: 10.1016/B978-0-444-52512-3.00157-1

[6] Zhang P, Barlow S, Krstic M, Herderich M, Fuentes S, Howell K. Within-vineyard, within-vine, and within-bunch variability of the rotundone concentration in berries of *Vitis vinifera* L. cv. Shiraz. Journal of Agricultural and Food Chemistry. 2015;**63**(17):4276-4283. DOI: 10.1021/acs. jafc.5b00590

[7] Costa JM, Vaz M, Escalona J, Egipto R, Lopes C, Medrano H, et al. Modern viticulture in southern Europe: Vulnerabilities and strategies for adaptation to water scarcity. Agricultural Water Management. 2016;**164**:5-18. DOI: 10.1016/j. agwat.2015.08.021

[8] de Oliveira AF, Nieddu G. Accumulation and partitioning of anthocyanins in two red grape cultivars under natural and reduced UV solar radiation. Australian Journal of Grape and Wine Research. 2016;**22**(1):96-104. DOI: 10.1111/ajgw.12174

[9] Martínez A, Gomez-Miguel V. Vegetation index cartography as a methodology complement to the terroir zoning for its use in precision viticulture. OENO One. 2017;**51**(3):289. DOI: 10.20870/oeno-one.2017.51.4.1589

[10] van Leeuwen C, Roby JP, de Resseguier L. Soil-related terroir factors: A review. OENO One. 2018;**52**(2):173-188. DOI: 10.20870/oeno-one.2018.52.2.2208

[11] Gómez-Miguel V. Terroir. In: Böhm J, editor. Atlas das Castas da Península Ibérica: História, Terroir, Ampelografia. Lisboa, Portugal: Dinalivro; 2011. pp. 104-153

[12] van Leeuwen C, Bois B. Update in unified terroir zoning methodologies. E3S Web of Conferences. 2018;**50**:01044

[13] van Leeuwen C, Roby J, Pernet D, Bois B. Methodology of soil-based zoning for viticultural terroirs. Bulletin de l'OIV. 2012;**83**:947-949

[14] Vaudour E. Los Terroirs Vitícolas: Definiciones, Caracterización y Protección. Zaragoza: Acribia, D.L.; 2010

[15] Gómez-Miguel V, Sotés V. Zonificación del terroir en España. In: Fregoni M, Schuster D, Paoletta A, editors. Terroir Zonazione Viticoltura: Tratatto Internazionale. Rivoli Veronese: Phytoline; 2003. pp. 187-226

[16] Hall A, Lamb D, Holzapfel B, Louis J. Optical remote sensing applications in viticulture—A review. Australian Journal of Grape and Wine Research. 2002;**8**(1):36-47. DOI: 10.1111/j.1755-0238.2002.tb00209.x

[17] Tisseyre B, Ojeda H, Taylor J. New technologies and methodologies for site-specific viticulture. Journal

International des Sciences de la Vigne et du Vin. 2007;**41**(2):63-76

[18] Bramley RGV, Ouzman J, Thornton C. Selective harvesting is a feasible and profitable strategy even when grape and wine production is geared towards large fermentation volumes. Australian Journal of Grape and Wine Research. 2011;**17**(3):298-305. DOI: 10.1111/j.1755-0238.2011.00151.x

[19] Bramley RGV, Ouzman J, Boss PK. Variation in vine vigour, grape yield and vineyard soils and topography as indicators of variation in the chemical composition of grapes, wine and wine sensory attributes. Australian Journal of Grape and Wine Research. 2011;**17**(2):217-229. DOI: 10.1111/j.1755-0238.2011.00136.x

[20] Rouse JWJ, Haas RH, Schell JA, Deering DW. Monitoring vegetation systems in the great plains with ERTS. In: Proceedings of the 3rd ERTS Symposium, NASA. SP-351 1. 1973. pp. 309-317

[21] Qi J, Chehbouni A, Huete AR, Kerr YH, Sorooshian S. A modified soil adjusted vegetation index. Remote Sensing of Environment 1994 5;48(2):119-126. DOI: 10.1016/0034-4257(94)90134-1

[22] Steele M, Gitelson AA, Rundquist D. Nondestructive estimation of leaf chlorophyll content in grapes. American Journal of Enology and Viticulture. 2008;**59**(3):299-305

[23] Zarco-Tejada PJ, Catalina A, Gonzalez MR, Martin P. Relationships between net photosynthesis and steady-state chlorophyll fluorescence retrieved from airborne hyperspectral imagery. Remote Sensing of Environment. 2013;**136**:247-258. DOI: 10.1016/j.rse.2013.05.011

[24] Gomez-Miguel VD, Sotes V, Martinez A, Gonzalez-SanJose ML. Use of remote sensing in zoning's studies for terroir and precision viticulture: Implementation in DO Ca Rioja (Spain). In: Proceedings of the 39th World Congress of Vine and Wine. Vol. 7. 2016. p. 01025. DOI: 10.1051/bioconf/20160701025

[25] Cook S, Bramley R. Precision agriculture—Opportunities, benefits and pitfalls of site-specific crop management in Australia. Australian Journal of Experimental Agriculture. 1998;**38**(7):753-763. DOI: 10.1071/EA97156

[26] Bramley RGV, Hamilton RP. Terroir and precision viticulture: Are they compatible? Journal International des Sciences de la Vigne et du Vin. 2007;**41**(1):1-8

[27] Matese A, Toscano P, Di Gennaro SF, Genesio L, Vaccari FP, Primicerio J, et al. Intercomparison of UAV, Aircraft and Satellite Remote Sensing Platforms for Precision Viticulture. Remote Sensing. 2015;**7**(3):2971-2990. DOI: 10.3390/rs70302971

[28] Tonietto J, Carbonneau A. A multicriteria climatic classification system for grape-growing regions worldwide. Agricultural and Forest Meteorology. 2004;**124**(1-2):81-97. DOI: 10.1016/j.agrformet.2003.06.001

[29] Gómez-Miguel V, Sotés V. Delimitación de zonas vitícolas de la DOCa Rioja. Madrid: Universidad Politécnica de Madrid; 1997

[30] Soil Survey Staff. Keys to Soil Taxonomy. 12th ed. Washington, D.C.: USDA, Natural Resources Conservation Service; 2014

[31] Instituto Geográfico Nacional. Plan Nacional de Ortofotografía. PNOA. Madrid, España: IGN-CNIG; 2009

[32] Nieves M, Forcada R, Gómez-Miguel V. Precisión, escala y densidad

de observaciones en los estudios de suelos. Boletín de la Estación Central de Ecología. 1985;**27**:46-56

[33] van Leeuwen C, Friant P, Chone X, Tregoat O, Koundouras S, Dubourdieu D. Influence of climate, soil, and cultivar on terroir. American Journal of Enology and Viticulture. 2004;**55**(3):207-217

[34] Lamb D, Weedon M, Bramley R. Using remote sensing to predict grape phenolics and colour at harvest in a Cabernet Sauvignon vineyard: Timing observations against vine phenology and optimising image resolution. Australian Journal of Grape and Wine Research. 2004;**10**(1):46-54

[35] Li H, Calder CA, Cressie N. Beyond Moran's I: Testing for spatial dependence based on the spatial autoregressive model. Geographical Analysis. 2007;**39**(4):357-375. DOI: 10.1111/j.1538-4632.2007.00708.x

[36] Memoria 2015. Logroño, España: CRDOCa Rioja; 2016:79

# 10

# Ecology of Plant Communities in Central Mexico

*Joaquín Sosa-Ramírez, Vicente Díaz-Núñez and Diego R. Pérez-Salicrup*

**Abstract**

In Central Mexico converge three biogeographic provinces: Altiplano sur, Sierra Madre Occidental and Costa del Pacífico. Each one of them is composed by different plant communities: Thorn Forest, Temperate Mountain Forest and Dry Tropical Forest respectively. Our objective is to show, through phytoecological analysis, the species richness, diversity and the structure of the plant communities from the Temperate Mountain Forest and from the Tropical Dry Forest. In the Temperate Mountain Forest, 50 forest species were recorded, with a Shannon Wiener diversity index H′ = 1.63 on altitudes from 2400 to 2600 m. The Whittaker β index is $B_w$ = 7.22. In the tropical dry forest, we identified 79 plants species with a mean diversity index H′ = 3.49 on altitudes from 1951 to 2100 m. In this ecosystem the $B_w$ index is 8.12. This study offers important information for the establishment of management practices, considering the protection status from the areas in which this vegetation type is distributed.

**Keywords:** Aguascalientes, Sierra Fria, Temperate Mountain Forest, Tropical Dry Forest, biogeographic provinces

## 1. Introduction

Mexico is one of the five countries with the greatest biological diversity in the world, due, in part to the confluence of the Neartic (North America) and Neotropical biogeographic zones (Mexico, Central and south America). As well as, the species evolutionary processes in its territory [1]. The Mexican territory represent only 1% of the earth's surface; nevertheless, Mexico belongs to the select group of the five countries considered megadiverse, along with Brazil, Colombia, China and Indonesia [2, 3]. Due to its geographic locations and its multiple landscapes, a large number and diversity of ecosystems converge in the national territory. For that reason, Mexico is ranked 12th in terms of global forest area [4]. Even though, multiple efforts have been made for the forest conservation during the last decade of the XXI century, on a global scale, forest have been transformed to other uses at a rate of $1.3x10^6$ million ha/yr. or they have been affected by natural disturbances that have partially or totally changed their structure. This amount represents a 19% decrease in comparison to the exchange rate registered in the last decade of the 20th century ($1.6X10^6$ million ha/yr) [5]. Temperate forests in Mexico are found mostly, although not exclusively, in the mountainous areas along the Sierra Madre Occidental (the area with the highest concentration of forest ecosystems in the country), the mountains of Sierra Madre Oriental, the Sierra Norte de Oaxaca and the Altos de Chiapas, as well as in different mountain ranges and isolated mountains in the Altiplano and intermingled in the tropical plains [6]. The conifer and oak forest in Mexico represent the most extensive vegetation cover in terms of vegetation types dominated by woody species, this species covers 16.4% of the total surface of the country, being only surpassed by the xeric shrubland which is the vegetation type which has the largest extension [5]. These ecosystems are important both economically and ecologically, since they support productive activities, harbor great biological diversity and serves as a refuge for wildlife. Likewise, forest provide

The State of Aguascalientes has a total extension of 555, 867.4 hectares, of which 291,792.4 hectares equivalent to 52.5% present some forest type [9]. According to the classifications issued by different sources [10, 11], the State of Aguascalientes is made up by three large ecoregions (biogeographic regions), the Temperate Mountain Forest, the Tropical Dry Forest (also known as lowland deciduous forest) and the thorn forest (including crasicaule shrubland and xeric shrubland). The first ecosystem type is mainly distributed in la Sierra Fria, Sierra del Laurel, Sierra de Tepezalá and Cerro de Juan el Grande in El Llano municipality (**Figure 1**). The largest area covered by Temperate Mountain Forest vegetation in Aguascalientes is located in an area locally known as Sierra Fria, this site is a Protected Natural Area by state and federal decree which covers close to 107,000 ha [12]. In the Temperate Mountain Forest, the plant communities the most common vegetation types are oak forests (*Quercus* spp.), pine trees (*Pinus* spp.), oak-pine, pine-oak, juniper (*Juniperus* spp.), manzanita shrubland (*Arcostaphyllos pungens*) and different associations of these genera. The vegetation that has mainly colonized the sites that had been disturbed are *Juniperus deppeana* and *A. pungens*, although there has also been an increase in conifer populations [13]. The second largest formation where this ecosystem is found is located in la Sierra del Laurel in the Southwest corner of the State occupying close to 17,000 ha. This area presents similar plant communities

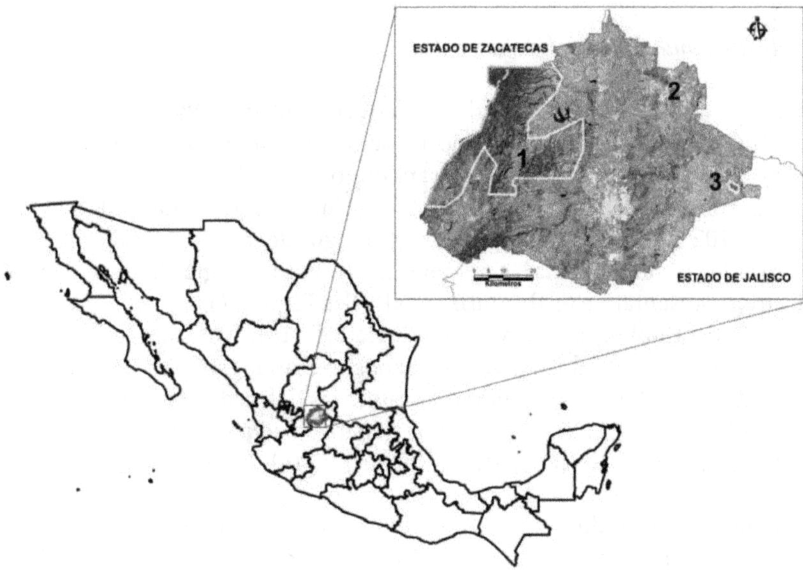


**Figure 1.**
*Distribution areas of the Temperate Mountain Forest in Aguascalientes state. (1) sierra Fria protected natural area (San José de Gracia, Pabellón de Arteaga, Rincón de Romos and Calvillo municipalities); (2) Mountain Hill of Tepezalá, and (3) Juan el Grande Mountain (El llano municipality).*

but with a greater dominance of oak populations (*Quercus* spp.) and lower density of manzanita (*A. pungens*).

The largest area occupied by the tropical dry forest is mainly located in the Calvillo municipality, although, there are relics of vegetation indicative of this ecosystem in the Jesus Maria, San Jose de Gracia and Aguascalientes municipalities, which suggests a larger presence of this vegetation type in the past. In the tropical dry forest, forest structures made up of shrubs and trees between 2 and 8 m high and some relics of medium tropical forest. In Aguascalientes, this is one of the ecosystems with the highest species richness [14]. The most representative vegetation in this ecoregion corresponds mainly to the *Lysiloma*, *Bursera*, *Ipomoea*, *Acacia*, *Eysenhardthia*, *Opuntia*, *Mimosa* and *Agave* genera.

Our objective was to provide an overview of some ecological aspects (species richness, diversity and distribution) of woody species natural communities in the most representative ecosystems of the State of Aguascalientes, assuming that there would be a high similarity degree with the vegetation of neighboring sites, considering both the environmental and physiographic characteristics from this State.

## 2. Materials and methods

Three studies were conducted individually. During 2008–2015, the natural communities of the temperate mountain forest in the area commonly known as Sierra Fria, in the northwest of the State of Aguascalientes, as well as the main disturbances that have affected them in the past and present were analyzed [10, 13]. Likewise, during the period 2011_2015 a study was carried out to determine the diversity, dynamics and functioning of the tropical dry forest in the Calvillo municipality [14, 15].

### 2.1 Temperate Mountain Forests

#### *2.1.1 Study area and sampling design*

This study was carried out in to the Sierra Fria Protected Natural Area (SF-PNA) which is 106,114.6 hectares in size and is located in the northwest of the Aguascalientes State. This area has an altitude ranging between 2,100 and 3,050 masl. The study area comprised 25 thousand hectares, in a polygon located between the coordinates 102°31′31″ to 102°37′44″ west longitude and 22°05′47″ at 22°14′03″ north latitude, assuming that the conditions both geographic, ecological and climatic are representative of the entire ANP (See **Figure 2**).

A stratified sampling strategy was developed [16]. The sampling strata were delimited based on the altitude, solar exposure, and geoform of the site (flat, concave and convex terrain). The first stratum was defined using a Digital Elevation Model (DEM) of the ANP SF, elaborating a spatial grid according to five altitudinal categories: i) 2,000-2,200, ii) 2,200-2,400, iii) 2,400-2,600, iv) 2,600-2,800, and v) >2 800 masl.

To stablish the altitudinal strata, the level curves from study site were defined using the DEM. The solar exposure was approached using an exposure map made with a SPOT 2010® imagine on which the DEM of the site was superimposed. Subsequently, a mesh map was prepared using the ArcGis 10.2. The geoform was obtained based on the slope, where flat terrain = sites with a slope ≤ 10%, concave t. = slope ≥ 10 and ≤ 25% and convex t. = slope ≥ 25%.

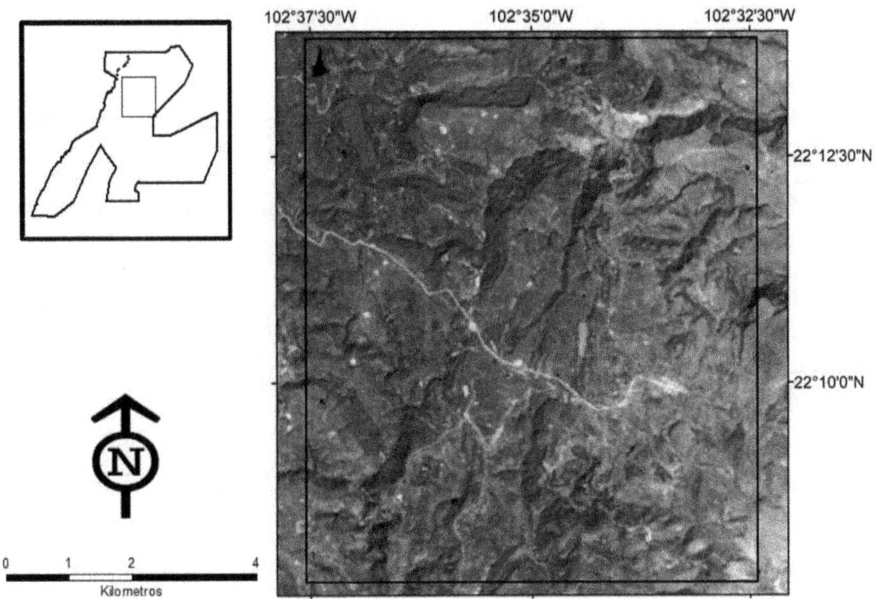


**Figure 2.**
*Location of the protected natural area sierra Fria, the study area of the Temperate Mountain Forest.*

### *2.1.2 Identification, distribution and abundance of forest species*

To identify the tree and shrub diversity in the study area, we conducted 60 phytoecological inventories in 60 different sites distributed randomly using the sampling scheme already described (**Table 1**).

The field samplings were performed in rectangular plots of 600 $m^2$, with a central line 100 m in length and two lateral lines with three m of separation. In each inventory, the frequency of the tree and shrub species present were determined, as well as the site environmental variables. Individuals with DBH ≥ 5 cm and height ≥ 1.50 m were considered as trees. Individuals below these categories were

| **Altitude levels** | **Topographic position** | | | | | | | | | |
|---|---|---|---|---|---|---|---|---|---|---|
| | Concave | | | | Convexe | | | | Flat | Total |
| | N | S | E | W | N | S | E | W | | |
| 2000–2200 | 0 | 0 | 0 | 0 | 1 | 1 | 0 | 0 | 1 | 3 |
| 2200–2400 | 0 | 1 | 0 | 0 | 3 | 2 | 0 | 0 | 0 | 6 |
| 2400–2600 | 8 | 4 | 1 | 1 | 8 | 3 | 0 | 0 | 3 | 28 |
| 2600–2800 | 2 | 0 | 0 | 1 | 4 | 3 | 0 | 0 | 11 | 21 |
| > 2800 | 2 | 0 | 0 | 0 | 0 | 0 | 0 | 0 | 0 | 2 |
| **Total** | **12** | **5** | **1** | **2** | **16** | **9** | **0** | **0** | **15** | **60** |
| **Total inventories** | | | | | | **60** | | | | |

*The intersections between lines and columns whose value is zero, indicate areas with little representativeness in the landscape and consequently an absence of samplings.*

**Table 1.**
*Number of samplings performed at different altitudinal levels, topographic positions and solar exposures, derived from the sampling system.*

considered as juveniles and shrubs. The variables recorded in the site were: altitude, slope (in %), solar exposure (N, S, E, W), physiography (flat land, hillock, plateau, middle slope, high slope, ravine bottom, creek), coverage (c1 = ≤10%; c2 = 11–30%; c3 = 31–50%; c4 = 51–70% y c5 = ≥ 70%) and geoform. Management variables related to land use (no use, forest exploitation, wildlife management, grazing, agriculture and conservation) were considered as well as intensity of use (null, moderate, over-exploited and not determinable). Each one of the sampling points were geographically located by Transverse Mercator Units (UTM).

In order to identify the oak and conifer species in the field, keys generated by De la Cerda [17] and Siqueiros [18], respectively, were used. The unknown species were collected in botanical presses and identified at the Autonomous University of Aguascalientes herbarium (HUAA). To leave evidence of the new species records in the ANP SF, specimens were deposited in the HUAA.

### *2.1.3 Distribution and abundance of species*

To estimate the distribution of tree and shrub forest species, the presence of each of the species found in each of the 60 sampling sites was quantified. In the case of species considered as restricted distribution (eg. *Quercus cocolobifolia, Pinus chihuahuana*, and *P. duranguensis* var. *quinquefoliata*), samples were taken at specific sites (n = 4), according to the information provided by De la Cerda [17] and Siquéiros [18]. Species with a wide distribution were those that occurred in the greatest number of sites.

The frequency of the species found was determined on 100 m transect at ground level, observing 100 separate points every meter. The species found at each point were recorded (when there was more than one vegetation layer), counting the number of times that each species appeared (absolute frequency) [16] over the whole transect. Relative frequency was calculated using the Equation [19]:

$$\text{Relative frequency} = \left(\frac{\text{Species frequency}}{\sum \text{Frequency values of all species}}\right) \tag{1}$$

Where:

Frequency of the species x = absolute frequency obtained from each site sampling.

Subsequently, an abundance index was calculated using the equation:

$$\text{Spp.ai} = \frac{\sum_{\text{relative frequencies}}}{\text{Number of sampled sites}} \tag{2}$$

Where:

Spp.ai = Identified Species abundance index.

With this data, distribution and abundance graphs of the main arboreal-shrub forest species were created. The phytoecological analysis was used to calculate the species richness and the Shannon index diversity ($H'$) and the beta Whittaker's (ßw) index respectively, the first were calculated as a function of the altitudinal level, the second also incorporating the geoform using the Species Diversity and Richness® (Pisces Conservation LTD) software. Pear calculate the indexes we used the equation:

$$H' = -\sum_{i=1}^{S} p_i \log_2 p_i \tag{3}$$

Where

S = species richness; Pi = proportion of the individuals of species i with respect to the total number of individuals; ni = number of individuals of species i

$$ß_w = \frac{S}{\bar{S}} \tag{4}$$

Where:

S = Species richness and S = mean richness of the site.

## 2.2 Dry Tropical Forest (DTF)

### *2.2.1 Study area*

Although there are some studies that suggest the existence of relics of Dry Tropical Forest (DTF) vegetation in some municipalities of the Aguascalientes State [15, 20], this ecosystem has a greater representation both in surface area and in its conservation status in Calvillo municipality. The study was conducted in 26 sites with DTF vegetation cover in Terrero de la Labor ejido, located within the Sierra Fria Protected Natural Area, in the Municipality of Calvillo, State of Aguascalientes, in Central Mexico. The ejido polygon comprises an area of 5,861 ha. [21], of which, the DTF occupies 45% of its total area (**Figure 3**). It is located within the extreme coordinates: 102°43'58.88" West Longitude and 22°6'4.78" North Latitude and at the Southeast end 102°41'24.95" West Longitude and 21°44'27.61" North Latitude.

### *2.2.2 Selection of the study sites and sampling design*

We used a stratified sampling design system [16]. Sampling strata were delimited based on geoforms, slope, exposure and altitude. To characterize geoforms, three criteria were used: concave, convex and flat terrain. A concave geoform was defined when the slope ranged between 10 and 25%, which usually corresponded to ravines and small depressions. When the sites had a slope between 25 and 60% they were characterized as convex sites. Flat terrains had slopes <10%.

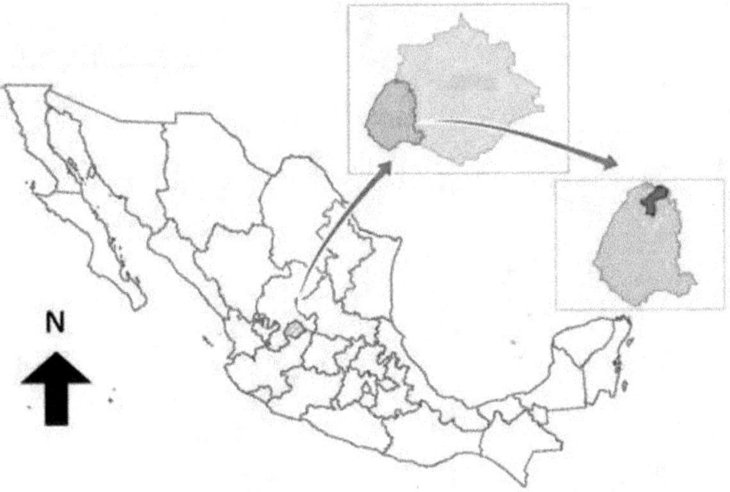


**Figure 3.**
*Location of the study area. (A) Mexico, (B) state of Aguascalientes, (C) municipality of Calvillo and (D) Terrero de la labor Ejido.*

Solar exposure was defined using an exposure map made with a Geographic Information System from a 2008 Spot® satellite image and a digital elevation model (MDE). Only the main cardinal points (North, South, East and West) were considered. To locate the altitudinal strata, the contours of the zone defined from the MDE were used. Subsequently, a grid map was developed for the identification of the sampling areas (See **Figure 4**).

### 2.2.3 Selection and characterization of sites to quantify of the composition and abundance of woody species

We established 26 sites to quantify phytoecological inventories, distributed in the landscape according to the above mentioned sampling system. At each point, the projected coordinates of the site were taken with GPS Garmin 48 XL line in UTM format, zone 13 North and with reference Datum WGS84 and with accuracies of 5 to 12 m with differential kinematic adjustment (WAAS). Subsequently, the points were placed on a SPOT 2010 satellite image (**Figure 5**). Site variables considered were the slope (%), solar exposure, physiography of the terrain, intensity and type of use and canopy coverage.

Slope at each sampling site was obtained by direct field measurement with a Bruntton clinometer with a precision of +/− 2° of variation for each 100 meters of length. This data in turn was contrasted with the data obtained from the digital elevation model with precision of 1 to 2 meters in the Z value. Five classes were used to define the slope: i) <10%, ii) 11–30, iii) 31–50, iv) 51–70 and v) > 70. Exposure to solar radiation was estimated considering the cardinal points North (N), South (S), East (E) and West (O), as well as their combinations.

The altitude of each site was obtained directly in the field with the support of a GPS with barometric adjustment to reduce the effect of mathematical variation of the Geoid model and with precision of 1 to 3 meters. This was compared with the data obtained from the prospecting of points against elevation level curves obtained

**Figure 4.**
*Geographic representation of dry tropical Forest and the sampling points in Terrero de la labor Ejido, in the municipality of Calvillo, Aguascalientes.*

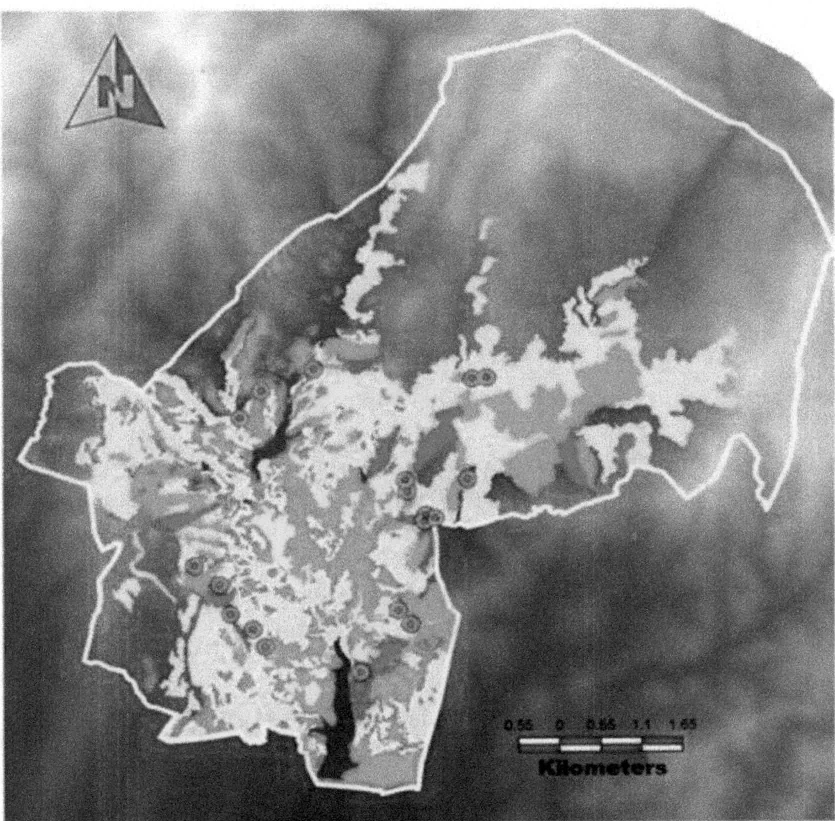


**Figure 5.**
*Ipsographic model of Ejido Terrero de la labor Ejido polygon, and distribution of the sampling points in the DTF.*

from the digital elevation model to reduce the potential errors of direct measurements.

The physiography of the terrain was characterized considering flat terrain (slope < 10%), steep (without slope), medium slope (10–25%) and high slope (>60%). The exposure of the sites was quantified with a compass and the magnetic north was taken as reference for its definition in the previously defined ranges. Exposure for each stand of the sampling site was also analyzed along with the digital model of exposures generated from the digital elevation model. The **Table 2** shown the sample points distributed in the landscape of the Dry Tropical Forest.

Other characteristics considered in the description of the sites were the degree of modification (i.e. transformation of geographical space, introduction of species), its intensity (light, medium and overexploited), as well as the type of use by local inhabitants (hunting, grazing, gathering, etc.).

### *2.2.4 Species richness*

To describe species composition, we used a sampling design based on nested plots in an area of 1024 $m^2$ in each inventory, using the criteria of the minimum area [16]. We started with a plot of 1 x 1 m in a direction perpendicular to the slope in which all present species were recorded, and subsequently, the plot. Subsequently, the plot was increased in size to 2 X 1, 2 X 2, 2 X 4, 4 X 4 m etc. registering the new species for each increment in the area of the squares until reaching the maximum

| Meters above sea level (masl) | Topographic position | | | | | | | | | | Total |
|---|---|---|---|---|---|---|---|---|---|---|---|
| | Concave coverage (%) | | | | | Convex coverage (%) | | | | | |
| | <10 | 11–30 | 31–50 | 51–70 | >71 | <10 | 11–30 | 31–50 | 51–70 | >71 | |
| 1800–1900 | 0 | 0 | 0 | 0 | 1 | 0 | 1 | 0 | 3 | 0 | 5 |
| 1901–2000 | 0 | 0 | 0 | 0 | 0 | 1 | 1 | 3 | 5 | 1 | 11 |
| 2001–2100 | 0 | 0 | 0 | 0 | 0 | 1 | 1 | 2 | 2 | 0 | 6 |
| >2100 | 0 | 0 | 0 | 0 | 0 | 1 | 1 | 0 | 4 | 0 | 4 |
| Total inventories | 0 | 0 | 0 | 0 | 1 | 0 | 0 | 5 | 14 | 1 | 26 |

**Table 2.**
*Distribution of samplings sites according to the proposed design.*

extension (i.e.: 32 x 32 m = 1024 m2), to obtain an area/species curve. We then identified the area in which the present species stabilized. This sampling method increased the probability of finding rare species as the area increased, an effect known as Rarefaction [22].

Identification of species was estimated in the field by morphological characters described in previous studies. Specimens that could not be identified in the field were collected and later identified in the Herbarium of the Autonomous University of Aguascalientes (HUAA).

We used the linear intercept survey method (Canfield line). A 100 m long line was perpendicular to the slope, starting at the GPS coordinates of the sampling site, then intersection lines were defined were individuals of DTF species were counted at constant intervals of 1 meter. Shrub and tree individuals were categorized into five heights classes 0–1 m, 1.1–2 m, 2.1–4 m , 4 –8 m , 8 –15 m and > 16 m. For each class we measured canopy cover of each species by measuring the perpendicular projection of the crown and the frequency of species. To estimate crown, cover the following formula was used:

Cover C()=Σ length of individuals of species i/total length of intersections X 100.

To estimate frequencies, we used the formula:

Frequency (F) = Σ of number of times that individuals of the species intercepted by the line/Σ total species intercepted X 100.

### 2.2.5 Data analysis

Species composition was estimated through the identification of the species found in each of the sampling plots. To find a limit on the number of samples and to reduce the possibility of under- or over-sampling, we conducted a rarefaction analysis. The Shannon-Wienner alpha diversity ($H'$) was calculated for each of the sites and for each altitudinal level using the Richness and diversity species® software, considering that there could be variation in diversity according to the change in environmental conditions in temperature and precipitation as mentioned in the Standard Atmospheric Index (decrease of 0.6°C/100 m altitude).

The formula of the Shannon index is:

$$H' = -\sum_{i=1}^{S} p_i \log_2 p_i \tag{5}$$

where:

- $S$– Total number of species (species richness)
- $p_i$– Proportion of individuals of species i in respect to total of individuals (i.e.: relative abundance of species i): $\frac{n_i}{N}$
- $n_i$– number of individuals of species i
- $N$– Total number of individuals of all species

The index considers the number of species present in the study area (species richness), and the relative number of individuals of each of those species (abundance).

To estimate replacement rates of species Whitakker's β diversity was computed, using the diversity found for each altitudinal level analyzed as reference.

$$\beta = \frac{S}{\alpha - 1} \tag{6}$$

Where: β = Whitakker's β diversity.
S = Total number of species in samples.
α = Mean number of species in samples.

# 3. Results

## 3.1 Temperate Mountain Forest

### *3.1.1 Richness and diversity species*

In the 60 sites, 50 species were recorded, corresponding to 20 families and 27 genera (**Table 3**), of which, due to their structure, 47% (n = 24) were considered trees (height ≥ 3.5 m) and 53% (n = 27) shrubs and juveniles. The best represented families were *Fagaceae* (11 species), *Pinaceae* (8 species) and *Ericaceae* (5 species). The species *Q. obtusata* Bonpl., *J. duranguensis* Martínez and *Crataegus* sp. are new reports in the Sierra Fria.

On average, the highest $H'$ diversity index is found in sites whose altitude ranges between 2400 and 2600 and 2600–2800 mamsl ($H'$ = 1.48 y 1.63, respectively), the former associated with ravines and difficult access places; the second index corresponds to places with higher moisture content and without use. The lowest indexes (H = 1.22 y 1.36) were found in altitudinal ranges of 2200–2400 and 2000–2200 m, respectively, located on flat lands with intensive management and high resource use rates. According to the geoform, the diversity Wittaker's *ß* was greater on the convex sites (*ßw* = 5.80), followed by the concave sites (*ßw* = 4.27) and flat lands (*ßw* = 4.04). According to the altitudinal level, the highest diversity was found in the sites whose altitude ranges between 2,400 and 2,600 m (*ßw* = 7.22), mainly in ravines and places hard to access. In contrast, the lowest indexes were found on site with an altitude lower than 2, 400 m (*ßw* = 4.52), located on flat lands, under intensive management and easy access.

In the **Figure 6**, we shown an example of dominant vegetation in Temperate Mountain Forest (in conifers, *Pinus leiophylla* and *P. teocote* in order) in the Sierra Fria Protected Natural Area.

| Species | Key | Common name* | Family | Forest classification** | Use¥ | Report*** |
|---|---|---|---|---|---|---|
| *Acacia farnesiana* | *Acafar* | Huizache | *Leguminosae* | Tr | Nu | Y |
| *Asclepias linearis* | *Aline* | Romerillo | *Apocynaceae* | Sh | Nu | Y |
| *Arbutus arizonica* | *Aariz* | Madroño | *Ericaceae* | Sh | Fe | Y |
| *Arbutus xalapensis* | *Axala* | Madroño rojo | *Ericaceae* | Sh | Fe | Y |
| *Arbutus glandulosa* | *Aglan* | Madroño blanco | *Ericaceae* | Sh | Fe | Y |
| *Arctostaphylos pungens* | *Apun* | Manzanita | *Ericaceae* | Sh | Fe | Y |
| *Budleia scordioides* | *Bsco* | Vara blanca | *Compositae* | Sh | Fr | Y |
| *Budleia cordata* | *Bcor* | Tepozan | *Compositae* | Tr | Fr | Y |
| *Bursera fagaroides* | *Burfaga* | Venadilla | *Burseraceae* | Sh | Nu | Y |
| *Comerostaphyllis spp.* | *Comesp* | Pacuato | *Ericaceae* | Sh | Med | Y |
| *Dalea bicolor* | *Dabic* | Engordacabra | *Fabaceae* | Sh | Fr | Y |
| *Dasylirion acotriche* | *Dasaco* | Sotol | *Agavaceae* | Sh | Nu | Y |
| *Dodonaea viscosa* | *Dovisc* | Jarilla | *Sapindaceae* | Sh | Med | Y |
| *Eucaliptus camaldulensis* | *Eucamal* | Eucalipto | *Myrtaceae* | Tr | Nu | Y |
| *Fraxinus uhdei* | *Frauhd* | Fresno | *Oleaceae* | Tr | Nu | Y |
| *Garria ovata* | *Garova* | planta peluda | *Garryaceae* | Sh | Nu | Y |
| *Pinus chihuahuana* | *Pinchi* | Pino Prieto | *Pinaceae* | Tr | Nu | Y |
| *Pinus duranguensis Mart.* | *PiduM* | Pino verde | *Pinaceae* | Tr | Ew | Y |
| *Pinus duranguensis f. quinquefoliata* | *PiduQ* | Pino verde | *Pinaceae* | Tr | Nu | Y |
| *Pinus leiophylla* | *Pile* | Pino Prieto | *Pinaceae* | Tr | Nu | Y |
| *Pinus lumholtzii* | *Pilum* | Pino llorón | *Pinaceae* | Tr | Nu | Y |
| *Pinus michoacana* | *Pimich* | Pino barbón | *Pinaceae* | Tr | Nu | Y |
| *Pinus cembroides* | *Picem* | Pino chaparro | *Pinaceae* | Tr | Nu | Y |
| *Prosopis laevigata* | *Prolae* | Mesquite | *Pinaceae* | Sh | Nu | Y |
| *Pinus teocote* | *Pinteo* | Pino | *Pinaceae* | Tr | Ew | Y |
| *Jatropha dioica* | *Jadio* | Sangre de grado | *Euphorbiaceae* | Sh | Nu | Y |
| *Juniperus flacida* | *Jufla* | Olmo triste | *Cupresaceae* | Sh | Nu | Y |
| *Juniperus deppeana* | *Judep* | Táscate | *Cupresaceae* | Tr | Fe-Tu | Y |
| *Juniperus duranguensis* | *Judur* | Cedro chino | *Cupresaceae* | Sh | Nu | New |
| *Opuntia leucotricha* | *Opuleu* | Nopal duraznillo | *Cactaceae* | Sh | Nu | Y |
| *Opuntia streptacantha* | *Opust* | Nopal cardón | *Cactaceae* | Sh | Nu | Y |
| *Prunus serotina* | *Pruser* | Cerezo negro | *Rosaceae* | Sh | Ft | Y |
| *Quercus cocolobifolia* | *Queco* | Palo manzano | *Fagaceae* | Tr | Fe | Y |
| *Quercus chihuahuensis* | *Quechih* | Palo blanco | *Fagaceae* | Tr | Fe | Y |

| Species | Key | Common name* | Family | Forest classification** | Use¥ | Report*** |
|---|---|---|---|---|---|---|
| *Quercus laeta* | *Quela* | Palo blanco | *Fagaceae* | Tr | Fe | Y |
| *Quercus grisea* | *Quegri* | Palo chino | *Fagaceae* | Tr | Fe | Y |
| *Quercus potosina* | *Quepo* | Palo chaparro | *Fagaceae* | Tr | Fe | Y |
| *Quercus microphylla* | *Quemic* | Chaparrito | *Fagaceae* | Sh | Nu | Y |
| *Quercus resinosa* | *Queres* | Encino hojudo | *Fagaceae* | Tr | Le | Y |
| *Quercus rugosa* | *Querug* | Palo blanco | *Fagaceae* | Tr | To | Y |
| *Quercus sideroxyla* | *Quersid* | Palo rojo | *Fagaceae* | Tr | Fe-le | Y |
| *Quercus eduardii* | *Queredu* | Palo rojo | *Fagaceae* | Tr | Fe-le | Y |
| *Quercus sp.* | *Encino 1* | Encino | *Fagaceae* | Sh | Fe | Y |
| *Quercus obtusata* | *Querobt* | Encino | *Fagaceae* | Sh | Fe | New |
| *Yucca filifera* | *Yufi* | Palma | *Agavaceae* | Sh | Nu | Y |
| *Odontotrichum amplum* | *Adoamp* | Vaquerilla | *Asteraceae* | Sh | Nu | Y |
| *Phytecellobium leptophyllum* | *Phylep* | Gatuño de la sierra | *Leguminosae* | Sh | Nu | Y |
| *Eisenhardtia polystachya* | *Eipol* | Varaduz | *Fabaceae* | Sh | Nu | Y |
| *Crataegus spp.* | *Crasp* | Tejocote | *Rosaceae* | Tr | Nu | New |
| *Quercus sp-2* | *Encino 2* | Encino | *Fagaceae* | Sh | Nu | Y |
| *Ipomoea stans* | *Ipost* | Galuza | *Convolvulaceae* | Sh | Nu | Y |

**The common names were provided by the habitants of "La Congoja" community and do not necessarily correspond to the common name in other localities where these species could be found.*
***Within the forest classification, Tr = Tree and Sh = Shrub.*
****The reports correspond to the flora identified previously, the new reports correspond to the individuals identified in this study.*
*¥The use of the forest species recorded, depends on the forest managers experience, in this way, Nu = no use; Fe = firewood extraction; Fr = use as forage plant; Med = Medicinal use; Pt = Timber use; for extraction as fence pole; Ew = extraction as wood stripe from* P. teocote*; le = leave extraction for ornamentals; To = tools.*

**Table 3.**
*Forest species identified in an area of the SF-natural protected area, Aguascalientes.*

### *3.1.2 Distribution and abundance of species*

The most widely distributed species belong to the genus *Juniperus* (locally known as cedros or táscates), *Quercus* (oaks) and *Arbutus* (locally known as madrone). *J. deppeana* is the most widely distributed species followed by *Quercus potosina* and by *Arctostaphyllos pungens.* The madrones (*Arbutus xalapensis* and *A. glandulosa*) appear in fourth and fifth place, respectively (**Figure** 7).

Out of 50 recorded species, 6 are the ones with the highest abundance indexes. *Q. potosina*, the species best represented in the landscape. This species presents the highest abundance index (ia = 0.1585), followed by *J. deppeana* (ia =0.1102) which also presents the widest distribution. Inside the genus *Pinus*, *P. leiophylla* is the most abundant, even above manzanita (*Arctostaphyllos pungens*) and red oaks (*Q. sideroxyla* y *Q. eduardii*; **Figure 8**).

There are species such as *Pinus chihuahuana*, *Pinus lumholtzii* and *Pinus duranguensis* that present restricted distribution, but are abundant in very specific

**Figure 6.**
*Typical vegetation of the Temperate Mountain Forest. (a) Landscape dominates by conifers in the sierra Fria, in this case, by* Pinus leiophylla*; (b) wild ash twig (*Fraxinus uhdei*); (c) Manzanita (*Arctostaphyllos pungens*) specimen in a plateau of the sierra Fria; (d) oak specimen locally known as Palo chino (*Quercus grisea*).*

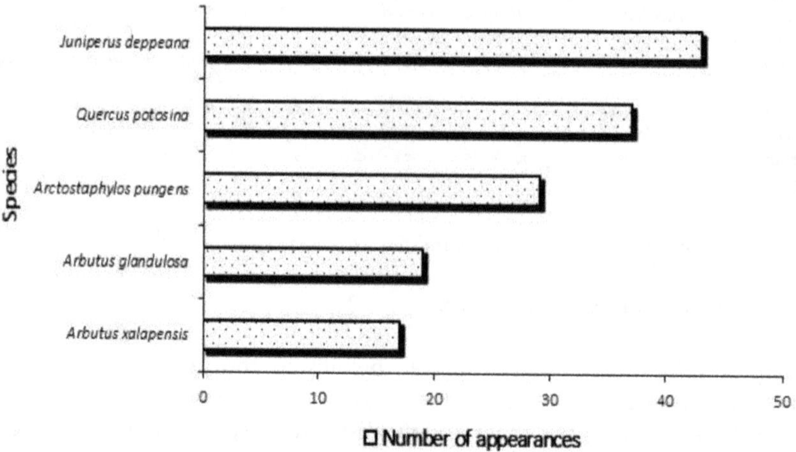


**Figure 7.**
*Forest species with wide distribution in the study area inside the SF-natural protected area.*

sites. The distribution analysis based on the altitudinal gradients and geoform suggests that the altitudinal stratum between 2 000 and 2 200 m is the one with the lowest tree and shrub species richness. The best represented species in this range belong to the xeric shrubland being three of them such as *Dodonaea viscosa*, *Phytecellobium leptophyllum*, and *Odontotrichum amplum*, considered as overgrazing indicator species [22]. From the second stratum (2 200 to 2 400 mamsl) *Pinus* and

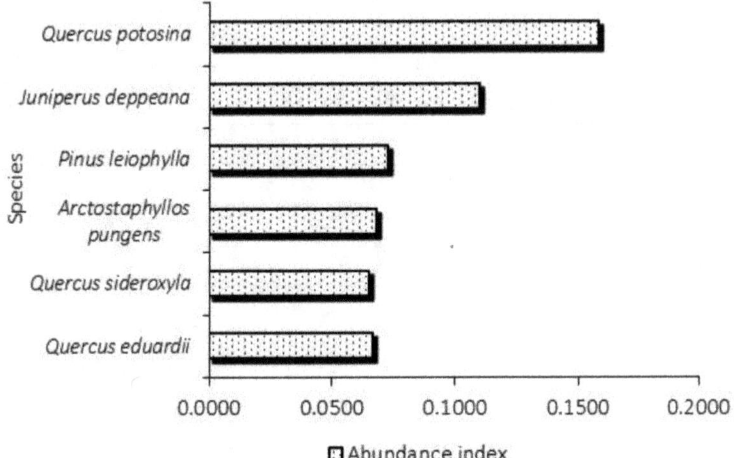


**Figure 8.**
*Abundancy indexes from the species best represented in the SF-natural protected area, the most representative ecosystem of Temperate Mountain Forest in Ags. The X axis represents the abundancy index which ranges between 0.0670 (*Quercus eduardii*) and 0.1585 (*Q. potosina*). The maximum value of the abundancy index could be 1.*

*Quercus* species begin to appear, although isolated *Quercus resinosa* individuals can be found at higher altitudes (**Table 4**).

Out of the dominant conifer species at the SF-Natural Protected Area, *Pinus leiophylla* and *P. teocote* are distributed at altitudes ranging from 2 400 to 2 600 masl. Between 2 600 and 2 800 mamsl these two species are more dispersed and located mainly in ravines. *P. leiophylla* is also located on plateaus at 2 700 m (e.g. Mesa del Águila and Mesa del Aserradero). Red oaks (*Q. eduardii* and *Q. sideroxyla*) are distributed at altitudes from 2 400 to 2 600 m, mainly along the ravines (**Table 3**).

In **Figure 9** we shown some species of *Pinus* genera dominants in the intermediate altitudinal strata of SF-Protected Natural Area.

## 3.2 Dry Tropical Forest

### *3.2.1 Richness and diversity woody species*

We identified 79 species of trees and shrubs, within 45 genera and 14 families (see **Table 5**). The best represented families were Fabaceae (13 genera), Asteraceae (11 genera) and Cactaceae (9 genera). The genera better represented were *Opuntia* (n = 4 spp.), *Acacia* (n = 4 spp.) and *Bursera* (n = 3 spp.). The genero *Salvia* is also important.

The *H′* diversity found in the DTF of the ejido Terrero de la Labor ejido is constant. The highest diversity index found was 3.49 in two of the 26 analyzed sites, which apparently are well conserved sites. On the contrary, three sites had the lowest *H′* diversity index with 2.77 (**Table 6**). Although there are apparently no differences, the highest diversity indexes are located mainly in ravines and north facing exposures, and in locations with difficult access (see **Table 7**).

### *3.2.2 Distribution and abundance of woody species in the DTF*

Of the 79 species identified, eight are distributed in more than 70% of the plots of Terrero de la Labor ejido. The species with the greater distribution are the

| | ALTITUDE (MASL)± | | | | | | | | | | | | | | |
|---|---|---|---|---|---|---|---|---|---|---|---|---|---|---|---|
| **Species** | **A1** | | | **A2** | | | **A3** | | | **A4** | | | **A5** | | |
| | **2** | **2.1** | **2.19** | **2.2** | **2.3** | **2.39** | **2.4** | **2.5** | **2.59** | **2.6** | **2.7** | **2.79** | **2.8** | **2.9** | **3** |
| *Arctostaphylos pungens* | | | | | | | | | | | | | | | |
| *Dodonaea viscosa* | | | | | | | | | | | | | | | |
| *Juniperus deppeana* | | | | | | | | | | | | | | | |
| *Quercus potosina* | | | | | | | | | | | | | | | |
| *Bursera fagaroides* | | | | | | | | | | | | | | | |
| *Eisenhardtia polystachya* | | | | | | | | | | | | | | | |
| *Juniperus flacida* | | | | | | | | | | | | | | | |
| *Acacia farnesiana* | | | | | | | | | | | | | | | |
| *Prosopis laevigata* | | | | | | | | | | | | | | | |
| *Arbutus glandulosa* | | | | | | | | | | | | | | | |
| *Quercus resinosa* | | | | | | | | | | | | | | | |
| *Yucca filifera* | | | | | | | | | | | | | | | |
| *Phytecellobium leptophyllum* | | | | | | | | | | | | | | | |
| *Asclepias linearis* | | | | | | | | | | | | | | | |
| *Quercus eduardii* | | | | | | | | | | | | | | | |
| *Odontotrichum amplum* | | | | | | | | | | | | | | | |
| *Pinus leiophylla* | | | | | | | | | | | | | | | |
| *Pinus teocote* | | | | | | | | | | | | | | | |
| *Quercus rugosa* | | | | | | | | | | | | | | | |
| *Quercus chihuahuensis* | | | | | | | | | | | | | | | |
| *Quercus sideroxyla* | | | | | | | | | | | | | | | |
| *Arbutus xalapensis* | | | | | | | | | | | | | | | |
| *Pinus lumholtzii* | | | | | | | | | | | | | | | |
| *Juniperus duranguensis* | | | | | | | | | | | | | | | |
| *Quercus cocolobifolia* | | | | | | | | | | | | | | | |
| *Quercus grisea* | | | | | | | | | | | | | | | |
| *Quercus laeta* | | | | | | | | | | | | | | | |

**The species distribution in different altitudinal gradients was as a function of the 10 dominant species (obtained from the frequency/site) at each altitudinal stratum.*
*±Altitudes (A1-A5) are calculated in m* 1000.*
*‡The bars with gray shades indicates that this species is abundant at the altitudinal gradient where it was found. In contrast, the black shades indicate that although this species is not abundant, it was found in.*

**Table 4.**
*Dominant species distribution by altitudinal strata.*

*Myrtillocactus geometrizans* (garambullo), *Ipomoea murucoides* (palo bobo), *Eysenhardtia polystachya* (varaduz), *Bursera fagaroides* (venadilla), and *Forestiera phillyreoides* (palo blanco) (**Figure 10**), which were located in 96, 92, 90, 88 and 86% of the plots respectively, assuming that the sampling sites are representative of the entire landscape.

On the other extreme, the rarest species were *Plumeria rubra*, *Ficus petiolaris* and *Fraxinus purpurea*. The first species was only located in one site, while the last two

**Figure 9.**
*Populations of* Pinus *(*Pinus *spp) at the SF-natural protected area. The photograph on the left side shows a* Pinus leiophylla *population at the Barranca de Piletas. The pothogragh in the right side shows an image of* Pinus duranguensis. *Photographs as courtesy of Clemente Villalobos llamas and Vicente Díaz Núñez.*

| Species | Family | Common name |
|---|---|---|
| *Acacia berlandieri* Benth. | Fabaceae | Carbonera |
| *Acacia farnesiana* (L.) Willd. | Fabaceae | Tepame |
| *Acacia pennatula* (Schltdl. & Cham.) Benth. | Fabaceae | Huizache o Cascalote |
| *Agave angustifolia* Haw. | Asparagaceae | Lechuguilla |
| *Albizia plurijuga* (Standl.) Britton & Rose | Fabaceae | Tepeguaje blanco |
| *Alnus acuminata* Kunth | Betulaceae | Aile |
| *Amelanchier denticulata* (Kunth) K. Koch | Rosaceae | Duraznillo |
| *Amphipterygium molle* (Hemsl.) Hemsl. & Rose | Anacardiaceae | Cuachalalate |
| *Asclepias linaria* Cav. | Apocynaceae | Algodoncillo |
| *Ayenia mexicana* Turcz. | Sterculioideae | |
| *Baccharis heterophylla* Kunth | Asteraceae | Escobilla |
| *Bouvardia multiflora* (Cav.) Schult. & Schult. f. | Rubiaceae | Clavelito |
| *Brickellia veronicifolia* (Kunth) A. Gray | Asteraceae | Orégano de monte |
| *Buddleja cordata* Kunth | Buddlejaceae | Tepozán blanco |
| *Buddleja sessiliflora* Kunth | Buddlejaceae | Tepozán verde |
| *Bursera bipinnata* Donn. Sm. | Burseraceae | Lantrisco |
| *Bursera fagaroides* (Kunth) Engl. | Burseraceae | Venadilla |
| *Bursera penicillata* (DC.) Engl. | Burseraceae | Arbol de chicle |
| *Calliandra eriophylla* Benth. | Fabaceae | Calandria |
| *Castilleja tenuifolia* M. Martens & Galeotti | Scrophulariaceae | Hierba del cancer |
| *Cedrela dugesii S. Watson* | Meliaceae | Cedro |

| Species | Family | Common name |
|---|---|---|
| *Ceiba aesculifolia* (Kunth) Britten & Baker f. | Malvaceae | Pochote |
| *Celtis caudata* Planch. | Ulmaceae | Capulincillo |
| *Celtis pallida* Torr. | Ulmaceae | Vara en cruz |
| *Chusquea sp* | Poaceae | Camalote |
| *Colubrina triflora* Brongn. Ex G. Don | Rhamnaceae | Algodoncillo |
| *Cordia sonorae* Rose | Boraginaceae | Amapa o Vara prieta |
| *Croton ciliatoglandulifer* Ortega | Euphorbiaceae | Algodoncillo |
| *Dasylirion acrotrichum* (Schiede) Zucc. | Asparagaceae | Sotol |
| *Dodonaea viscosa* Jacq. | Sapindaceae | Jarilla |
| *Erythrina flabelliformis* Kearney | Fabaceae | Colorín |
| *Eupatorium sp* | Asteraceae | Copalillo |
| *Eysenhardtia polystachya* (Ortega) Sarg. | Fabaceae | Palo azulo o Varaduz |
| *Eysenhardtia punctata* Pennell | Fabaceae | Palo cuate |
| *Ferocactus histrix* Lindsay | Cactaceae | Biznaga costillona |
| *Ficus petiolaris* Kunth | Moraceae | Ficus silvestre |
| *Forestiera phillyreoides* (Benth.) Torr. | Oleaceae | Palo blanco |
| *Fraxinus purpusii* Brandegee | Oleaceae | Saucillo |
| *Gymnosperma glutinosum* (Spreng.) Less. | Asteraceae | Cola de zorra |
| *Heliocarpus terebinthinaceus* (DC.) Hochr. | Meliaceae | Cicuito o Cuero de indio |
| *Ipomoea murucoides* Roem. & Schult. | Convolvulaceae | Palo bobo |
| *Iresine sp.* | Amaranthaceae | Cola de zorra |
| *Jatropha dioica* Sessé | Euphorbiaceae | Sangregrado |
| *Koanophyllon solidaginifolium* (A. Gray) R. M. King & H. Rob. | Asteraceae | Caballito |
| *Karwinskia humboldtiana* (Schult.) Zucc. | Rhamnaceae | Coyotillo |
| *Leucaena esculenta* (Moc. & Sessé ex DC.) Benth. | Fabaceae | Guaje rojo |
| *Lippia inopinata* Moldenke | Verbenaceae | Palo oloroso |
| *Lysiloma acapulcense* (Kunth) Benth. | Fabaceae | Ébano o Palo fierro Tepeguaje |
| *Lysiloma microphyllum* Benth. | Fabaceae | Tepeguaje |
| *Mammillaria bombycina* Quehl | Cactaceae | Biznaga de seda |
| *Mammillaria sp.* | Cactaceae | Biznaga |
| *Manihot caudata* Greenm. | Euphorbiaceae | Pata de gallo |
| *Mimosa monancistra* Benth. | Fabaceae | Gatuño o Uña de gato |
| *Mimosa sp.* | Fabaceae | Huizache |
| *Myrtillocactus geometrizans* (Mart. ex Pfeiff.) Console | Cactaceae | Garambullo |
| *Montanoa leucantha* (Lag.) S.F. Blake | Asteraceae | Talacao o Vara blanca |
| *Opuntia leucotricha* DC. | Cactaceae | Nopal chaveño o duraznillo |
| *Opuntia robusta* J.C. Wendl. | Cactaceae | Tuna tapona |
| *Opuntia sp.* | Cactaceae | Nopal |
| *Opuntia streptacantha* Lem. | Cactaceae | Nopal cardón |

| Species | Family | Common name |
|---|---|---|
| *Perymenium mendezii* DC. | Asteraceae | |
| *Pistacia mexicana* Kunth | Anacardiaceae | Lantrisco |
| *Pittocaulon filare* (McVaugh) H. Rob. & Brettell | Asteraceae | Palo loco |
| *Plumbago pulchella* Boiss | Plumbaginaceae | Chilillo medicinal |
| *Plumeria rubra* L. | Apocynaceae | Flor de mayo |
| *Prosopis laevigata* (Humb. & Bonpl. ex Willd.) M.C. Johnst. | Fabaceae | Mezquite |
| *Ptelea trifoliata* L. | Rutaceae | Naranjo agrio o Zorrillo |
| *Quercus laeta* Liebm. | Fagaceae | Roble blanco |
| *Salvia mexicana* L. | Labiatae | Tlacote |
| *Salvia sp.* | Labiatae | Salvias |
| *Stachys coccínea* Ortega | Labiatae | Mirto |
| *Stenocereus queretaroensis* (F. A. C. Weber) Buxb. | Cactaceae | Pitahaya |
| *Tecoma stans* (L.) Juss. ex Kunth | Bignoniaceae | Tronadora |
| *Trixis angustifolia* DC. | Asteraceae | Vara verde |
| *Verbesina serrata* Cav. | Asteraceae | Vara blanca |
| *Viguiera quinqueradiata* (Cav.) A. Gray ex S. Watson | Asteraceae | Vara amarilla |
| *Wimmeria confusa* Hemsl. | Celastraceae | Algodoncillo |
| *Yucca filifera* Chabaud | Asparagaceae | Palma |
| *Zanthoxylum fagara* (L.) Sarg. | Rutaceae | Rabo lagarto |

**Table 5.**
*List of species identified in the dry tropical Forest of Terrero de la labor Ejido, Calvillo, Ags.*

| Altitud level (masl) | Sampled sites | *H´* |
|---|---|---|
| 1851–1900 | 7 | 3.08 |
| 101–1950 | 5 | 2.57 |
| 1951–2000 | 7 | 3.60 |
| 2001–2100 | 5 | 3.14 |
| >2100 | 2 | 3.25 |

**Table 6.**
*Average H′ diversity indices associated to different altitudinal ranges in the DTF of Terrero de la labor Ejido.*

were only found in two and three sampling sites, respectively. Their low frequency could be associated to their presence in mid statured forests. The most abundant species are those that, even though they are not those with a wide distribution in the landscape, in the places where they are located their frequency is higher than the rest of the identified species. In the DTF of the Terrero de la Labor, the most abundant species belonged to five different genera, of which the most important are *Lysiloma microphylla* (tepeguaje), *Ipomoea murucoides* (palo bobo), and *Bursera fagaroides* (locally known as venadilla) (**Figure** 7). In the case of *Ipomoea murucoides*, it occupies the second place in both distribution and abundance (see **Figure 11**).

| Slope range (%) | Sampled sites | $H'$ |
|---|---|---|
| 0–9 | 6 | 2.94 |
| 9–25 | 3 | 2.88 |
| 26–37 | 5 | 2.92 |
| 37–49 | 4 | 3.26 |
| 49–64 | 5 | 3.34 |
| >65 | 3 | 3.39 |

**Table 7.**
*Diversity indexes associated to different slopes of the sites.*

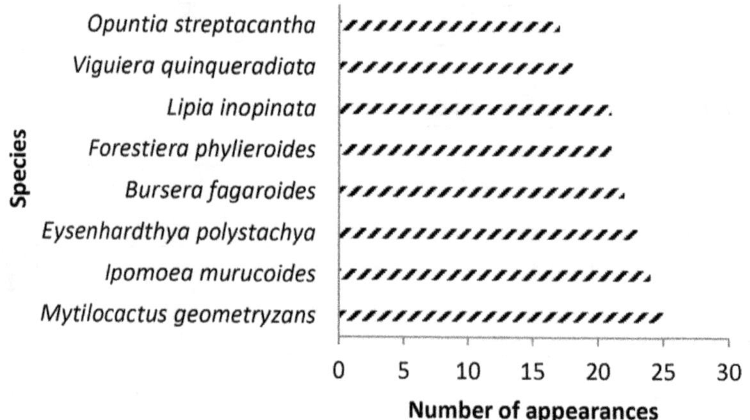


**Figure 10.**
*Species best represented in the DTF Terrero de la labor Ejido.*

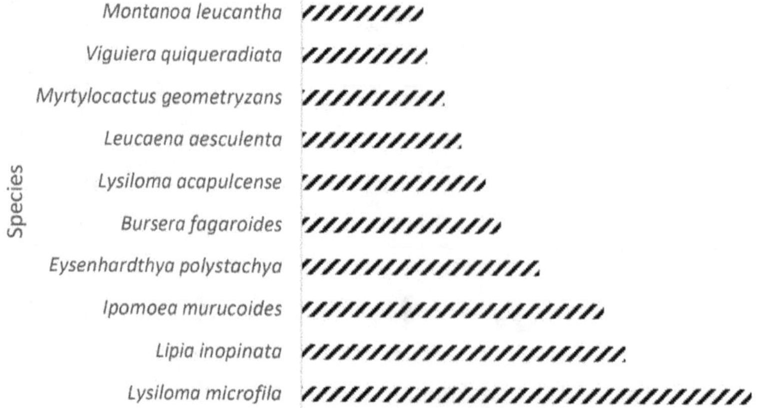


**Figure 11.**
*List of species with the highest abundance in Terrero de la labor Ejido.*

The **Figure 12** shown some species of the dominant vegetation in tropical dry forest, in this case, of the Terrero de la labor and las Moras ejidos in the Municipality of Calvillo, Aguascalientes State.

## 4. Discussion

The loss of biodiversity is one of the environmental problems that has managed to arouse broad global interest in the last two decades [4, 23]. Some of the main

**Figure12.**
*Diversity of forest species in the tropical dry Forest. (a) Landscape of the tropical dry Forest in the Terrero de la labor and las Moras Ejido; (b) an example of* Manihot caudata, *locally known as jaboncillo; (c) specimen of* Bursera fagaroides *(locally known as venadilla or papelillo). Photographs courtesy of Vicente Díaz Núñez, Joaquín Sosa-Ramírez and Jesús Argumedo-Espinoza.*

causes are related to human activities, either directly (overexploitation) or indirectly (habitat alteration), although there is generally an interaction between them. The communication systems have impacted in such a way that both the government and the private sector, as well as society in general, consider a priority to direct greater efforts towards conservation programs. The basis for an objective analysis of biodiversity and its change lies in its correct evaluation and monitoring.

In the Temperate Mountain Forest, the 50 woody species identified show a high species richness in comparison with other mountain regions. The best represented genera correspond to oak trees (*Quercus spp.*) and pines (*Pinus spp.*). The studied area harbors a small portion (6.8%) of the oak species that inhabit Mexico (161 species) [24]; although, this percentage is lower than those reported in areas with a greater territorial surface and higher rainfall, such as the case of San Luis Potosi and Jalisco States, which has identified 45 and 51 oak species respectively, that represent 27.95 and 36.9% of the total oak species registered in Mexico [25], the SF-NPA represent less 5% of the territorial surface in the mentioned states. In relation to pines, the studied area has about 17% of the species identified in Mexico [26]. This proportion is similar to that reported by Márquez-Linares et al., [27] in an area of pine-oak forest, in Durango, Mexico, where they recorded 8 pine species. In relation to "Las Joyas" scientific station, in the Sierra de Manantlán Biosphere Reserve, the *Quercus* diversity (16 spp.) is similar to the one in the Sierra Fria, although the area of las Joyas is smaller (Ca. 3600 ha.).

In the Sierra Fría, the most widely distributed and abundant species are the potosine oak (*Q. potosina*) and alligator juniper (*J. deppeana*). In the case of *Q. potosina*, its distribution and abundance may be related to the dominant physiography in this area, as well as to the mean annual precipitation (650 mm). The appearance of *J. deppeana* is possibly related to the disturbances that occurred in the Sierra Fria during the period between 1920 and 1940 [28]. This species has probably been a pioneer in the recovery of the vegetation cover, although the presence of manzanita (*A. pungens*) has also been documented colonizing sites where disturbances occurred, either natural, as in the case of fires or, anthropogenic, such as forest clearance and harvesting. Pines population is restricted to the Sierra Fría and the Sierra del Laurel. In the Sierra Fria, *Pinus teocote* (locally known as pino ocote) and *Pinus leiophylla* (locally known as pino prieto) are the two most abundant pine species. Its population is abundant in humid places and altitudes higher than 2,500 masl. *P. leiophylla* isolated specimens have been found on flat lands, which suggests that in the past this species had a greater distribution. In the Sierra del Laurel only two pine species have been identified, the pino triste (*Pinus lumholtzii*) and the pino piñonero (*Pinus cembroides var. cembroides*) in isolated populations, which suggests that in the past they were more abundant; however, the existing information is incipient.

The *H'*diversity indexes for each altitudinal stratum suggest that, between 2,400 and 2,600 mamsl, the plant richness of the SF-Natural Protected Area is similar to temperate forests, similar to what the β Whitakker index showed.

The distribution of species such as *J. deppeana* and *Q. potosina*, the most abundant and widely distributed, are influenced by flat sites and canopy covers that vary between 30 and 50%. One explanation is that *Q. potosina* tolerates high drought rates and *J. deppeana* is a pioneer species in disturbed sites, as suggested by Minnich et al. (1994) [28]. On the other hand, the presence or absence of the species may also be dictated due to their dispersal capacity or to the presence or absence of dispersers [19]. The results obtained contribute to describe the habitat of the species, which is an essential factor in programs aiming the restoration and management of temperate climate forests [8, 29], actions that, at least in the case of Mexico, have shown few results.

The species richness in BTS is generally lower than in humid tropical forests [30], although higher than in Temperate Mountain Forests [25]. The BTS is dominated by relatively short trees, most of which lose all their foliage during the dry season. In this community, herbaceous life form, thin woody species, and vines are common, but epiphytes and thick lianas are less abundant and diverse than in humid forests [31]. Diversity is generally higher without a clear dominance of any species, to the point that many of them are rare [32]. In this type of ecosystem, it is common to identify some genera such as *Bursera, Lonchocarpus, Lysiloma* and *Jatropha*, as well as emerging columnar cacti [33].

The species richness found at the Terrero de la labor Ejido BTS (N = 79) is similar to that reported by Trejo (2005) [33], where he points out that on average the tropical dry forest in Mexico harbors around 74 species with a DBH ≥ 1 cm in 0.1 ha. However, in the study site, some species considered «rare» which are indicators of medium forest (e.g *Amphipterygium molle*) were found in ravines and better preserved sites, suggesting that at some point this ecosystem had a greater presence in the landscape.

The analysis of the diversity, distribution and abundance associated with the Tropical Dry Forest in Aguascalientes has been little addressed, so the study conducted in the BTS of the Municipality of Calvillo represents one of the first efforts to understand this ecosystem natural heritage [14]. Previously, partial floristic studies had been carried out, studies which mainly referred to the dominant vegetation types and some important species, however on these studies there were gaps in relation to the ecology of the plant communities [20]. On the other hand, other studies mention some factors related to the mortality of these natural communities [15], but there is no information on vegetation diversity which reflects the real tropical dry forest importance.

This work contributes directly to the management of the ecosystems analyzed. Knowledge about species richness and their distribution provides an overview of the territory's conservation state, considering that both the Temperate Mountain Forest and the Tropical Dry Forest studied are part of the Sierra Fria Protected Natural Area, which is the protected area with the biggest extension in the State. On the other hand, the bases are established for the restoration of degraded ecosystems, either through active restoration or through mechanisms of ecological succession (passive restoration) [29].

## Acknowledgements

The authors acknowledge the participation of Jesus Argumedo-Espinoza for his cartographic support. Likewise, we thank the facilities provided of the owners of the Sierra Fria, as well as Jesus Velasco Serna of the Terrero de la Labor ejido for in the gathering of field information.

## Author details

Joaquín Sosa-Ramírez[1*], Vicente Díaz-Núñez[2] and Diego R. Pérez-Salicrup[3]

1 Agricultural Sciences Center, Autonomous University of Aguascalientes, Av. Universidad, Ciudad Universitaria, Aguascalientes, Ags., México

2 Scientific Research Invited by the Agricultural Sciences Center, Autonomous University of Aguascalientes, Avenida Universidad, Ciudad Universitaria, Aguascalientes, Ags., México

3 Ecosystem Research Institute, National Autonomous University of Mexico, Morelia, Mich, Mexico

*Address all correspondence to: jsosar@correo.uaa.mx

## References


[1] Sarukhán, J., P. Koleff, J. Carabias, J. Soberón, R. Dirzo, J. Llorrente-Busquets, G. Halfter, R. González, I. March, A. Mohar, S. Anta, J. de la Maza. 2009. Natural Capital of Mexico: current knowledge, evaluation and perspectives of sustainability. Synthesis/Capital Natural de México: conocimiento actual, evaluación y perspectivas de sustentabilidad. Síntesis. México: CONABIO. 100 p.

[2] Llorente-Bousquets, J., y S. Ocegueda. 2008. State of knowledge of biota. In: CONABIO (2008). Natural Capital of Mexico, vol. I: Current knowledge of biodiversity/Estado del conocimiento de la biota. En: CONABIO (2008). Capital natural de México, vol. I: Conocimiento actual de la biodiversidad. CONABIO, México, 2008.

[3] Mora, F. 2019. The use of ecological integrity indicators within the natural capital index framework: The ecological and economic value of the remnant natural capital of Mexico. Journal for Nature Conservation 47:77–92

[4] Organización de las Naciones Unidas para la Agricultura y la Alimentación (FAO). 2016. Global Forest Resources Assessment 2015: How are the world's changing? Second edition. FAO. 54 p. Available in: http://www.fao.org/3/a-i4793e.pdf

[5] Organizaciòn de las Naciones Unidas para la Agricultura y la Alimentación (FAO). 2010. Evaluación de los Recursos Forestales Mundiales. Informe Principal/Global Forest Resources Assessment. Main Report. FAO. Roma. 381 p.

[6] Challenger, A. y J. Soberón. 2008. Los ecosistemas terrestres de México. En: Capital natural de México, Vol. I: Conocimiento actual de la biodiversidad, J. Soberón, G. Halfter y J. Llorente (eds.)/Terrestrial ecosystems of Mexico. In: Natural Capital of Mexico, Vol. I: Current knowledge of biodiversity, J. Soberón, G. Halfter and J. Llorente (eds.). Comisión Nacional para el Conocimiento de la Biodiversidad –CONABIO-, México. ISBN 978–607-7607-03-8. p. 87–108.

[7] Balvanera, P. 2012. Los servicios ecosistèmicos que ofrecen los bosques tropicales/ The ecosystem services offered by tropical forests. Ecosistemas 21: (1). http://www.revistaecosistemas.net/articulo.asp?Id=709

[8] Baskent, E.Z. 2020. A framework for characterizing and regulating ecosystem services in a management planning context. Forests 11: 102; doi:10.3390/f11010102

[9] Comisión Nacional Forestal (CONAFOR). 2014. National Forest and Soil Inventory 2014–2018. Main report/ Inventario Nacional Forestal y de Suelos 2014–2018. Informe de resultados. SEMARNAT-CONAFOR. 200 p.

[10] Díaz, V., J. Sosa-Ramírez y D.R. Pérez-Salicrup. 2012. Trees and shrubs Distribution and abundance in Sierra Fria, Aguascalientes, Mexico/ Distribución y abundancia de las especies arbóreas y arbustivas en la Sierra Fría, Aguascalientes, México/. Polibotánica 34: 99–126.

[11] Inventario Estatal Forestal y de Suelos (IEFyS). 2012. State Forest and Soil Inventory of Aguascalientes/ Inventario Estatal Forestal y de Suelos de Aguascalientes. Gobierno del Estado de Aguascalientes-SEMARNAT-CONAFOR. 122 P.

[12] Gobierno del Estado de Aguascalientes. 2015. Decree of the Protected Natural Area "Área Silvestre Estatal Sierra Frìa"/Decreto del Área Natural Protegida "Área Silvestre Estatal

Sierra Fría". Diario Oficial del Estado de Aguascalientes. Disponible en: http://eservicios2.aguascalientes.gob.mx/NormatecaAdministrador/archivos/EDO-12-47.pdf

[13] Díaz-Núñez, V., J. Sosa-Ramírez, and D.R. Pérez-Salicrup. 2016. Vegetation patch dynamics and tree diversity in a conifer and oak forests in Central Mexico. Botanical Sciences 94: 229–240.

[14] Argumedo-Espinoza, J., J. Sosa-Ramírez, V. Díaz-Núñez, D.R. Pérez-Salicrup, and M.E. Siqueiros-Delgado. 2017. Diversity, distribution and abundance of woody plants in a dry tropical forest: recomendation for its management. In: Ortega-Rubio (Ed). 2017. Mexican Natural Resources Management and Biodiversity Conservation. Springer. Pp. 479–500

[15] Díaz-Núñez, V., J. Sosa-Ramírez e I. P. Macías-Medina. 2014. Phytosanitary diagnosis of vegetation in priority ecosystems of Aguascalientes, Mexico/ Diagnóstico fitosanitario de la vegetación en ecosistemas prioritarios de Aguascalientes, México. Comisión Nacional Forestal-Secretaría de Medio Ambiente del Estado de Aguascalientes. 84 p.

[16] Daget Ph. y M. Godron. 1982. Analysis of the ecology of species in communities/ Analyse de l'ecologie des espéces dans les communautés. Masson, Paris. 163p.

[17] De la Cerda, M. E. 1999. Oaks of Aguascalientes/Encinos de Aguascalientes. Universidad Autónoma de Aguascalientes. Segunda Edición. 77 p.

[18] Siquéiros, D.M.E. 1989. Conifers of Aguascalientes/Coníferas de Aguascalientes. Universidad Autónoma de Aguascalientes, 68 p.

[19] Krebs, C. K. 1993. Factors that limit distributions: Dispersal. In Ecology, C. K. Krebs. John Whiley and Sons. p. 41–56.

[20] Siqueiros-Delgado, M.E., J.A. Rodríguez-Ávalos, J. Martínez-Ramírez y J.C. Sierra-Muñoz. 2016. Current status of the vegetation of Aguascalientes, Mexico/ Situación actual de la vegetación del Estado de Aguascalientes/. Botanical Scienses 94: 455–470

[21] INEGI. 2007. Agrarian nuclei. Municipality basic tables. Certification Program for Ejido Rights and Land Titling/Núcleos agrarios. Tabulados básicos por municipio. Programa de Certificación de Derechos Ejidales y Titulación de Solares. PROCEDE. México.

[22] Begon, M., C. T. Towsend y J. L. Harper. 2006. Ecology: from individuals to ecosystems. Blackwell Publishing, Oxford. 738 p.

[23] Brawman, K.A., and G.C. Daily. 2008. Ecosystem services. Human ecology. Elsevier. Pp. 1148–1154

[24] Valencia, A. S. 2004. Diversity of *Quercus* genera (Fagaceae) in Mexico/ Diversidad del género *Quercus* (Fagaceae) en México. Boletín de la Sociedad Botánica de México 75: 33–53.

[25] Sabás-Rosales, J.L., J. Sosa-Ramírez, and J.J. Luna-Ruiz. 2015. Diversity, distribution and basic habitat characterization of the Oaks (*Quercus*: Fagaceae) of San Luis Potosí, México/ Diversidad, distribución y caracterización básica del hábitat de los encinos (*Quercus*: Fagaceae) del Estado de San Luis Potosí, México. Botanical Sciences 93: 881–897. DOI 10.17129/botsci.205

[26] Saénz-Romero, C., A. E. Snively y R. Lindig-Cisneros 2003. Conservation and restoration of pine forest genetic resources in Mexico. Silvae Genetica 52: 233–237.

[27] Márquez-Linares, M. A., S. González-Elizondo y R. Álvarez-Zagoya. 1999. Components of tree diversity in pine-oak forests of Durango, Mex./ Componentes de la diversidad arbórea en bosques de pino encino de Durango, Mex. Madera y Bosques 5: 67–78.

[28] Minnich, R. A., J. Sosa-Ramírez, V. E. Franco, W. J. Barry y M. E. Siqueiros. 1994. Preliminar recognizing of vegetation and human impacts in Sierra Fria, Aguascalientes/Reconocimiento preliminar de la vegetación y de los impactos de las actividades humanas en la Sierra Fría, Aguascalientes. Investigación y Ciencia 12: 23–29.

[29] Rey-Benayas, J. M., J. M. Bullock y A. C. Newton. 2008. Creating woodland islets to reconcile ecological restoration, conservation, and agricultural land use. Frontiers in Ecology and the Environment 6: 329–336.

[30] Gentry, A. H. 1995. Diversity and floristic composition of neotropical dry forests. In Seasonally dry tropical forests, S. H. Bullock, H. A. Mooney y E. Medina (eds.) Cambridge University Press, Cambridge. p. 146–190.

[31] Pineda-García, F., L. Arredondo-Amezcua y G. Ibarra-Manríquez. 2007. Richness and diversity of wood species in deciduous tropical forest of El Tarimo, basin of del Balsas, Guerrero/ Riqueza y diversidad de especies leñosas del bosque tropical caducifolio El Tarimo, Cuenca del Balsas, Guerrero/Richness and diversity of Woody species in the tropical dry forests of El Tarimo, Cuenca del Balsas, Guerrero. Revista Mexicana de Biodiversidad 78: 129–139.

[32] Durán, Z.V.H., M.J.R. Francia, P.C. R. Rodríguez, R.A. Martínez y R.B. Cárceles. 2006. Soil-erosion and runoff prevention by plant covers in a mountainous area (SE Spain): Implications for sustainable agriculture. The Environmentalist 26:309–31.

[33] Trejo, I. 2005. Analysis of the diversity of the deciduous tropical forest in Mexico. In: Halfter et al., (2005). On biological diversity: the meaning of alpha, beta and gamma diversities/ Análisis de la diversidad de la selva baja caducifolia en México. En: Halfter et al., (2005). Sobre diversidad biológica: el significado de las diversidades alfa, beta y gama. CONABIO, SEA, CONACyT. ISBN 84–932807-7-1. Pp. 111–122.

# 11

# Assessment of the State of Forest Plant Communities of Scots Pine (*Pinus sylvestris* L.) in the Conditions of Urban Ecosystems

*Elena Runova, Vera Savchenkova,*
*Ekaterina Demina-Moskovskaya and Anastasia Baranenkova*

**Abstract**

Siberian cities are characterized by one feature: many of them have preserved natural woodlands during construction, which on the one hand give a completely unusual, unique appearance to cities, on the other hand, trees suffer from recreational load, high levels of pollution and other anthropogenic factors. To assess the condition of pine stands, 3 test areas (0.5 ha, 0.1 ha and 1.9 ha) were laid. All considered plantings of natural origin are areas of woodland that were preserved during the construction of the city and are subject to recreational and industrial pollution. The test sites belong to areas with a high anthropogenic load, as they are located along highways and in close proximity to residential and public buildings and are part of parks with a high recreational load. The average age of trees is 70–80 years. The sanitary condition of the massif and its landscape characteristics are also determined. The critical condition of the massif is established, requiring sanitary logging and other forestry measures that could reduce recreational and anthropogenic loads.

**Keywords:** Scots pine (*Pinus sylvestris* L.), dendrometric characteristics, sanitary condition, damage to the trunk and crown, care

## 1. Introduction

Place of research: Irkutsk region, Bratsk city, residential district Energetik. On the territory of Bratsk and the Bratsk region, southern taiga and taiga natural complexes of Central Siberia prevail [1, 2]. In the vegetation of the city territory, forests of natural origin and urban plantations stand out. The dominant breed in forests of natural origin is Scots pine (*Pinus sylvestris* L.) - 57% of the total composition of the woodland, hanging birch (*Betula pendula* Roth.) And fluffy birch (*Betula pubescens* Ehrh.) Make up 17%, Siberian larch (*Larix sibirica* Ledeb.) - 6%, aspen (*Populus tremula* L.) - 16%, in much smaller quantities there are ordinary spruce (*Picea abies* L.), Siberian spruce (*Picea obovata* Ledeb.), goat willow (*Salix caprea* L.), shrub alder (*Duschekia fruticosa* Rupr.), common grouse (*Sorbus aucuparia* L.) and Siberian grouse (*Sorbus sibirica* Hedl.).

Inner-city vegetation is artificially created communities that are not self-regulating systems, they need constant care, which in most cases they do not receive. The predominant breed in urban planting is balsamic poplar (*Populus balsamifera* L.), In much smaller quantities are represented hanging birch (*Betula pendula* L.) and fluffy birch (*Betula pubescens* Ehrh.) - 11%, tree-shaped caragana (*Caragana arborescens* Lam.) - 6%, squat elm (*Ulmus pumila* L.) - 4%, Siberian mountain ash (*Sorbus sibirica* Hedl.) - 4%, Siberian larch (*Larix sibirica* Ledeb.) - 3%, apple berry (*Malus baccata* L.) - 3% of the total. The remaining representatives of trees and shrubs make up 2% or less of the total.

The city of Bratsk is located on the banks of the Bratsk and Ust-Ilimsk water reservoirs formed on the Angara River during the construction of hydroelectric power plants and is an agglomeration of dispersed residential areas separated by significantly vast forests and water spaces. Residential areas, various in size and degree of improvement are former villages that arose near industrial enterprises under construction. The area of the city is 428 km$^2$.

Bratsk is located in the north-west of the Irkutsk region in the central part of the Angara ridge. The city arose in 1955, in connection with the construction of the Bratsk hydroelectric station, north of the old village of Bratsk (Bratsk, Bratskoye), founded as an ostrog in 1631.

Climatic conditions are not favorable for diverse vegetation in urban plantations and forest areas. The climate is sharply continental with long harsh winters (down to - 35-45 ° C) and short hot summers (up to +25–30 ° C). During the year and day, the temperature here can vary within large limits. The cold period lasts an average of six months (from the second decade of October to the third decade of April). The average long-term frost-free period in the central part of the city is 94 days. The first frosts are recorded on September 8, the last - on June 5. About 370 mm of precipitation falls per year. The characteristics of the soils of the region include their fineness due to the large dissection of the relief and the diversity of the lithological composition of the rocks, the low temperature regime due to deep seasonal freezing and slow thawing, insufficient moisture due to the small amount of precipitation and spring waters that roll down the still thawing soils and dirt. Soils are subject to wind and water erosion, which reduces the content of humus and reduces fertility. Soils lack organic and mineral fertilizers and need new agricultural techniques [2].

## 2. Research methodology and methods

To conduct studies on the state of forest woody vegetation, 3 test areas (0.5 ha, 0.1 ha and 1.9 ha) were laid with a number of trees from 150 to 664 plants. All considered plantations of natural origin are areas of the forest preserved during the construction of the city, subject to recreational effects and industrial pollution. Test areas belong to territories with a high anthropogenic load, as they are located along roads and in close proximity to residential and public buildings and are part of parks with a high recreational load. The average age of trees is 70–80 years.

The purpose of this research work is to study the conditions for the growth and development of Scots pine masses (*Pinus sylvestris* L.). in the center of the village of Energetik in the city of Bratsk with high attendance. The massif was preserved during the construction of the village of Energetik, the age of the trees is approximately the same - 70-80 years. Studies were carried out according to generally accepted methods adopted in forest dendrometry [3, 4] (**Figure 1**).

At the test areas, a continuous inventory of trees was carried out with measuring diameters at the height of the chest with a measuring fork, the height of each tree using a altimeter - an eclimeter, the condition of the crown, the state of the roots

**Figure 1.**
*Test area №1.*

and trunk, the curvature of the trunk was determined visually. Kraft classes and a sanitary assessment class were also defined. The class of sanitary assessment is as follows: 1 - healthy (without signs of weakening); 2 - weakened; 3 - severely weakened; 4 - drying; 5 - fresh dry; 5 (a) - fresh wind; 5 (b) - fresh brown; 6 - old dry; 6 (a) - old wind; 6 (b) - old brown; 7 - emergency trees.

Next, the area of the entire project area is defined using the compass and measuring ribbon. For instrumental evaluation, 10 model trees were selected for each trial area. To assess the internal state of the wood, the following were used: Arbotom ® pulsed tomograph and Resistograph ® micro-drilling device of the German company Rinntech. For instrumental evaluation, 10 model trees were selected for each trial area. The principle of operation of Arbotom ® is based on determining the velocity of the pulse through the wood between sensors. Vibration sensors are arranged in series to measure trunk circumference. Measurements were taken at chest height. After the sensors are connected to the power supply unit and the manufacturer's specialized software, a number of shocks are applied to each sensor, after which information about the pulse speed between them is recorded by the computer. The standard deviation limit was set at 10%. The software presents the results in the form of matrix values, linear graphs and plane graphs (tomograms) (**Figure 2**).

The principle of operation of the Resistograph ® device is based on the determination of wood resistance to drilling. The Resistograph ® design uses an ultra-thin drill (1 mm) to reduce damage to the study object. To level the error taking into account the point profile, 2 perpendicular measurements were made on each model tree, the results of which were averaged [5–9].

Processing of the obtained data with isolation of destruction zones is carried out by visual evaluation of the obtained graphs taking into account the averages for the tree in question and for the total sample, as well as the distribution pattern of the

**Figure 2.**
*Arbotom® wood condition measurement.*

**Figure 3.**
*Test area №3.*

| Test Area Number | Breed | Percentage of total trees,% |
|---|---|---|
| 1 | Scots pine (*Pinus sylvestris* L.) | 91 |
| | Siberian larch (*Larix sibirica* Ldb.) | 5 |
| | Birch fluffy (*Betula pubescens* Ehrh.) | 4 |
| 2 | Scots pine (*Pinus sylvestris* L.) | 100 |
| 3 | Scots pine (*Pinus sylvestris* L.) | 100 |

**Table 1.**
*Breed composition of wood stands on test areas.*

relative density of wood. Direct correlation of instrument readings with wood density is not possible due to the absence of fixed graded scales.

An agglomerative hierarchical clustering method was chosen for statistical processing of results. Given the nature of the sampling, the Euclidean distance was adopted as the distance between objects:

$$\rho(x, x') = \sqrt{\sum_i^n (x_i - x'_i)^2}, \quad (1)$$

where i are signs; n - number of characteristics.

Clustering was carried out using a complete communication algorithm (far neighbor method), according to which the degree of proximity is estimated by the degree of proximity between distant objects of the clusters (**Figure 3**).

Data is entered in Excel and processed using statistics methods [4]. All test areas are pure or almost pure single-tier pine trees (**Table 1**). In this regard, further evaluation was carried out only on Scots pine (*Pinus sylvestris* L.).

As can be seen from **Table 1**, common pine, as the most flexible in the biological and ecological aspect, tree species predominate in the composition of woodlands.

## 3. Research result

The results of the studies and treatment of the experimental materials obtained are as follows: Average biometric parameters of wood stands on trial areas are given in **Table 2.**

As can be seen from **Table 2**, the trees in all test areas are severely weakened, there are stunted and drying trees and even old dry. Since the trees are single-age and single-tier, the classification by the degree of growth and dominance by Kraft classes was used. The distribution by Kraft class is shown in **Table 2**. Weakened trees according to the Kraft classification are (IV, Va, Vb) in trial areas of more than 20 percent, which indicates the depressed state of the tree stand, which is also confirmed by sanitary assessment points, which average 2.8 points in trial area No. 1; 3.2 points on trial area No. 2 and 2.5 points on trial area No. 3.

Dynamics of average height of trees by thickness stages is shown in **Figures 4–6.**

As can be seen from **Table 2** and **Figures 4–6**, all woodlands have a clearly underdeveloped height and at the age of 70–80 years belong to the woodlands of 4–5 bonitet classes (on the scale of Professor Orlov), which indicates low productivity of woodlands and very unfavorable growing conditions. Productivity (Orlov bonitet classes are determined by the ratio of height and diameter at a certain age). The low bonitet class indicates poor growing conditions (soil or climatic). If we compare the average height of trees of the same age of class III bonitet (through which most forests of the Angara region grow), then the studied trees lack by –3 9

| Test Area Number | Average age, years | Average diameter, cm | Average height, m | Vearage height of crown base, m | Root condition | Sanitary grade,% | Kraft grade, % |
|---|---|---|---|---|---|---|---|
| 1 | 70–80 | 30.21 ± 1.2 | 17.2 ± 0.9 | 7.2 ± 0.3 | Not visible | 1–14<br>2–11<br>3–55<br>4–17<br>5–3 | I – 18<br>II – 12<br>III – 42<br>IV – 26<br>V – 2 |
| 2 | 70–80 | 26.24 ± 0.8 | 16.4 ± 0.7 | 7.0 ± 0.2 | Not visible | 1–3<br>2–17<br>3–58<br>4–7<br>5–14<br>6–1 | I – 15<br>II – 28<br>III – 36<br>IV – 10<br>Va – 10<br>Vб – 1 |
| 3 | 70–80 | 21.6 ± 1.1 | 10.1 ± 0.6 | 5.3 ± 0.5 | Visible – 87<br>Not visible – 575 | 1–0<br>2–82<br>3–1<br>4–10<br>5–0<br>6–7 | I – 12<br>II – 8<br>III – 39<br>IV – 21<br>Va – 19<br>Vб – 1 |

**Table 2.**
*Biometric indicators of the studied Scots pine tree* (Pinus sylvestris *L.).*

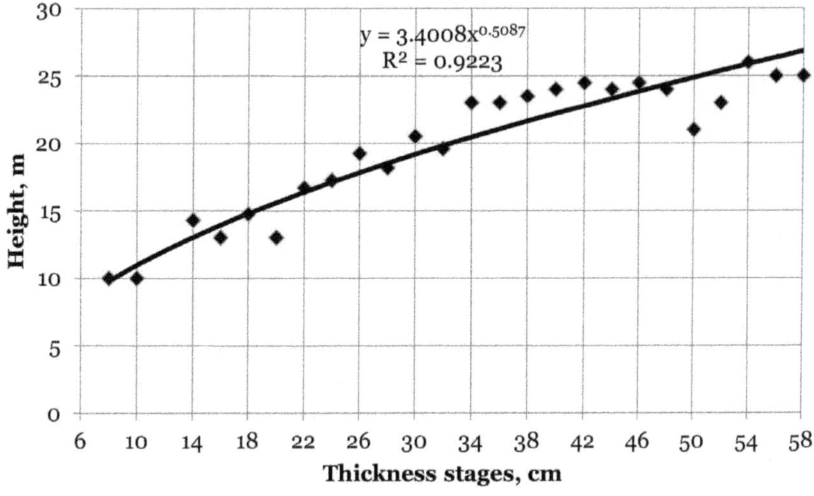


**Figure 4.**
*Height distribution by thickness stages for site No. 1. The figure shows the relationship between the thickness of trees determined at a height of 1.3 meters (abscissa axis) and the height of trees in meters (axis of ordinates) on the first test area.*

meters in height, which is very significant, especially on test area No. 3, where the lowest height and the highest recreational load.

When analyzing the condition of tree crowns, it is worth noting that deviations in crown shapes from the norm (flag-shaped, compressed or cut crown) are found in 50% of trees on site No. 1, 35% on site No. 2 and 53% on site No. 3. The percentage of crown condition is shown in **Figures 7–9**.

During measurements, the presence of trunk defects was noted in 95% of cases at site No. 1; 99% - at site No. 2; 96% - at site No. 3. At the same time, it should be noted that in most cases there were 2 or more types of external wood defects on the trunk (**Figures 9** and **10**; **Table 3**).

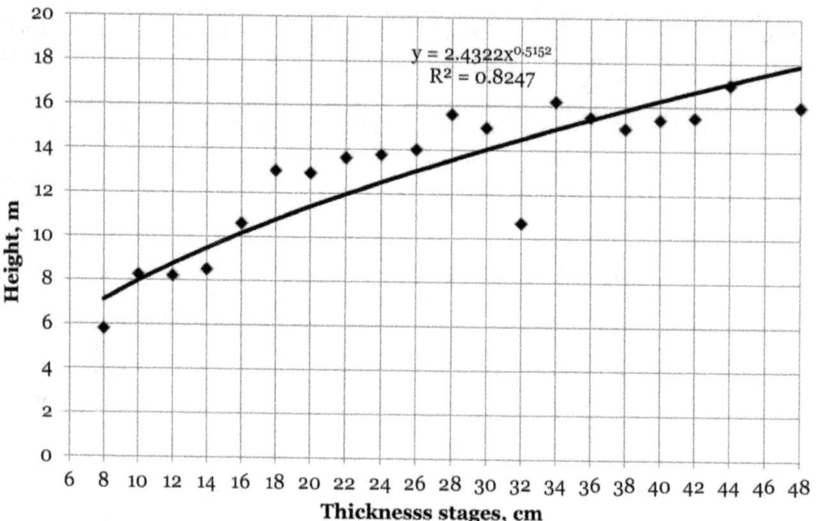


**Figure 5.**
*Height distribution by thickness stages for site No. 2. The figure shows the relationship between the thickness of trees determined at a height of 1.3 meters (abscissa axis) and the height of trees in meters (axis of ordinates) on the second test area.*

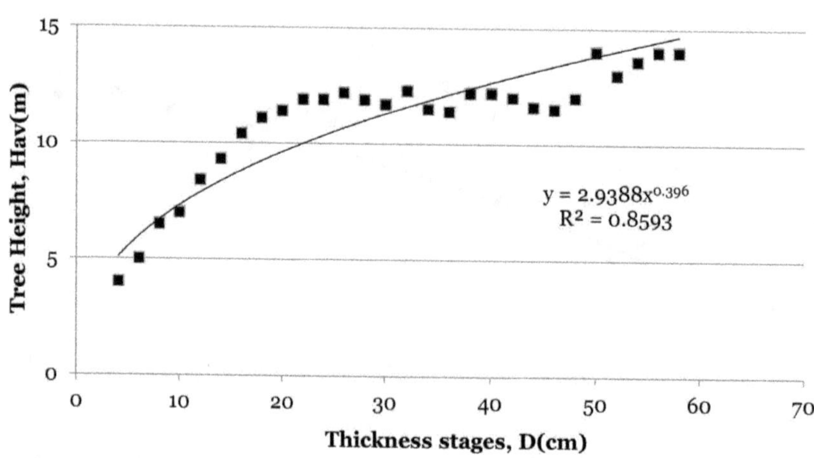


*The distribution of heights by thickness stages for site No. 3. The figures show the correlation between the thickness of trees determined at a height of 1.3 meters (abscissa axis) and the height of trees in meters (ordinate axis) on the third test area.*

All model trees studied are ripe and restless. Most trees have a slope or curvature of the trunk and crown asymmetry. Visible wood defects are often observed - dryness, mechanical damage. The trunk height is from 12 to 19 meters, the trunk diameter is from 30 to 70 cm. **Table 4** shows the taxation indicators of the three trees most characteristic of the study object.

The nature of the tomograms obtained indicates a heterogeneous distribution of the wood density of the studied model trees. The speed varies from 912 m/s to 2018 m/s. The maximum frequency of occurrence falls in the range of 1003–1349 m/s. The degraded wood content ranges from 12 to 79% (average content is 30%). According to Resistograph ®, the average drilling resistance of the sample model trees was 121. The degraded wood content ranges from 26% to 85% (average content is 50%). In the vast majority of cases (96% of the sample), resistogram

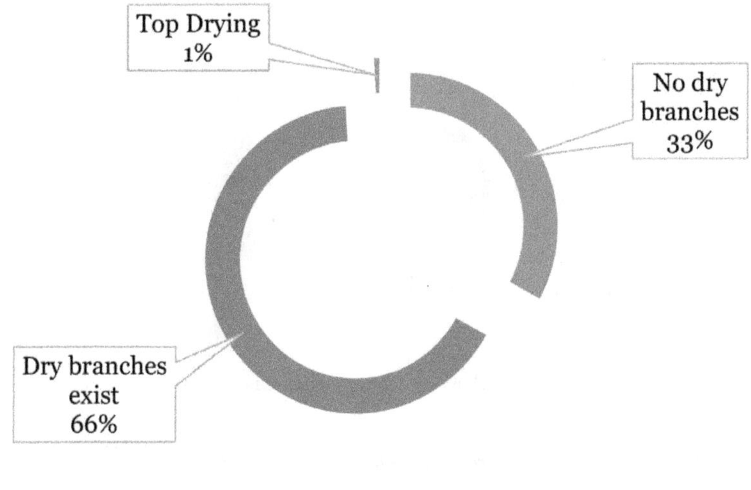


**Figure 7.**
*Presence of signs of drying of the crown of trees of site No. 1, % of the sample.*

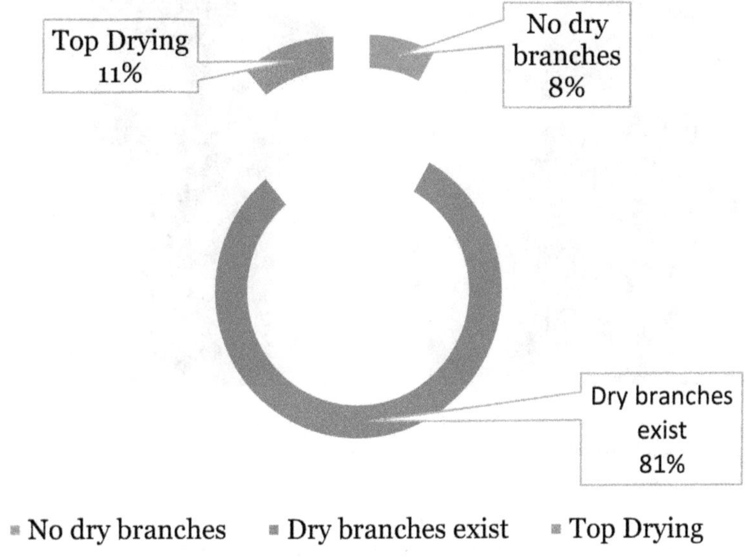


**Figure 8.**
*Presence of signs of drying of the crown of trees of site No. 2, % of the sample.*

readings reflect a significantly higher percentage of trunk destruction. This is a consequence of incomplete accounting of the area of peripheral areas of sickness during profile analysis by drilling.

Particularly strong oppression of trees can be traced on test area No. 3. Therefore, its characteristics should be discussed separately.

As can be seen from **Table 5**, the trees on this test area develop accordingly to the dwarf type, the bonitet class (woodland productivity) is only V, that is, the lowest at a given diameter. The average class of sanitary assessment is 3.6, which indicates almost the decay of the tree. **Table 5** shows the main forms of tree crowns and their number in pieces and percentages. It should be noted that on the 1.9 hectare site there are quite a large number of trees, namely 662 plants, in some areas the fullness of the tree stand significantly exceeds the maximum equal to 1.0.

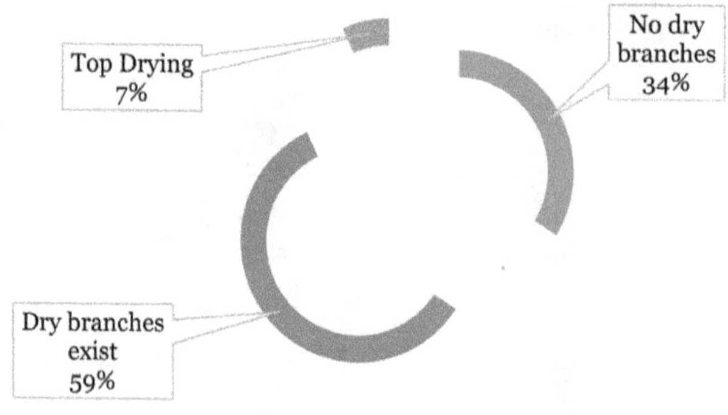


**Figure 9.**
*Presence of signs of drying of the crown of trees of site No. 3, % of the sample.*

**Figure 10.**
*Mechanical damage to tree trunks.*

| Test Area Number | Number of kinds of trunk defects | Share in sample,% |
|---|---|---|
| 1 | No defects | 5 |
| | 1 kind of defect | 14 |
| | 2 or more kinds of defects | 81 |
| 2 | No defects | 1 |
| | 1 kind of defect | 26 |
| | 2 or more kinds of defects | 73 |
| 3 | No defects | 4 |
| | 1 kind of defect | 30 |
| | 2 or more kinds of defects | 66 |

**Table 3.**
*Number of trunk defects in test areas.*

| № of the tree | 6 | 7 | 8 |
|---|---|---|---|
| $D_{main}$ cm | 38 | 44 | 52 |
| $D_{1,3}$ cm | 32 | 38 | 44 |
| $H_{tr}$, m | 12.5 | 18.8 | 19.0 |
| Tree age | 68 | 72 | 78 |
| Visible defects of the trunk | Drywall, trunk slope | Drywall, trunk slope | Drywall, trunk slope |
| Average pulse speed, m/s | 936 | 966 | 1011 |
| Average drilling resistance, relative units. | 137 | 134 | 142 |
| Disturbed wood content according to Arbotom ®,% | 78 | 52 | 15 |
| Disturbed wood content according to Resistograph ®,% | 50 | 57 | 69 |

**Table 4.**
*Taxation characteristics of model trees.*

| Type | Quantity | % |
|---|---|---|
| Flag-shaped | 326 | 49.10 |
| Oval | 311 | 46.84 |
| Round shaped | 3 | 0.45 |
| No crown or dried up | 24 | 3.61 |
| Total | 664 | 100 |

**Table 5.**
*Distribution of the number of trees by crown forms in test area No. 3.*

As the table shows, almost all crowns, namely 95.94% of the total number of trees, have either a flag-shaped crown or a strongly compressed oval crown. **Figure 11** shows in more detail the distribution of crowns in form.

**Table 6** and its results indicate extreme soil compaction as a result of long-term recreational load, there is no living soil cover characteristic of the forest, so it is impossible to determine the type of forest and type of forest conditions. More than 13% of trees have a highly bare root system, which has mechanical damage and does

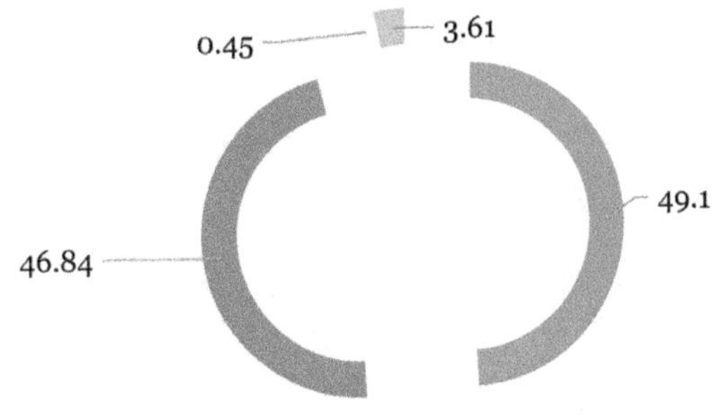


**Figure 11.**
*Crown form distribution chart, %.*

| Type | Quantity | % |
|---|---|---|
| Bare rooted | 88 | 13,11 |
| Not visible | 576 | 86,59 |

**Table 6.**
*Conditions of tree roots on test area No. 3.*

| Type | Quantity | % |
|---|---|---|
| partial drywall, rot | 268 | 40,36 |
| partial drywall | 91 | 13,70 |
| partial drywall, mechanical defects | 101 | 15,21 |
| partial drywall, mechanical defects, rot | 15 | 2,26 |
| partial drywall, rot, cancerous object | 1 | 0,15 |
| healthy | 28 | 4,22 |
| rot | 85 | 12,80 |
| rot, cancerous object | 2 | 0,30 |
| mechanical defects | 28 | 4,226 |
| mechanical defects, cancerous object | 1 | 0,15 |
| mechanical defects, rot | 43 | 6,48 |
| cancerous object | 1 | 0,15 |
| Total | 664 | 100 |

**Table 7.**
*Identification of the quality status of the tree trunk in test area No. 3.*

not allow plants to develop fully. The state of trees in terms of the quality of trunk wood is especially manifested, absolutely all trees have numerous mechanical and mushroom damages (**Table 7**).

The trunks have a drywall to a greater extent, which was formed from mechanical action on the trunks, drywall with open damaged wood led to the presence of mushroom lesions and rotting, up to the hollow. **Figure 12** shows the distribution of the number of trees by type of stem injury.

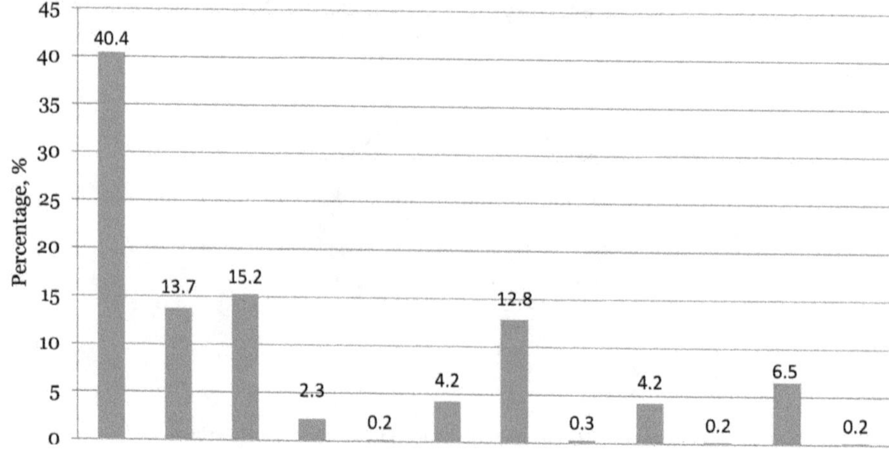


**Figures 12.**
*Diagram of stem damage distribution by number of trees, %.*

In addition to damage to the trunk wood, the trees under study had a different degree of crown damage, which was expressed by the presence of a large number of dry branches (**Table 8**).

**Figure 13** shows the percentage distribution of trees by crown condition. On the axis of ordinates are represented percentages, on the axis of abscissa - categories of trees as per crown.

| Type | Quantity | % |
|---|---|---|
| No dry branches | 223 | 33.58 |
| Dry branches exist | 394 | 59.34 |
| Half dried tree, top drying | 1 | 0.15 |
| Dry tree | 41 | 6.17 |
| Top drying | 3 | 0.45 |
| Half dried tree | 2 | 0.30 |
| Total | 664 | 100 |

**Table 8.**
*Presence of dry branches of investigated trees on test area No. 3.*

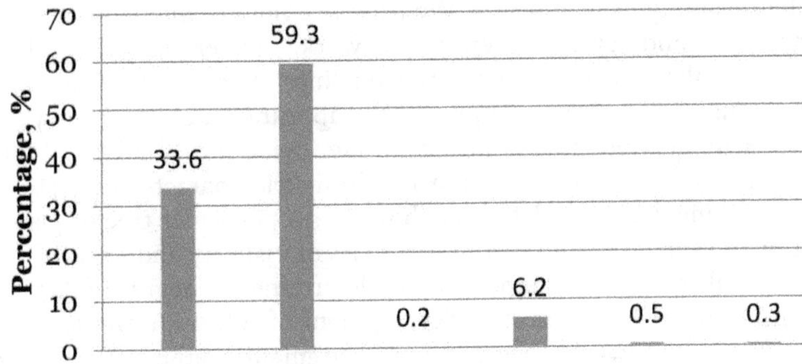


**Figure 13.**
*Distribution of the number of trees by the degree of damage to the crown (presence of dry branches), % on test area No. 3.*

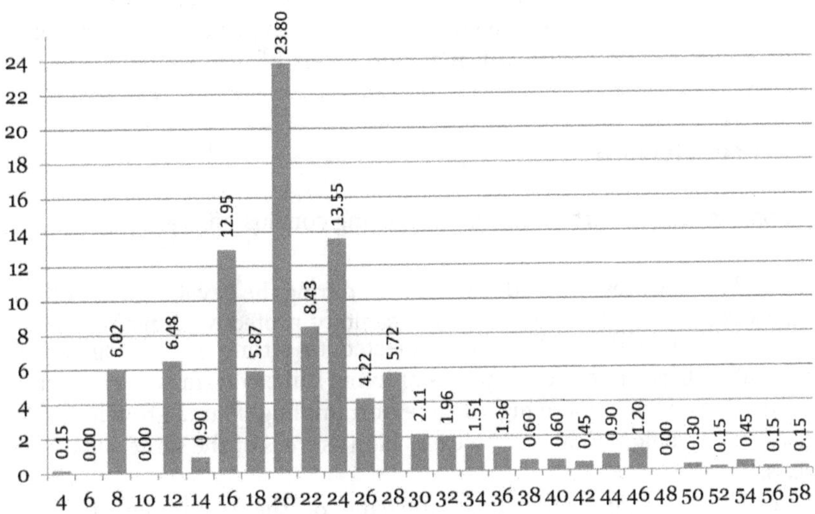


**Figure 14.**
*Distribution of number of trees in% by diameters in test area No. 3.*

Another evidence of the extreme oppression of trees in the pine massif is the nature of the distribution of trees along the thickness stages. In normal, healthy wood, this distribution is close to the normal distribution curve. Let us look at **Figure 14**, which shows a diagram of the distribution of trees by diameter at a height of 1.3 meters.

The **Figure 12** shows the percentage of trees by diameter on test area No. 3. Percentages are represented along the ordinate axis, along the abscissa axis - the thickness of trees at a height of 1.3 meters. It can be seen what a large range of diameters in the woodland is from 4 centimeters to 58 at a uniform age, which indicates strong intraspecific competition of trees.

## 4. Discussion of the results

Based on the conducted studies, it can be concluded that the results of studies of the Arbotom ® and Resistograph ® devices of the German company Rinntech [7–17] are quite often found in scientific publications. However, in most studies, the assess-ment of the state of the stem wood using appropriate instruments is performed separately [5, 7, 8, 11, 14]. The data are comparable with the data of well-known scientists in terms of quantitative indicators [7, 8, 10, 12–14], which confirms the reliability of our studies. We tried to compare the data of the readings of the two devices and compare them in our work. Cluster analysis methods were also used to simultaneously compare the readings in order to more accurately quantify the condi-tion of trees and predict the appearance of emergency trees that are dangerous to human life in the city under wind loads. Most often, urban plantings were described using visual or dendrometric characteristics [16–18], but only visual assessment did not determine how long a particular tree or plant species could exist in an urban environment without loss of viability and signs of accidents. That is, the visual assessment method cannot determine the degree of damage to the tree by internal rot and the degree of development of rot, up to the formation of a hollow [16_18]. In this work, visual and measurement methods for assessing the state of forest areas included in the urban environment were also carried out, which is unique, since it is typical only for the northern regions of Russia, where cities are relatively poorly built (50–70 years). A fairly high correlation was found between dendrometric parame-ters, the state of the roots and tops of trees, and the presence of internal trunk defects. The reduction in the life expectancy of trees in urban conditions under recreational loads and high levels of atmospheric pollution has been proven.

## 5. Chapter conclusions

Based on the studies carried out, the following conclusions can be drawn:

1. Natural forests preserved 60–65 years ago during the development of the city of Bratsk certainly perform esthetic and sanitary protective functions. But at the same time they themselves are subjected to strong anthropogenic effects. Typically, there are no forest plants of living soil cover, in some cases the trees are dead cover due to the high degree of trampling. Plants grow in 4–5 bonitet class, have a height of almost twice as high as trees of the same age of 1 bonitet class.
2. Trees have a large percentage of shape of the trunk defects, primarily drywalls and prophecy, which is associated with mechanical effects on tree trunks.

3. Tree crowns are cut, often have a flag-shaped crown, a large number of dry branches.

4. Studies have proved the presence of internal defects in all studied model trees. It is possible to conclude the general oppression of woody vegetation of the test areas under consideration.

5. Among the studied model trees, trial area No. 2 is noticeably distinguished in terms of wood hardness. However, in the general picture of the distribution of internal defects, significant selectivity between sites is not observed. Thus, one can conclude that the conditions for the growth of the woodland are relatively equal and the green spaces of various areas of the urban ecosystem of the city are evenly oppressed.

6. Under conditions of industrial pollution and increased recreational loads, processes of earlier aging of trees occur up to their natural death. In the forest environment, the life expectancy of common pine is from 350 to 600 years in Russia [10], and in the urban environment without proper care and with a high level of anthropogenic load, already at the age of 70–80 years there are pronounced signs of tree aging, which are manifested in the presence of dry branches, dry trees, the presence of internal rots, stem pests.

7. When compiling an assessment of the state of the plantation as a whole, it is advisable to recommend comparing the data of the two devices both according to the parameters of the expert assessment (proportion of disturbed wood) and according to the parameters of automated measurements (instrument data) to compile the most complete picture of the state of dendrocenosis.

8. The work performed is of great practical importance, as it allows to identify emergency trees that can be exposed to wind and windbreak under heavy wind loads, and to carry out timely replacement with younger and healthier trees.

9. All these signs indicate the need for additional studies of the internal condition of the wood of model trees of the sites considered by instrumental methods. In view of the high value of urban forests, non-destructive testing methods are recommended. Woodlands require a whole range of forestry measures to preserve these unique objects of the urban environment - such as sanitary cutting of dead trees, cutting off dry branches, introducing a fertile layer of land, and treating mechanical damage to the trunk.

10. The developed proposals should be used in the care of green spaces in urban urban ecosystems, especially in the care of areas of natural forests that are located inside urban development. Regular monitoring of the condition of such plantings is planned to be carried out not only for Pinus sylvestris L., but also for other tree and shrub species growing under anthropogenic loads.

## 6. Grainude

I express my gratitude to my colleagues in the work who took part in the collection of experimental material I.A. Garus, A.N. Mukhacheva.

## Author details

Elena Runova[1*], Vera Savchenkova[2], Ekaterina Demina-Moskovskaya[2] and Anastasia Baranenkova[2]

1 Bratsk State University, Bratsk, Russian Federation

2 Mytischi Branch of Bauman Moscow State Technical University, Moscow, Russian Federation

*Address all correspondence to: runova0710@mail.ru

## References

[1] Runova E. M., Anoshkina L. V., Gavrilin I. I. State of woody vegetation in urban ecosystems on the example of Bratsk: monograph.- BrSU publishing house, 2017. – 80 p.

[2] Geographical encyclopedia of the Irkutsk region: commonly. Essay/ed. L. M. Korytny; Irkut Government. Region, Institute of Geography named after VB Sochava SB RAS, Irkut. State un-t. –Irkutsk: Institute of Geography. V.B. Sochavy SB RAS, 2017.–336 p.

[3] OST 56–69-83. Trial areas, forest inventory. Bookmarking methods. Enter. For the first time 1984.01.01. M. Publishing house of standards, 1984.61 p.

[4] Runova E.M., Zhang S.A., Puzanova O.A., Savchenkova V.A. Dendrometry: Textbook for the ACT.-St. Petersburg: Lan Publishing House, 2020. -160 p. ISBN 978–5–8114-5916-2.

[5] Johnstone D., Moore G., Tausz M., Nicolas M. The measurement of wood decay in landscape trees//Arboriculture & Urban Forestry 36(3). 2010. P. 121–127.

[6] Chubinsky A.N., Tambi A.A., Fedyaev A.A., Fedyaeva N. Yu., Kulkov A.M. Directions of using physical methods for controlling the structure and properties of wood//Systems. Methods. Technology. 2015.2 (26). S. 152–158.

[7] Tyukavina O.N. The speed of passage of a sound pulse in pine wood//Bulletin of the Northern (Arctic) Federal University. Series: Natural Sciences. – Arkhangelsk, 2014 .–S. 78–85.

[8] Lavrov M.F. Determination of quality indicators of wood by drilling method/M.F. Lavrov, D.K. Chakhov, I. A. Doctors//Bulletin of the Moscow State Forest University. – Forest Bulletin. – 2014. No. 5. – S. 196–201.

[9] Runova E.M., Garus I.A., Mukhacheva A.N. The use of instrumental methods in assessing the state of the trunks Pinus sylvestris L. Forest engineering journal, No. 3, 2020, p. 72–85. Bibliography: p. 83–84 (15 names). - DOI: 10.34220/issn.2222-7962/2020.3/8.

[10] Rinn F. Eine neue Bohrmethode zur Holzuntersuchung//Holz-Zentralblatt. 1989. №15 (34). s. 529–530.

[11] Literature review of acoustic and ultrasonic tomography in standing trees/Arciniegas A., Prieto F., Brancheriau L., Lasaygues P. //Trees. 2014. № 28(6). p. 1559–1567.

[12] Rinn F. Holzanatomische Grundlagen der Schall-Tomographie an Bäumen//Neue Landschaft. 2004. № 7/04. s. 44–47.

[13] Rinn F. Statische Hinweise im Schall-Tomogramm von Bäumen//Stadt und Grün. 2004. № 7/2004. s. 41–45.

[14] Melnichuk I. A., Yasin N., Cherdantseva O. A. Diagnostics of the internal state of *TILIA CORDATA* MILL trees. with the use of the complex of acoustic ultrasound tomography equipment "ARBOTOM °" / /Bulletin of the Russian University of Friendship of Peoples. Series: Agronomy and animal husbandry. 2012.№ S5. P. 25–32.

[15] Palchikov S. B., Antsiferov A. B. Assessment of the state of trees affected by xylotrophic fungi using Resistograph and Arbotom devices. assessment of the condition of trees affected by xylotrophic fungi using Resistograph and Arbotom devices. Eurasian Union of Scientists (EUU) # 4 (25), 2016. Biological Sciences series, pp. 121–125

[16] Kolomyts E. G., Rosenberg G. S., Glebova O. V. The natural complex of the big city: landscape-ecological analysis. - Moscow: Nauka, 2000 – - 286 p.

[17] Mozolevskaya E. G. Methodological recommendations for assessing the viability of trees and the rules for their selection and assignment to felling and transplanting. Moscow: MGUL, 2003. 28s.

[18] Petrovskaya P. A., Stolyarova A. G. Basic principles of urbanized territory improvement / / Vestnik Rossiyskogo universiteta druzhby narodov. Series: Agronomy and animal husbandry. 2013.№ 5. S. 86–92.

# PERMISSIONS

All chapters in this book were first published by InTech Open; hereby published with permission under the Creative Commons Attribution License or equivalent. Every chapter published in this book has been scrutinized by our experts. Their significance has been extensively debated. The topics covered herein carry significant findings which will fuel the growth of the discipline. They may even be implemented as practical applications or may be referred to as a beginning point for another development.

The contributors of this book come from diverse backgrounds, making this book a truly international effort. This book will bring forth new frontiers with its revolutionizing research information and detailed analysis of the nascent developments around the world.

We would like to thank all the contributing authors for lending their expertise to make the book truly unique. They have played a crucial role in the development of this book. Without their invaluable contributions this book wouldn't have been possible. They have made vital efforts to compile up to date information on the varied aspects of this subject to make this book a valuable addition to the collection of many professionals and students.

This book was conceptualized with the vision of imparting up-to-date information and advanced data in this field. To ensure the same, a matchless editorial board was set up. Every individual on the board went through rigorous rounds of assessment to prove their worth. After which they invested a large part of their time researching and compiling the most relevant data for our readers.

The editorial board has been involved in producing this book since its inception. They have spent rigorous hours researching and exploring the diverse topics which have resulted in the successful publishing of this book. They have passed on their knowledge of decades through this book. To expedite this challenging task, the publisher supported the team at every step. A small team of assistant editors was also appointed to further simplify the editing procedure and attain best results for the readers.

Apart from the editorial board, the designing team has also invested a significant amount of their time in understanding the subject and creating the most relevant covers. They scrutinized every image to scout for the most suitable representation of the subject and create an appropriate cover for the book.

The publishing team has been an ardent support to the editorial, designing and production team. Their endless efforts to recruit the best for this project, has resulted in the accomplishment of this book. They are a veteran in the field of academics and their pool of knowledge is as vast as their experience in printing. Their expertise and guidance has proved useful at every step. Their uncompromising quality standards have made this book an exceptional effort. Their encouragement from time to time has been an inspiration for everyone.

The publisher and the editorial board hope that this book will prove to be a valuable piece of knowledge for researchers, students, practitioners and scholars across the globe.

# LIST OF CONTRIBUTORS

**Amber L. Hauvermale**
Department of Crop and Soil Sciences, Washington State University, Pullman, WA, USA

**Marwa N.M.E. Sanad**
Department of Genetics and Cytology, National Research Centre, Giza, Egypt

**Feyza Candan**
Biology Department, Arts and Science Faculty, Manisa Celal Bayar University, Manisa, Turkey

**Herminia García-Mozo**
Departamento de Botánica, Ecología y Fisiología Vegetal, Universidad de Córdoba, Córdoba, España

**Dean G. Fitzgerald and David R. Wade**
Premier Environmental Services Inc., Cambridge, Canada

**Patrick Fox**
Wiikwemkoong Unceded Territory, Department of Lands and Resources, Mide Kaaning, Wikwemikong, Canada

**Jimin Cheng, Wei Li, Liang Guo, Jingwei Jin and Chengcheng Gang**
Institute of Soil and Water Conservation, Northwest A&F University, Yangling, China
Institute of Soil and Water Conservation, Chinese Academy of Sciences and Ministry of Water Resources, Yangling, China

**Jishuai Su**
State Key Laboratory of Vegetation and Environmental Change, Institute of Botany, Chinese Academy of Sciences, Beijing, China

**Erhard Schulz**
Institut für Geographie und Geologie, Universität Würzburg, Germany

**Aboubacar Adamou, Sani Ibrahim and Issa Ousseini**
Département de Géographie, Université Abdou Moumouni de Niamey, Niger

**Ludger Herrmann**
Institut für Bodenkunde und Standortslehre, Universität Hohenheim, Germany

**Ilknur Babahan**
Department of Chemistry, Faculty of Arts and Sciences, Adnan Menderes University, Aydin, Turkey

**Birsen Kirim**
Department of Aquaculture Engineering, Faculty of Agriculture, Adnan Menderes University, Aydin, Turkey

**Hamideh Mehr**
Department of Polymer Engineering, University of Akron, Akron, Ohio, USA

**Sarwan Kumar**
Department of Plant Breeding and Genetics, Punjab Agricultural University, Ludhiana, India

**Álvaro Martínez and Vicente D. Gómez-Miguel**
Universidad Politécnica de Madrid, Madrid, Spain

**Joaquín Sosa-Ramírez**
Agricultural Sciences Center, Autonomous University of Aguascalientes, Av. Universidad, Ciudad Universitaria, Aguascalientes, Ags., México

**Vicente Díaz-Núñez**
Scientific Research Invited by the Agricultural Sciences Center, Autonomous University of Aguascalientes, Avenida Universidad, Ciudad Universitaria, Aguascalientes, Ags., México

**Diego R. Pérez-Salicrup**
Ecosystem Research Institute, National Autonomous University of Mexico, Morelia, Mich, Mexico

**Elena Runova**
Bratsk State University, Bratsk, Russian Federation

**Vera Savchenkova, Ekaterina Demina-Moskovskaya and Anastasia Baranenkova**
Mytischi Branch of Bauman Moscow State Technical University, Mosco, Russian Federation

# Index

Printed in the USA
CPSIA information can be obtained
at www.ICGtesting.com
JSHW051258021023
49509JS00006B/53